Assessing Science Learning

Perspectives From Research and Practice

Assessing Science Learning

Perspectives From Research and Practice

**Edited by Janet Coffey,
Rowena Douglas, and
Carole Stearns**

National Science Teachers Association

Arlington, VA

National Science Teachers Association

Claire Reinburg, Director
Jennifer Horak, Managing Editor
Judy Cusick, Senior Editor
Andrew Cocke, Associate Editor
Betty Smith, Associate Editor

ART AND DESIGN
Will Thomas, Jr., Director

PRINTING AND PRODUCTION
Catherine Lorrain, Director
Nguyet Tran, Assistant Production Manager

NATIONAL SCIENCE TEACHERS ASSOCIATION
Gerald F. Wheeler, Executive Director
David Beacom, Publisher

LIBRARY OF CONGRESS CATALOGING-IN-PUBLICATION DATA

Assessing science learning : perspectives from research and practice / edited by Janet E. Coffey,
Rowena Douglas, and Carole Stearns.
    p. cm.
  Includes index.
  ISBN 978-1-93353-140-3
  1. Science—Study and teaching—Evaluation. 2. Science—Ability testing. I. Coffey, Janet. II.
Douglas, Rowena. III. Stearns, Carole.
  LB1585.A777 2008
  507.1'073—dc22
                          2008018485

NSTA is committed to publishing material that promotes the best in inquiry-based science
education. However, conditions of actual use may vary, and the safety procedures and practices
described in this book are intended to serve only as a guide. Additional precautionary measures
may be required. NSTA and the authors do not warrant or represent that the procedures and
practices in this book meet any safety code or standard of federal, state, or local regulations. NSTA
and the authors disclaim any liability for personal injury or damage to property arising out of or
relating to the use of this book, including any of the recommendations, instructions, or materials
contained therein.

PERMISSIONS
You may photocopy, print, or e-mail up to five copies of an NSTA book chapter for personal
use only; this does not include display or promotional use. Elementary, middle, and high school
teachers *only* may reproduce a single NSTA book chapter for classroom- or noncommercial,
professional-development use only. For permission to photocopy or use material electronically from
this NSTA Press book, please contact the Copyright Clearance Center (CCC) (*www.copyright.
com*; 978-750-8400). Please access *www.nsta.org/permission* for further information about NSTA's
rights and permissions policies.

Contents

Section 3

High-Stakes Assessment: Test Items and Formats..................227

Section 4

Professional Development: Helping Teachers Link Assessment, Teaching, and Learning..337

Foreword

Elizabeth Stage
Director, Lawrence Hall of Science
University of California, Berkeley

I t is all too common to pick up a newspaper and see an article about student achievement (usually declining test scores) or district testing policies and the effects of No Child Left Behind on the allocation of instructional time. All around the country, annual testing for the purpose of accountability is dominating public conversations about education. This focus on accountability testing is just one of many assessment responsibilities teachers juggle daily, and probably the least important for supporting student learning. As the essays in this book attest, teachers also need to assess students to guide daily instructional decisions, to promote their further learning, and to assign grades. In a more perfect world, assessment for accountability and assessment for student learning would align, reinforcing one another. Unfortunately, more often than not, such synergy remains elusive.

In 2005, NSTA invited a distinguished group of researchers and teacher educators to share their current research and perspectives on assessment with an audience of teachers. As the conference demonstrated, a rich body of research on what works and what does not is available to inform teachers' assessment practices. The conference also demonstrated the value of an open dialogue among researchers and teachers on practical applications of assessment research to practice. The goal of this book, with chapters by the conference presenters, is to share these research-based insights with a larger audience and to help teachers bring together different assessment priorities and purposes in ways that ultimately support student learning.

This book is also a call for greater teacher involvement in assessment discussions, particularly at the state and local levels. Just as we know from classroom-based research that teachers can gain great insight by listening carefully to their students, so too researchers and policy makers will be better informed by listening to teachers—to the questions they have, the realities they face, and the dilemmas with which they struggle. Teachers should actively engage in conversations, participate in test design and item development, and help improve the assessment literacy of students and parents. Indeed, teachers' voices are prominent in many of the research efforts described in this book; teachers co-authored many of the chapters. Insights from teachers will help generate strands of research that contribute to richer understandings of assessment practice and its ultimate influence on student learning. While no simple fixes exist for the seemingly divergent assessment purposes, by working together, teachers and researchers can design powerful assessment contexts that help all students reach deep levels of conceptual understanding and high levels of science achievement.

Introduction

Janet Coffey and Carole Stearns

In an era of accountability, talk of assessment often conjures up images of large-scale testing. Although it dominates attention, annual testing is only a small corner of what occurs in the classroom in the name of assessment. Assessment is the chapter test, the weekly quiz, the checking of nightly homework assignments. Assessment can be the observations made as students engage in an activity or the sense-making of student talk as they offer explanations. It is the teacher feedback offered on the lab report, provided to students as they complete an investigation or after they have completed a journal entry. As all of these things and more, assessment is a central dimension of teaching practice.

As the multiple images of assessment suggest, within any classroom, assessment takes on many forms, and must serve multiple purposes. These purposes include accountability and grading. Another important purpose that has received increasing attention is assessment that supports student learning, rather than solely documenting achievement. Different ways to talk about assessment have emerged. We can talk about its *purpose*, as we just did above. We can talk about the *form* assessment takes—the multiple-choice test, the portfolio, the alternative assessment, the written comments or oral feedback, or the piece of student work. Different *uses* of information gleaned from assessment have led us to talk about assessment *of* learning and assessment *for* learning, or, in assessment terminology, summative and formative assessment. All of these purposes, forms, and functions are important; all are at play in the classroom.

Over the past decade, the National Science Foundation (NSF) has funded numerous research efforts that seek to better understand assessment in science and math education at all levels; the various strategies and systems; and the variety of forms, roles, and contexts for assessment *of* and *for* student learning. NSF has also funded assessment-centered teacher professional development efforts and creation of models for assessment systems that seek synergy among different purposes. In 2005–2006, the National Science Teachers Association convened two full-day conferences to help disseminate these NSF-funded research findings to practitioners. Many of the recipients of those grants shared their work at the conferences and have prepared chapters for this book in an effort to build connections between research and practice and to facilitate meaningful conversation.

Conversations between research and practice are not commonplace, yet greater exchange is essential. Practitioners help researchers better understand the terrain, including the practitioners' underlying rationales for their everyday decision making. These insights from those "on the ground" can inform research and contribute to generative lines of questioning. Although starting points and perspectives may differ, ultimately the assessment research and practitioner communities are working toward the same goal: to better understand the relationships between assessment and learning in order to create classroom environments that support our students' learning.

Researchers are afforded the luxury of stepping back; they can extract a part from the whole—the formative from the summative, for example. They can focus on particular strategies or activities, such as use of notebooks or assessment of lab reports. Teachers, on the other hand, need to make sense of assessment in all its complexity and juggle what may seem like competing priorities and purposes. There may even be times when the different roles teachers take on with respect to assessment appear to conflict: They are, at once, judge and juror, coach and referee. Teachers are asked to figure out ways to navigate these different roles and to align strategies to priorities. They are asked to implement assessment activities and strategies in such a way that a variety of purposes is served, and served well, while mitigating tensions that appear unavoidable.

Research does not hold all the answers. The research community still wrestles with very real and difficult issues that teachers face every day, such as equitable assessments, challenges associated with wide-scale professional development, and assessment designs that capture the complexity of disciplinary reasoning and understanding. As the education community makes progress on these fronts, new challenges and questions arise. No silver bullet exists, nor does a one-size-fits-all fix. However, research can offer insights into strategies and features that are particularly productive, and into frameworks that are particularly compelling.

The essays in this collection will introduce readers to some of the many voices in the national discourse on science assessment, a field currently at the crossroads of education and politics. The essays present authors' deeply held values and perspectives about the roles of assessment and how assessment must not only provide accountability data but also support the learning of students from different backgrounds. Readers will notice that many of the research studies are grounded in classroom practices and involve teachers as collaborators or in professional development settings. Practitioners' expertise in understanding the complexity of classrooms is crucial to realizing the importance of assessment in deep science learning for all students.

You will not hear a message of consensus here. The research community does not speak in a unitary voice—beyond the claims that there exists a tight coupling between assessment and student learning and that events and interactions that occur in classrooms in the name of assessment do matter. This is not a "how-to" manual. You will not find polished strategies or assessments to try tomorrow in your classroom. Research cannot offer assistance in that form. Strategies, approaches, and frameworks will need modification and accommodation in order to be meaningfully integrated into your classroom and school. As you read, we encourage you to reflect on your own practice, consider your own priorities, and make sense of what you are learning in light of your own school community

Organization of the Book

The chapters in this book are grouped into four sections: (1) formative assessment in the service of learning and teaching; (2) classroom-based strategies for assessing students' science understanding; (3) high-stakes tests; and (4) assessment-focused professional development.

Each section begins with a brief introduction and overview of the included chapters. The section introductions also offer a set of framing questions intended to help readers identify important themes and construct take-home messages that are relevant to their own teaching environment and needs.

The first section, "Formative Assessment: Assessment for Learning," introduces three perspectives on formative assessment: its role in improving student learning; research examining connections between a sequence of formative assessments and their impact on teaching and learning; and the importance of probing how students learn and their misconceptions. Many of the book's central ideas are introduced in this section:

- Roles of assessment in teaching and learning,
- Characteristics of meaningful assessment items,
- Need for research to validate assessment practices,
- Significance of assessing both the knowledge and misconceptions of students,

- Value of assessing students' ability to apply their knowledge, and
- Importance of assessment-focused professional development.

The opening chapter defines classroom-based formative assessment as an ongoing activity informing daily instructional decisions and accompanied by meaningful feedback to students. The author asserts that an essential precursor to raising student achievement in science is providing professional development that will help teachers improve their assessment practices, a topic addressed in many of the chapters and explored in great detail in Section 4.

A research study on correlations between use of embedded formative assessments, teacher practice, and student achievement is the subject of Chapter 2. The focus of the third chapter is the importance of knowing *how* students learn and the nature of their misconceptions. Readers will learn about tools the authors developed to gather and analyze this information.

The chapters in Section 2, "Probing Students' Understanding Through Classroom-Based Assessment," present specific classroom-based strategies for assessing students' science knowledge and understanding. Several of these strategies are closely linked with students' literacy and communication skills, primarily writing, but also drawing, reading, and oral communication. These chapters address the day-to-day issues that teachers confront, such as "How much do my students understand?" "What still confuses them?" "How can I encourage them to communicate more clearly?" and "What constitutes a good formative assessment?"

Several authors write about using familiar classroom artifacts such as students' drawings and notebook entries for assessment purposes. There is a chapter on teaching students to construct reasoned scientific explanations based on their own observations and analysis of data. Secondary teachers may be particularly interested in the chapter on assessing laboratory work. One chapter reports a research study on the use of science notebooks to assess English language learners. (Chapters in later sections also address the needs of English language learners, one in the context of eliminating bias in test items [Chapter 12] and another in a large-scale study of correlations between the science achievement of non-native speakers and the amount of assessment-based professional development their teachers receive [Chapter 17].)

Many of the chapters in this section consider assessments based on familiar classroom routines and artifacts (e.g., science notebooks, lab reports, conversations with and among students) that, when observed through an assessment lens, reveal valuable information about what and how students are learning. Other chapters in this section describe classroom-based assessment formats and items that were developed by researchers and subjected to field testing in multiple classroom settings. A team of developers describes a suite of formative assessment tasks designed to monitor student learning at several points during a multi-week unit of study. Another chapter introduces a technology-based assessment system developed to track students' problem-solving skills as they interact with a computer simulation. This section concludes with a chapter offering teachers guidelines on constructing standards-based formative assessment probes.

Section 3, "High-Stakes Assessment: Test Items and Formats," begins with an examination of the cognitive demands of several high-stakes test item formats. Authors focus on what students must know and be able to do to succeed on high-stakes tests and how teachers' own classroom assessment can help students meet these challenges. The opening chapter takes readers through the process of designing and field testing items that are closely linked to specific standards-based learning goals. The next chapter analyzes constructed-response test items, a format commonly used in national and international tests, such as TIMSS and NAEP. The authors present sample items and detailed scoring guides to help teachers better understand how such items are scored. Another chapter provides teachers with guidelines for analyzing the content and format of high-stakes test items and creating closely aligned questions to use in their own classrooms.

Section 3 continues with a chapter summarizing the National Research Council's (NRC) report on design principles for state-level science assessment systems. The authors discuss the goals of state-level assessment, calling attention to the distinct differences between these tests and the classroom-based assessments described in Section 2. The concluding chapter offers reflections by a literacy expert on high-stakes testing practices and test items in his field. He summarizes the lessons learned and offers some suggestions to science test developers.

In Section 4, "Professional Development: Helping Teachers Link Assessment, Teaching, and Learning," authors describe several large-scale professional development initiatives that emphasize building assessment expertise. Programs in Seattle, Washington, Toledo, Ohio, Miami, Florida, and Colorado Springs, Colorado are discussed. While each had a different approach to professional development design, all included a research component investigating potential correlations between the teachers' experiences and their student performance on high-stakes tests. Each study reports compelling data showing a positive correlation between teachers' participation in the professional development efforts and student achievement on high-stakes science tests.

A chapter on a classroom observation research tool titled the Reform Teacher Observation Protocol (RTOP) offers another approach to professional development. The authors discuss the use of this tool by secondary teachers to self-evaluate their classroom assessment practices. The final chapter describes strategies that school teams can use to analyze assessment data from multiple sources; including high-stakes tests, classroom-based assessments, and teacher observations, for the purposes of program evaluation and guiding instructional decisions.

* * *

This brief summary does little justice to the richness of the essays herein and to the multiple examples of meaningful science assessment practices they explore. The collection reflects work with socioeconomically and ethnically diverse populations to better understand the attributes of equitable assessment practices. While the authors may describe an assessment study conducted within a narrow context (science teachers will recognize the constraints required by a controlled experiment), the findings and recommendations are broadly applicable. For example, professional development programs in Seattle, Washington, offer many ideas equally relevant for schools and districts in other parts of the United States. Similarly the assessment potential of student notebooks extends far beyond classrooms in El Centro, California.

We hope that this book can be used to fuel the conversations about assessment sparked in the initial NSTA conference. From the informal interactions that occur among students and teachers to more formal exchanges, from item design to grading, and from classroom systems of reporting on progress to large-scale external state tests, fodder exists for deep and provocative discussion. In the essays that follow, readers have an opportunity to consider the issues closely and to reflect on the ways in which assessment can be better coordinated. We hope that, eventually, the entire system will become more synergistic in order to meet the many purposes of assessment while not neglecting or undermining any single one.

The editors are grateful to the researchers who contributed to this volume for their commitment to communicating their work to practitioners, the ultimate consumers of science assessment knowledge. We hope that readers will find many ideas that enrich their own understanding of the assessment landscape and help them better serve their students. We encourage teachers to actively engage in the national assessment conversation and to share insights they develop in their own classrooms.

Formative Assessment: Assessment for Learning

Section 1 focuses on the nature of formative assessment and its multiple roles in the classroom: monitoring student learning, guiding instruction, providing meaningful feedback to students, engaging them in self assessment, and revealing their misconceptions.

Discussion Questions
- What does formative assessment look like in the science classroom?
- In what ways can formative assessment impact student learning?
- What is the relationship between formative assessments and high-stakes tests?
- What can formative assessments teach us about how students learn and how to help them become better learners?

Chapter Summaries
The essay by Dylan Wiliam, Institute of Education, University of London, is based on his extensive experience with classroom teachers. Wiliam asserts that an essential step to raising student achievement in science is to provide professional development that will help teachers improve their formative assessment practices. He envisions classroom assessment as an ongoing activity informing daily instructional decisions and accompanied by meaningful feedback to students. Wiliam emphasizes his thesis with a

sports analogy. Coaches observe their athletes and then provide both diagnostic and formative feedback to improve their performance. The diagnostic helps the athlete identify what is wrong and the formative is a guide to making it better.

Richard Shavelson and his colleagues at Stanford University and the Curriculum Research and Development Group at the University of Hawaii report on their recent study with middle school physical science teachers. After developing a series of embedded formative assessments closely aligned with the local curriculum, they conducted controlled experiments investigating the impact of the assessments and the information these assessments provided on students' knowledge and motivation to learn science. The chapter offers readers a candid view of a multifaceted research project, the complexity of controlling the research variables in a real-world situation, the findings (many of them unexpected), and the challenges associated with classroom-based research.

Jim Minstrell recounts his transition from science teacher to researcher. He describes his early career as a high school physics teacher and his gradually turning his attention from *what* his students were learning to *how* they were learning. Now a teacher-researcher, Minstrell talks about the Diagnoser, a tool he and his FACET Innovations colleagues—Ruth Anderson, Pamela Kraus, and James E. Minstrell—developed to explore the nature of students' commonly held misconceptions and to extend his research on how students develop scientific understanding.

Improving Learning in Science With Formative Assessment

Dylan Wiliam
Institute of Education, University of London

In recent years, the No Child Left Behind law has focused attention on student achievement in science across the United States, but there are more important reasons for being concerned with student achievement. Increasing student achievement brings substantial benefits both to the individual and to society. For the individual, improved school achievement increases career earnings, improves health and well-being, and actually lengthens life (Wiliam and Thompson 2007; Lleras-Muney 2005). For society, the benefits include increased tax revenues, savings in public health costs, reduced law-enforcement and prison costs, as well as savings in welfare budgets (Hoff 2007).

Achievement in science in particular is likely to be increasingly important in the future for the needs of employment, but it will also be essential for democratic citizenship. Without an understanding of what science can do (and what it can't) and how science does what it does, public policies about issues such as genetically modified foods, assisted reproductive technologies, and human cloning are likely to be set on the basis of populist journalism rather than scientific evidence.

The focus of this chapter, therefore, is about how we can raise achievement in science in the United States. In this chapter I will argue that if we are serious about raising achievement for science, then we need to look beyond "what works" in education to notions of cost-benefit—not just whether a particular initiative raises achievement, but by how much, and at what cost. I will show that the evidence currently at hand suggests that this is done by investing in teacher professional development, but of a sort very different from what occurs in most school districts. I will also show that this professional development needs to be focused specifically on changing what teachers do in the classroom and, in particular, needs to be aimed at changing teachers' minute-by-minute and day-by-day use of assessment to modify instruction, sometimes called formative assessment or assessment for learning.

Value for Money in Education Reform

Many policy makers have focused on "what works" in education, but such a focus is misguided, since what is most important is the size of effect relative to cost. An intervention might "work" but the effects might be too small to be worth bothering with, or it might produce substantial effects but be too expensive to implement. The focus in school improvement needs to be on the ratio of the size of the benefit to the cost incurred in bringing it about.

When we adopt this perspective, we find that some effective interventions are too costly, and some interventions that have only a small impact on student learning nevertheless turn out to be implementable at a modest cost. To take one example, the research shows clearly that reducing class size raises student achievement. So what? The important questions should be how much improvement, and at what cost? And here the data are depressing.

Jepsen and Rivkin (2002) found that reducing elementary school class size by 10 students would increase the proportion of students passing typical mathematics and reading tests by 4% and 3% respectively. While there is evidence that the effects are larger for the early years, beyond this point, class-size reduction appears to be a very expensive way of increasing student achievement. To make this more precise, consider the cost of this intervention per classroom of 30 students. Reducing class-size from (say) 30 to 20 would increase the salary costs by approximately 50%, because we would need three teachers instead of two for a group of 60 students. Assuming an average teacher salary cost of $60,000 per year, this would cost $30,000 per year to implement.

Using tests such as those used in international comparisons such as TIMSS (Rodriguez 2004) as a benchmark, we find that the effect of these class-size reductions would be equivalent to students in the smaller classes learning in three months what it would take the students in the larger classes four months to learn; a 33% increase in the speed of learning.

A more recent study by Jenkins, Levacic, and Vignoles (2006) in England found that additional resources, if used to reduce class size, might have a larger increase on student achievement in middle and high school science (up to four times the effect size found by Jepsen and Rivkin), but this model assumed that additional science teachers of equivalent quality to those already on the job could be found, which is at the very least questionable (one of the key factors in the modest results in the Jepsen and Rivkin research cited above was that the extra teachers were not as good as those already in place). Nevertheless, it may be that there are special cases, such as early years, and secondary science, where class-size reduction may have a substantial effect on student achievement. However, in general, class-size reduction would appear to offer only modest increases in student achievement at very high cost.

How else could we obtain the same increase in student achievement? One obvious candidate is to increase teacher subject knowledge. Hill, Rowan, and Ball (2005) found that an increase of two standard deviations in what they termed "mathematical knowledge for teaching" was associated with an increase of up to 0.1 standard deviations in student achievement in mathematics. This was not a direct experiment, and so we cannot infer a causal relationship. However, it suggests that the same improvement in achievement that would be gained by reducing class size from 30 to 20 might be secured by increasing teacher subject-matter knowledge by two standard deviations. Unfortunately, there is currently little evidence about how to do this.

In contrast, supporting teachers in developing the use of assessment for learning has been shown to roughly double the speed of learning (Wiliam et al. 2004; Hayes 2003). In other words, students learned in six months what would have taken a year to learn in other classrooms. The cost of this intervention was around $8,000 per teacher, but of course, unlike class-size reduction, the cost of which has to be found annually, investing in teachers' professional development is a one-off expenditure, which can be depreciated over the teacher's remaining career. Against this, some cost of annual renewal needs to be allowed. Assuming that a new teacher will continue to teach for at least five years, and allowing time for four hours of meetings with colleagues per month for renewal, the cost of teacher professional development focused on assessment for learning would appear to be around $3,000 per teacher (and therefore per classroom) per year.

Compared to class-size reduction, therefore, improving teachers' use of assessment for learning would appear to promise two or three times the increase in student learning, for around one-tenth the cost. Even in the special case of secondary school science discussed above, assessment for learning produces roughly the same size of benefit as reducing class size by 30%, at less than one-fifth the cost.

Assessment for Learning

When teachers are asked how they assess their students, they typically talk about tests, examinations, quizzes, and other formal methods. When they are asked how they know whether their students have learned what they have taught, the answers are very different. They talk about homework, classwork, the things students say in classroom discussions, and even the expressions on their faces. This is the great disconnect in education worldwide. Assessment that serves the needs of teachers and their students is seen as completely separate from, and indeed, incompatible with, assessment that serves the needs of parents, administrators, policy makers, and other stakeholders.

At one extreme we have a teacher questioning a student, trying to elicit evidence of (mis)conceptions that are likely to impede future learning. At the other extreme we have the use of Advanced Placement tests, used both to give students college-level credit and, by some universities, to decide which students to admit. The obvious conclusion is that the latter kind of assessment hinders learning while the former helps, but things are not that simple.

The teacher's questioning of the student can cause damage, possibly irreparable, to the student's sense of self if undertaken in a humiliating way. And at the other extreme, Advanced Placement tests provide clear guidance to teachers and students about what the vague words in examination syllabuses mean. Furthermore, when used as "trial" examinations, sample papers allow students to benchmark themselves against the standard established by the College Board and to help each other rectify deficiencies. If we are to design assessment systems that help rather than hinder learning, we must go beyond looking at the assessments themselves and look at deeper issues about how the assessments help learners and their teachers know where the learners are in their learning, where they are going, and how to get there.

CHAPTER

1

Through extensive reviews of the available research evidence, and through extensive fieldwork with teachers, both in the UK and in the United States (see the suggestions for further reading at the end of this chapter), we have identified five "key strategies," which, when implemented appropriately, allow assessment to help, rather than hinder, learning.

1. Engineer effective classroom discussions, questions, activities, and tasks that elicit evidence of student learning.

The first step in using assessment to help learning is to collect the right sort of evidence, and here it is clear that the tools that teachers use to find out where students are in their learning are given too little attention.

Few teachers plan the kinds of tasks, activities, and questions that they use with their students specifically to elicit the right kind of evidence of student learning. As an example of a good question, consider the science question shown in Figure 1.1.

Figure 1.1 Science Item

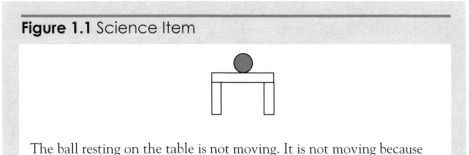

The ball resting on the table is not moving. It is not moving because
A. no forces are pushing or pulling on the ball.
B. gravity is pulling down, but the table is in the way.
C. the table pushes up with the same force that gravity pulls down.
D. gravity is holding it onto the table.
E. there is a force inside the ball keeping it from rolling off the table.

Source: Adapted by the author from Wilson, M., and K. Draney. 2004. Some links between large-scale and classroom assessments: The case of the BEAR assessment system. In M. Wilson (Ed.), *Towards coherence between classroom assessment and accountability: 103rd Yearbook of the National Society for the Study of Education.* Chicago, IL: University of Chicago Press.

Response A clearly relates to the well-known misconception that if there is no movement, then there are no forces acting. Response B is more complex. In one sense there is nothing wrong with B. After all, gravity *is* pulling the ball down, and the table *is* in the way. The reason that B is a less preferable response to C is not that it is incorrect; it is that it is not physics. Science provides us with a powerful, but unnatural, way of thinking about the world; after all, if it were natural, we would not need to teach it. Students who choose B rather than C have not understood the important idea that lack of movement denotes forces in equilibrium, not the absence of forces.

By crafting questions that explicitly build in the under- and overgeneralizations that we know students make, we can get far more useful information about what to do next. By equipping each student in the class with a set of cards with A, B, C, D, and E on them, and by requiring all students to respond simultaneously with their answers, the teacher can generate a very solid evidence base for deciding whether the class is ready to move on. If every student responds with C, then the teacher can move on with confidence that the students have understood. If everyone simply responds with A, then the teacher may choose to re-teach some part of the topic. The most likely response, however, is for some students to respond correctly and for others to respond incorrectly. This provides the teacher with an opportunity to conduct a classroom discussion in which students with different views can be asked to justify their selections. Moreover, because the teacher knows which student gave which response, the discussion can be better orchestrated (e.g., "Shane, you also chose B. Did you choose it for the same reason that Alicia gave, or a different reason?").

Of course, planning such questions takes time, but by investing the time before the lesson, teachers are able to address students' confusion during the lesson, with the students still in front of them. Teachers who do not plan such questions are forced to put children's thinking back on track by giving extended comments while grading, thus dealing with the students one at a time, after they have left the classroom.

2. Provide feedback that moves learning forward.
The research on feedback shows that much of the feedback that students receive has, at best, no impact on learning, and can actually be counterproductive. One extraordinary study (Kluger and DeNisi 1996) reviewed

more than 3,000 research reports on the effects of feedback in schools, colleges, and workplaces. They then rejected studies that failed to reach the highest standards of methodological rigor and were left with just 131 studies. Across these 131 studies, they found that, on average, feedback did increase achievement, but that in 50 of the studies (i.e., almost two in five), feedback actually made people's performance worse than it would have been without feedback. The key feature of these studies was that feedback was, in the psychological jargon, "ego-involving." In other words, the feedback focused attention on the person rather than the quality of the work—for example, by giving scores, grades, or other forms of report that encouraged comparison with others. For the 81 studies that found a positive impact on performance, Kluger and DeNisi found that the biggest impacts occurred when feedback told not just what to do to improve, but also how to go about it.

An example from athletics may be helpful here. If a young fast-pitch softballer has an ERA (earned run average) of 10, we know that she is not doing well. This is the *monitoring* assessment. The monitoring assessment identifies that there is a problem, but doesn't identify what it is. By looking at her pitching, we might see that the reason that her ERA is so high is that she is trying, unsuccessfully, to pitch a rising fastball. This is a pitch that is thrown with enough backspin so that as it reaches the batter, it rises sharply, making it almost unhittable. The problem is that if it does not rise, then the result is a fastball over the center of the plate, which is very easy to hit, and this is exactly what is happening to our pitcher. Her rising fastball is not rising. This is the *diagnostic* assessment. The diagnostic assessment identifies where the problem is, but by itself, doesn't give the athlete any clue about how to go about making improvements. However, if the pitching coach can see that the reason that the pitcher is struggling to pitch the rising fastball is because she is not dropping the pitching shoulder low enough to deliver the pitch from below the knee, then this gives the athlete something to work with. This is the *formative* assessment. Just as we use the term *formative* to describe the experiences that shape us as we grow up, a formative assessment is one that shapes learning. Much of the feedback that students get while learning science is no more helpful than telling our fast-pitch softballer to make sure her rising fastball rises or telling a bad comedian that he needs to be funnier. If feedback is to impact learning it must focus on what needs to happen next; in other words, it must be a *guide* to action and not just a demand for it. For an example of a feedback system focused on middle school science, see Clymer and Wiliam (2006/2007).

3. Clarify and share learning intentions and success criteria with learners.

In an article titled "The View From the Student's Desk," written more than 35 years ago, Mary Alice White (1971) said:

> *The analogy that might make the student's view more comprehensible to adults is to imagine oneself on a ship sailing across an unknown sea, to an unknown destination. An adult would be desperate to know where he [sic] is going. But a child only knows he is going to school.... The chart is neither available nor understandable to him.... Very quickly, the daily life on board ship becomes all important.... The daily chores, the demands, the inspections, become the reality, not the voyage, nor the destination.* (p. 340)

In a similar vein, I have walked into many science classrooms and asked students what they were doing, only to be told something like "page 34,"as if that were all I needed to know. For many students, school is just a series of tasks whose purposes are unclear and even what counts as success is mysterious, especially for students from less-advantaged backgrounds.

One study that is particularly notable for reducing the achievement gap between lower- and higher-achieving students was conducted by White and Frederiksen (1998). The study involved three teachers, each of whom taught four parallel seventh-grade classes in two U.S. schools. The average size of the classes was 31 students. In order to assess the representativeness of the sample, all the students in the study were given the Iowa Test of Basic Skills (ITBS), and their scores were close to the national average. All 12 classes followed a novel curriculum (called ThinkerTools) for 14 weeks. The curriculum had been designed to promote thinking in the science classroom through a focus on a series of seven scientific investigations (approximately 2 weeks each).

Each investigation incorporated a series of evaluation activities. In two of each teacher's four classes, these evaluation episodes took the form of a discussion about what they liked and disliked about the topic. For the other two classes, they engaged in a process of "reflective assessment." Through a series of small-group and individual activities, the students were introduced to the nine assessment criteria (each of which was assessed on a 5-point scale) that the teacher would use in evaluating their work. At the end of each episode within an investigation, the students were asked to assess their performance against two of the criteria. At the end of the investigation, students had to assess their performance against all nine. Whenever they assessed themselves, they had to write a brief statement showing which aspects of their work formed the basis for their rating. At the end of each investigation, students presented their work to the class, and the students used the criteria to give one another feedback.

When the researchers analyzed the achievement of the students, the weakest students in the "reflective assessment" group performed as well as the best students in the control group, and the other students did even better, the result of which was to reduce by half the achievement gaps in the "reflective assessment" classes.

4. Activate students as owners of their own learning.

One of the great traps of teaching is the belief that teachers create learning. This is particularly important when teachers are under pressure to improve student results, because studies have shown that when teachers are told they are responsible for making sure that their students do well, the quality of their teaching deteriorates, as does their students' learning (Deci et al. 1982); hence the old joke about schools being places where children go to watch teachers work.

Only learners create learning, and so, when we look at the role that assessment plays in promoting learning, the crucial feature is not the validity of the assessment, or its reliability, but its impact on the student. No matter how reliable or valid the assessment is, if it communicates to students that they cannot learn, it will hinder learning. Particularly important here is the work of Carol Dweck, who over a 30-year period has examined the way that students make sense of their successes and failures in school (see Dweck 2000, for a very readable summary of this huge volume of work). As a result of their experiences, some students come to believe that ability is fixed. The reason that this is so injurious to future learning is that every time students with this belief are faced with a challenging task, their first reaction is to engage in a calculation about whether they are likely to succeed or not. If they feel confident that they will succeed, or if they feel confident that the task is so hard that many others will also fail, they will attempt the task. However, if they feel that there is a danger that they will fail while others will succeed, they will disengage in order to protect their sense of self. Put simply, they are deciding that they would rather be thought lazy than dumb. It is the same choice that most adults would make.

There are other students, who, for a variety of reasons, have come to regard ability as incremental rather than fixed. They believe that "smart" is not something you *are* but something you *get*. For these students, challenging tasks are opportunities to increase their abilities, so whether their beliefs in their chances of success are high or low, they engage with a task in order to grow. What is particularly interesting is that the same student can believe the ability in science is fixed, while seeing ability in sports or music as incremental. Most students believe that ability in, for example, triple jump, throwing the javelin, or guitar playing can be improved by practice. We need to inculcate the same beliefs about science.

In general, we need to activate students as owners of their own learning, so that they see challenge as a spur to personal growth, rather than as a threat to self-image. We need students who own their learning to the extent that they can self-manage both their emotional and their cognitive responses to challenge, so that all their energies are spent on developing capability rather than disguising its absence (see Wiliam 2007 for a summary of research in this area).

5. Activate students as learning resources for one another.

The research on collaborative learning is one of the success stories of education research. Research in many areas of education produces ambiguous or contradictory findings whereas the research on collaborative learning, most notably the work of Robert Slavin (Slavin, Hurley, and Chamberlain 2003), has shown that activating students as learning resources for one another produces some of the largest gains seen in any educational interventions, provided two conditions are met. The first is that the learning environment must provide for group goals, so that students are working *as* a group, rather than just working *in* a group. The second is individual accountability, so that each student is responsible for his or her contribution to the whole, so there can be no "passengers."

With regard to assessment, then, a crucial feature is that the assessment encourages collaboration among students while they are learning. To achieve this, the learning intentions and success criteria must be accessible to the students (see above), and the teacher must support the students as they learn how to help each other improve their work. One particularly successful format for doing this has been the idea of "two stars and a wish." The idea is that when students are commenting on each others' work, they do not give evaluative feedback, but instead have to identify two positive features of the work (two "stars") and one feature that they feel merits further attention (the "wish").

Teachers who have used this technique with students as young as five years old have been astonished to see how appropriate the comments are, and, because the feedback comes from a peer rather than someone in authority over them, the recipient of the feedback appears to be more able to accept the feedback (in other words, students focus on growth rather than preserving well-being). In fact, teachers have told us that the feedback that students give each other, while accurate, is far more hard-hitting and direct than they would themselves feel able to provide. Furthermore, the research shows that the person providing the feedback benefits just as much as the recipient, because he or she is forced to internalize the learning intentions and success criteria in the context of someone else's work, which is less emotionally charged than doing so in the context of one's own work.

The "Big Idea": Keeping Learning on Track

The "big idea" that ties these strategies together is that assessment should be used to provide information to be used by students and teachers that is then used to modify instruction in real time in order to better meet student needs. In other words, assessment is used to keep learning on track.

That this is not common practice can be seen by imagining what would happen if an airline pilot navigated the way that most teachers teach. The pilot would set a course from the starting point (say New York) to the destination (say San Francisco). The pilot would then fly on this heading for the calculated time of travel, and then, when that time had elapsed, would land the plane at the nearest airport, and upon landing ask, "Is this San Francisco?" Worse, even if the plane had actually landed in Sacramento, the pilot would require all the passengers to leave, because he had to get on to his next job.

This would be absurd, and yet, this is how many teachers teach. They teach a unit for two or three weeks, and at the end of that teaching, they assess their students. And whatever the result of that assessment, the teacher is then on to the next unit, because of the district's pacing guide. If we are to keep learning on track, assessment cannot wait until the end of the unit. Instead, like the pilot, the teacher plans a course but then takes frequent readings along the way, adjusting the course as conditions dictate.

Substantial increases in student achievement are possible if we can increase the amount of assessment for learning in classrooms, but achieving this is no easy task.

Putting It Into Practice: The Case for Teacher Learning Communities

The fact that we know what needs to be done does not, of course, mean that we know how to do it. While the work of Black et al. (2003) has shown what we can achieve, the track record of professional development in producing significant effects on a large scale is rather unimpressive. However, this should not worry us unduly because very little of the professional development that teachers have received in the past is consistent with what we know makes for effective teacher change (Wiliam and Thompson 2007).

Why this is the case is complex, and beyond the scope of this article (see Wiliam 2003 for an extended discussion). What is clear is that in general, researchers have underestimated the complexity of what it is that teachers do; in particular, researchers have failed to understand how great an impact context has on teachers' practices. That is why "What works?" is not the right question, because everything works somewhere, and nothing works everywhere. The right question is "Under what conditions will this work?" And even if we might be able to answer this question scientifically at some point in the future, we are so far away from an answer now that we have to rely on the professional judgment of teachers.

This is why we cannot tell teachers what to do. This conclusion does not stem from a desire to be nice to teachers. Indeed, if I could identify a way of telling teachers what to do that would raise student achievement, I would have no hesitation in mandating it if I had the power to do so. Schools exist for the students, not to provide employment opportunities for teachers. The reason that we cannot tell teachers what to do is that we cannot provide them with reliable guides to action. The situations that they face in their classrooms are just too varied for us to predict. That is why professional judgment is important; we have to develop the ability of teachers to react appropriately to situations for which they have not been specifically prepared.

The specific changes that I am arguing for here appear to be quite difficult for teachers to implement because they involve changing habits. A teacher with 20 years experience may well have asked something like half a million questions in her career. And when you've done something the same way half a million times, it's quite difficult to start doing it in any other way. But there is a deeper reason why change is difficult, even for inexperienced teachers. Teachers learn most of what they know about teaching before their 18th birthday. In the same way that most of us learn what we know about parenting through being parented, teachers internalize the "scripts" of school as students. Even the best four-year teacher education programs will find it hard to overcome the models of practice that student teachers will have learned in the 13 or 14 years that they spent in school.

This is why, if we are to have any chance of really changing teacher practice, we have to take seriously how hard this is going to be. We are asking teachers to change the routines and the practices that get them through the day; in the transition, they will get worse before they get better. Indeed, many of the teachers we have worked with described making these changes as "scary." They saw involving the students more in their learning as requiring them to "give up control" of their classrooms. A year later, the same teachers described the process as one of sharing responsibility for learning with the learners—the same process, but viewed in a radically different perspective.

The process that I believe provides the best mechanism for supporting teachers in making these changes is through the use of teacher learning communities (TLCs). I say this not out of some ideological commitment to the benefits of teachers talking to each other, but because of the nature of the changes we are seeking to produce. If we were trying to increase teacher subject knowledge, then TLCs would not be a very sensible approach—it would be far better to arrange for high-quality direct instruction. But when we are trying to change deeply ingrained, routinized practices or habits, then it seems that TLCs offer the best hope, and indeed, the results we have achieved in the United States have been very encouraging.

Over the past three years, we have tried a number of different approaches to establishing and sustaining TLCs, and as a result of this experimentation, it appears to us that five principles appear to be particularly important: gradualism, flexibility, choice, accountability, and support.

Gradualism

Asking teachers to make wholesale changes in their practices is a little like asking a golfer to change her swing during a tournament. Teachers have to maintain the fluency of their classroom routines, while at the same time disrupting them. Teachers should develop an action plan that specifies a small number of changes—ideally two or three—that they will make in their teaching. As teachers establish new techniques in their practices, they can take on additional ones. For administrators, there will be a temptation to push teachers to change faster than they might otherwise do, but the result will be only a shallow adoption of the new practices as long as the teachers are being monitored. As soon as the supervision is relaxed, the teachers will revert to their earlier practices, and nothing will have been achieved. Even districts in a hurry will have to hasten slowly.

Flexibility

Teachers will need to modify techniques to make them work in their classrooms; in the process of adapting techniques, teachers often refine and improve them. One high school mathematics teacher heard about the "traffic light" technique in which at the end of a piece of work, students indicate their confidence in their understanding of a piece of work with a green, yellow, or red circle, representing complete, partial, or little understanding. She decided that she would not wait until the end of the lesson to engage students in this self-assessment and gave each student a disk, green on one side and red on the other. At the start of each lesson students place the disk on their desk with the green side showing. A student can indicate confusion at any time by turning the disk over to show red. The teacher found that students who had never asked a question all year in class were prepared to signal their confusion in this way.

Another teacher tried this approach, but found it difficult to see the disks from the front of the class so she provided each student with three paper cups—one green, one yellow, and one red—nested inside each other on the students' desks. Students used these cups to indicate whether they were following the teacher's explanation (green), wanted the teacher to slow down (yellow), or wanted her to stop in order to ask a question (red). The teacher made students accountable for signaling correctly by establishing a rule that whenever one student showed red, a student who was showing green or yellow would be chosen at random to come to the front of the class to answer the question posed by the student showing red. In this classroom there is nowhere to hide!

Choice

As noted above, teachers often describe the process of changing their practices as "scary," but when they are responsible for choosing what they will change about their practices they feel empowered, especially when they can choose from a range of techniques that appeal to them. This choice lies, however, within a framework of accountability. While teachers are free to choose what they change, they are accountable for changing something.

Accountability

Most professionals involved in teacher development will have had the experience of generating considerable enthusiasm for, and commitment to, change during a workshop, only to find that all the good intentions seem to be erased once the teachers return to the classroom. Teachers should be held accountable for making changes by colleagues at monthly meetings of their teacher learning community. Each teacher describes what he or she tried and how it went. Teachers repeatedly tell us that having to face their colleagues helps them move their "change" task to the top of their in-box.

Support

Along with ideas for what to change and the support of a teacher learning community, two elements are highly desirable, if not essential, for teacher learning. The first is training for those who will lead the learning communities. The person leading the learning community must be clear about his or her role. The role of the leader is not to create teacher change but to engineer situations in which the teacher change can take place. Those in supervisory roles often find this more difficult than do teachers, because those teaching every day understand how difficult it is to change practice.

The second element is peer observation. Collaborative planning in the monthly TLC meetings can help teachers focus on what they want to develop in their practices, but teachers need support in carrying out these resolutions. To distinguish these observations clearly from those routinely carried out to manage performance, these observations should be done by peers rather than supervisors. The teacher being observed must set the agenda for the observation and spell out for the observer what should count as evidence. By defining the observer's role, both in terms of what is to be looked for and what counts as evidence, the observer's own prejudices are minimized, and the difference between this and supervisory observation is emphasized. (For further details on setting up and sustaining learning communities for formative assessment see Wiliam [2007/2008].)

Integrated Assessment

The available research evidence, as well as our experience of working with many school districts in the United States, suggests that the use of teacher learning communities, focused on the use of minute-to-minute and day-to-day assessment to adjust instruction to meet student needs, represents the most powerful single approach to improving student achievement. However, if we are to maximize the impact on student learning, other parts of the system need to be "in sync." In addition to the minute-by-minute and day-by-day assessments that allow teachers to keep learning on track, teachers also need a range of more formal assessment tasks and activities that support valid and reliable conclusions about the extent of student learning. Our experience is that teacher-made assessments often focus on shallow aspects of learning, rather than the "big ideas." Developing high-quality assessments that involve students and motivate them to improve takes time, but there are good examples of how to go about this (Stiggins, Arter, and Chappius 2004). On a longer timescale, quarterly assessments that are paced to the curriculum can provide school leadership teams with valuable information about the progress—or lack of it—that is being made by students, and annual diagnostic analyses of high-stakes tests can provide important insights into the alignment between the teaching and the national curriculum. None of these different kinds of assessment is in conflict with any of the others. Each represents an important part of a complex machine, providing information at the right level of specificity for the decision that needs to be made. Together they form a balanced assessment system that can produce unprecedented increases in science achievement, benefiting both the individual and society as a whole.

References
Black, P., C. Harrison, C. Lee, B. Marshall, and D. Wiliam. 2003. *Assessment for learning: Putting it into practice*. Buckingham, UK: Open University Press.

Clymer, J. B., and D. Wiliam. 2006/2007. What's wrong with the way we grade science? *Educational Leadership* 64(4): 36–42.

Deci, E. L., N. H. Speigel, R. M. Ryan, R. Koestner, and M. Kauffman. 1982. The effects of performance standards on teaching styles: The behavior of controlling teachers. *Journal of Educational Psychology* 74: 852–859.

Dweck, C. S. 2000. *Self-theories: Their role in motivation, personality and development.* Philadelphia: Psychology Press.

Hayes, V. P. 2003. *Using pupil self-evaluation within the formative assessment paradigm as a pedagogical tool.* EdD thesis, University of London.

Hill, H. C., B. Rowan, and D. L. Ball. 2005. Effects of teachers' mathematical knowledge for teaching on student achievement. *American Educational Research Journal* 42(2): 317–406.

Hoff, D. J. 2007. Economists tout value of reducing dropouts. *Education Week* 26(Feb. 14): 5, 15.

Jenkins, A., R. Levacic, and A. Vignoles. 2006. *Estimating the relationship between school resources and pupil attainment at GCSE* (Vol. RR727). London, UK: Department for Education and Skills.

Jepsen, C., and S. G. Rivkin. 2002. *What is the tradeoff between smaller classes and teacher quality?* Cambridge, MA: National Bureau of Economic Research.

Kluger, A. N., and A. DeNisi. 1996. The effects of feedback interventions on performance: A historical review, a meta-analysis, and a preliminary feedback intervention theory. *Psychological Bulletin* 119(2): 254–284.

Lleras-Muney, A. 2005. The relationship between education and adult mortality in the United States. *Review of Economic Studies* 72(1): 189–221.

Rodriguez, M. C. 2004. The role of classroom assessment in student performance on TIMSS. *Applied Measurement in Education* 17(1): 1–24.

Slavin, R. E., E. A. Hurley, and A. M. Chamberlain. 2003. Cooperative learning and achievement. In W. M. Reynolds and G. J. Miller (Eds.), *Handbook of psychology, volume 7: Educational psychology* (pp. 177–198). Hoboken, NJ: Wiley.

Stiggins, R. J., J. A. Arter, and S. Chappius. 2004. *Classroom assessment for student learning: Doing it right—using it well.* Portland, OR: Assessment Training Institute.

White, M. A. 1971. The view from the student's desk. In M. L. Silberman (Ed.), *The experience of schooling* (pp. 337–345). New York: Rinehart and Winston.

White, B. Y., and J. R. Frederiksen. 1998. Inquiry, modeling, and metacognition: Making science accessible to all students. *Cognition and Instruction* 16(1): 3–118.

Wiliam, D. 2003. The impact of educational research on mathematics education. In A. Bishop, M. A. Clements, C. Keitel, J. Kilpatrick, and F. K. S. Leung (Eds.), *Second international handbook of mathematics education* (pp. 469–488). Dordrecht, Netherlands: Kluwer Academic Publishers.

Wiliam, D. 2007. Keeping learning on track: Formative assessment and the regulation of learning. In F. K. Lester Jr. (Ed.), *Second handbook of mathematics teaching and learning* (pp. 1053–1098). Greenwich, CT: Information Age Publishing.

Wiliam, D. 2007/2008. Changing classroom practice. *Educational Leadership* 65(4): 36–42.

Wiliam, D., and M. Thompson. 2007. Integrating assessment with instruction: What will it take to make it work? In C. A. Dwyer (Ed.), *The future of assessment: Shaping teaching and learning*. Mahwah, NJ: Lawrence Erlbaum.

Wiliam, D., C. Lee, C. Harrison, and P. J. Black. 2004. Teachers developing assessment for learning: Impact on student achievement. *Assessment in Education: Principles Policy and Practice* 11(1): 49–65.

Wilson, M., and K. Draney. 2004. Some links between large-scale and classroom assessments: The case of the BEAR assessment system. In M. Wilson (Ed.), *Towards coherence between classroom assessment and accountability: 103rd Yearbook of the National Society for the Study of Education* (Part II, pp. 132–154). Chicago, IL: University of Chicago Press.

Further Reading

Black, P. J., and D. Wiliam. 1998. Inside the black box: Raising standards through classroom assessment. *Phi Delta Kappan* 80(2): 139–148.

Black, P., and C. Harrison. 2002. *Science inside the black box: Assessment for learning in the science classroom*. London, UK: NFER-Nelson.

Black, P., C. Harrison, C. Lee, B. Marshall, and D. Wiliam. 2004. Working inside the black box: Assessment for learning in the classroom. *Phi Delta Kappan* 86(1): 8–21.

Clymer, J. B., and D. Wiliam. 2006. Improving the way we grade science. *Educational Leadership* 64(4): 36–42.

Leahy, S., C. Lyon, M. Thompson, and D. Wiliam. 2005. Classroom assessment: Minute-by-minute and day-by-day. *Educational Leadership* 63(3): 18–24.

Wiliam, D. 2006. Assessment: Learning communities can use it to engineer a bridge connecting teaching and learning. *Journal of Staff Development* 27(1): 16–20.

Wiliam, D. 2006. Assessment for learning: Why, what and how. *Orbit: OISE/UT's magazine for schools* 36(3): 2–6.

## CHAPTER 2

On the Role and Impact of Formative Assessment on Science Inquiry Teaching and Learning

Richard J. Shavelson, Yue Yin, Erin M. Furtak, Maria Araceli Ruiz-Primo, Carlos C. Ayala
Stanford Educational Assessment Laboratory

Donald B. Young, Miki K. Tomita, Paul R. Brandon, Francis M. Pottenger III
Curriculum Research & Development Group

S cience education researchers, like science teachers, are committed to finding ways to help students learn science. Like teachers, we researchers start with an informed hunch about something that we think will improve teaching. Then we work with teachers and try out our hunch in real classrooms. If we get positive results, we share them with a wide range of educators. Sometimes we find out that our hunch does not work, and we try to figure out what went wrong so that we can improve it the next time. In other cases, we find that while the idea may have been good, the technique will not work in practice. In those cases, we continue our search for other ways to help improve students' learning of science.

In reviewing the literature on assessment, Paul Black and Dylan Wiliam found strong evidence that embedding assessments in science curricula would lead to improved student learning and motivation (Black and Wiliam 1998; see also Wiliam, Chapter 1 in this book). Based on this finding, our team of teachers, curriculum and assessment developers, and science education researchers developed a series of *formative assessments* to embed in a middle school physical-science unit on sinking and floating. We wanted to see if this kind of assessment, which helps teachers to determine the status of students' learning while a unit is still in progress, would improve sixth- and seventh-grade students' knowledge and motivation to learn science. If it worked, we knew we might have a large-scale impact on teaching and learning.

In this chapter, we begin by describing what we mean by formative assessment and outline the potential and challenges of trying to implement and study this promising technique for scientific inquiry teaching. We then describe our study on formative assessment in middle schools, including some mistakes and wrong turns, and what we found when we tested our ideas experimentally. We conclude with future challenges in improving science education with formative assessment.

What Is Formative Assessment?

Formative assessment is a process by which teachers gather information about what students know and can do, interpret and compare this information with their goals for what they would like their students to know and be able to do, and take action to close the gap by giving students suggestions as to how to improve their performance. In this way, formative assessment is carried out for the purpose of improving teaching and learning while instruction is still in progress.

To clarify what we mean by *formative assessment*, consider the large-scale, high-stakes assessments that are carried out in all U.S. schools today. These types of assessments are summative in nature; that is, they provide a summary judgment about, for example, students' learning over some period of time. The goal of summative assessment is to inform external audiences primarily for evaluation, certification, and accountability purposes. Since the federal No Child Left Behind legislation was passed in 2001, summative assessment has certainly received a great deal of publicity in the popular media and has, to a certain degree, swamped the important formative function of assessment.

By focusing on formative assessment, we hope to put assessment back into its rightful place as an integral part of the teaching-learning process. Formative assessment takes place on a continuous basis, is conducted by the teacher, and is intended to inform the teacher and students, rather than an external audience (Shavelson 2006). We view classroom formative assessment as a continuum ranging from informal formative assessment to formal formative assessment. The position of a particular formative assessment technique on the continuum depends on the amount of planning involved, the formality of technique used, and the nature of the feedback given to students by the teacher. We focus on three important formative assessment techniques—(1) "on-the-fly," (2) planned-for-interaction, and (3) embedded in the curriculum (Figure 2.1) and describe each in turn.

Figure 2.1 Variation in Formative Assessment Practices

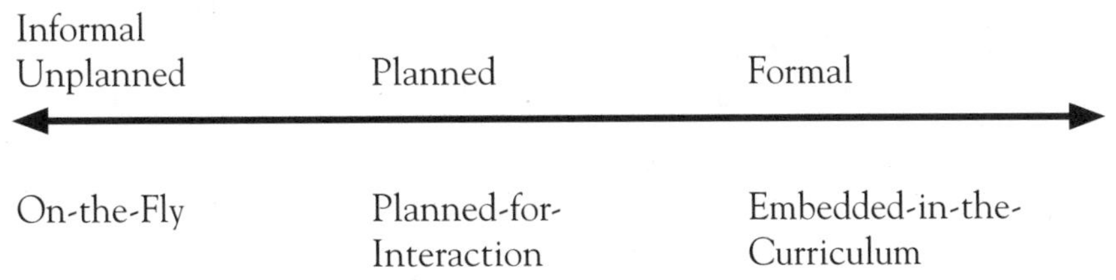

Informal Unplanned	Planned	Formal
On-the-Fly	Planned-for-Interaction	Embedded-in-the-Curriculum

On-the-Fly Formative Assessment. On-the-fly formative assessment occurs when "teachable moments" unexpectedly arise in the classroom. For example, teachers circulate between groups to listen in on conversations and make suggestions that give students new ideas to think about. A teacher might overhear a student in a small group investigating sinking and floating say that, as a consequence of an experiment just completed, "Density is a property of the plastic block. It doesn't matter what the mass or volume is, the density stays the same for that kind of plastic." The teacher recognizes that the student has a grasp of what density means for that block, and presents the student with other materials to see if she and her group-mates can generalize the density idea to a new situation. In this way, the teacher challenges the student to test her new idea by having her and her group measure the mass/volume relationships of a new material. Moreover, when satisfied that the students are onto something, the teacher calls for other students to hear what this group found out.

This vision of taking advantage of the "teachable moment" sounds a lot like good teaching, not necessarily *assessment*. This is exactly our point: Teaching and assessment are and should be considered as one and the same.

Rather than teachers planning assessment as a separate event during the class period, on-the-fly assessment is seamless with instruction and is based on the teacher capitalizing on opportunities as they arise to help students to move forward in reaching learning goals.

However, as we learned from our research, such on-the-fly formative assessment and action ("feedback") may be natural for some teachers but difficult for others. Identification of these moments is initially intuitive and then later based on cumulative wisdom of practice. Moreover, even if teachers can identify the moment, they may not have the confidence, techniques, or content knowledge to sufficiently challenge and respond to students.

Planned-for-Interaction Formative Assessment. In contrast, planned-for-interaction formative assessment is deliberate. Teachers plan for and craft ways to get information about the gap between what students know and need to know, rather than use questions just to "keep the show going" during an investigation or whole-class discussion. Consider, for example, teacher questioning—a ubiquitous classroom event. While developing a lesson plan, a teacher can prepare a set of "central questions" that get at the heart of the learning goals for that day's lesson and that have the potential to elicit a wide range of student ideas. For example, these questions may be general ("Why do things sink and float?") or more specific ("What is the relationship between mass and volume in floating objects?" "Can you give me an example of something really heavy that floats? Why do you think it floats?"). At the right moment during class, the teacher poses these questions to the class, and through a discussion the teacher learns what students know and allows different ideas to be presented and discussed. In this example, the teacher planned the assessment prompt in advance rather than waiting for unexpected opportunities to arise. Although not every student in class may respond to each question, the information gained from the students' responses allows the teacher to act on the information collected by fine-tuning instruction or intervening with individual students.

Embedded-in-the-Curriculum Formative Assessment. Alternatively, teachers or curriculum developers may embed more formal assessments ahead of time in the ongoing curriculum to intentionally create "teachable moments." These assessments are embedded after junctures or joints in a unit where an important goal should have been reached before going on to the next lesson. Embedded assessments inform the teacher about what students currently know and what they still need to learn (i.e., "the gap") so that teachers can provide timely feedback to students.

In their simplest forms, formal formative assessments are designed to provide information on important goals that students should have reached at critical joints in a unit before going onto the next lesson. In their advanced forms, formative assessments are based on a developmental progression of the ideas students have about a particular topic (such as why things sink and float). In contrast to the other two types of formative assessment, embedded assessments are more sophisticated because they are designed to collect critical information about student learning at the same time. The main difference between planned-for and embedded formative assessment is in the designer. Whereas planned-for assessment is usually done by the teacher as a part of the lesson-planning process, embedded assessments are usually designed by curriculum and assessment developers working with experienced teachers.

Embedded formative assessments are valuable teaching tools for at least four reasons. First, they are consistent with curriculum developers' understanding of the curriculum and are therefore consistent with instructional goals. Second, assessment developers contribute technical expertise that increases the quality of the assessments. Third, the involvement of experienced teachers in developing embedded assessments means that they are practical and based on the wisdom of practice. And fourth, embedded assessments provide thoughtful, curriculum-aligned, and valid ways of determining what students know, rather than leave the burden of planning on the teacher.

Formal embedded assessments come "ready-to-use" as part of a preexisting curriculum, and instructional decisions made from them may improve students' learning. Therefore, in our study, we sought to learn whether embedded formative assessments actually helped teachers close the learning gaps in their classrooms.

Potential and Challenges

Formative assessment is a potentially powerful teaching idea embodying knowledge and skills for creating and capitalizing on teachable moments. In the context of science education, formative assessment links teaching and learning in the service of building students' understanding of the natural world and of how the methods of science justify knowledge claims. In using formative assessments, we sought to move students from naive conceptions of the natural world to scientifically justifiable conceptions ("conceptual change"). To change their conceptions, students need to link what they find out through inquiry investigations to their current conceptions of the natural world and to change those conceptions when their evidence does not fit their "theory." Formative assessment's critical characteristic, then, lies in identifying learning gaps and providing immediate feedback to students that helps them close gaps.

This said, many teachers are in some ways skeptical about incorporating formative assessment substantively into their teaching practice, even when they know that it is important. Teachers have many questions about their role in formative assessment, and for good reason. For example, formative assessment creates a conflict with the teacher's traditional grade-giving role in summative assessment. How can the teacher on the one hand ask students to lay bare their understanding of a concept and at the same time have the responsibility for giving the student a grade? In other cases, teachers may have only experienced *summative* assessment when they were students themselves, or in their teacher education programs. Consequently, they may not have personal experience with the ways that *formative* assessment can improve the quality of teaching and learning. Other questions arise as well. Should teachers really change their beliefs about their role as assessors? Why should teachers change their practices to accommodate a yet unproven teaching technique? Will our emphasis on formative assessment eventually fade away as have other reform techniques?

Clearly, teachers' skepticism is appropriate; part of the science education researcher's role is to test out new (or not so new) techniques to see if they stand up to scientific scrutiny. To this end, our team designed and conducted a study that put formative embedded assessment to the test.

Embedding Formative Assessment in a Science Curriculum

Our study of formative embedded assessment addressed two central research purposes: first, to learn how to build and embed formative assessments in science curricula and, second, to examine the impact of formative assessments on students' learning, motivation, and conceptual change.

Building and Embedding Formative Assessments in Science Curricula
As noted above, we sought to move students from naive conceptions of
the natural world to scientifically justifiable ones. To this end, we wanted
students to link what they were finding out through investigations to their
conceptions about the natural world. The intent was for students to change
those conceptions when their evidence didn't fit their "theory."

We embedded formative assessments in the Foundational Approaches
in Science Teaching (FAST) curriculum unit on the properties of matter—
more specifically, buoyancy (Pottenger and Young 1992). As a first step, we
identified the *goals* for the unit. The main goal was for students to develop,
through a series of inquiry investigations, a relative density-based explanation for sinking and floating (or, as we came to call it during the study,
"Why things sink and float" or "WTSF"). We then worked from the goals
backward to the beginning of the unit, identifying key junctures between
lessons ("investigations") where important goals needed to be met. We then
inserted assessments to provide information about student performance.

Despite our well-conceived plans, in the end, a seemingly straightforward
process of developing formative assessments was anything but straightforward. We made some wrong turns and learned from our mistakes.

Pilot Study: From Embedded Formative Assessments to
Reflective Lessons
Our basic idea was to develop and embed formative assessments where the
"rubber hit the road"—that is, at critical curricular joints where students'
conceptual understanding was expected to develop from a simple level to
a more sophisticated one. In this way, teachers would know whether students were advancing in their knowledge as the curriculum progressed. We
expected that assessments embedded at the critical joints would provide
timely information to (a) help teachers and students locate the levels of
students' understanding, (b) determine whether students had reached the
desired level, (c) diagnose what students still needed to improve, and (d)
help students move to the next level.

At each critical joint, we created a set of assessments designed to tap
different kinds of knowledge that students should construct in learning
about sinking and floating. There were facts (e.g., density is mass per unit
volume—*declarative knowledge*) and procedures (e.g., using a balance scale
to measure the mass of an object—*procedural knowledge*). But most impor-

tant, and often implicit in curricula, was the use of this declarative and procedural knowledge in inquiry science to build a model or mini-theory of *why* things sink and float (e.g., a model of relative densities—*schematic knowledge*). Consequently, we embedded assessments of these types of knowledge at four natural joints in a 10-week unit on buoyancy. The assessments served to focus teaching on different aspects of learning about mass, volume, density, and relative density. Feedback on performance focused on problematic areas revealed by the assessments.

In order to embed assessments that were based on research and that could identify in a valid and reliable way what students know, we created four extensive assessment "suites" (combinations of individual assessments—graphing, short answer, POE [predict-observe-explain], and PO [predict and observe]). These assessments covered the declarative, procedural, and schematic knowledge underlying buoyancy. Each suite included multiple-choice (with space for students to justify their selections) and short-answer questions that tapped all three types of knowledge. We also included a substantial combination of concept maps (structure of declarative knowledge), performance assessments (procedural and schematic knowledge), predict-observe-explain assessments based on lab demonstrations (schematic knowledge), and/or "passports" verifying hands-on procedural skills (e.g., measuring an object's mass).

Three brave teachers volunteered to try out this extensive battery of embedded assessments in a pilot study. After the completion of the pilot study, the teachers warned us that the original formative assessments were too time-consuming and the amount of information obtained from them was overwhelming. Our lead pilot-study teacher, who was also a member of our assessment team, gently pointed out the problems that pilot-study teachers faced using our assessment suites. She suggested that perhaps there could be only a few assessments that directly led to a single, coherent goal, such as knowing *why things sink and float*. She pointed out that FAST provided ample opportunity for teachers to observe and provide feedback to students on their declarative and procedural knowledge. She urged us to focus on schematic knowledge and on students' developing an accurate mental model of why things sink and float in the assessment suite.

Moreover, Lucks (2003) viewed and analyzed videotapes of the pilot study teachers using the assessment suites. She found that our teachers were treating the "embedded assessments" more as external tests that were something apart from the curriculum—in other words, as *summative* assessment—rather than using the formative assessments as a way to find out what the students were learning. Thus, the teachers treated the new assessments like any other test that they were required to give to the students during the year, rather than as opportunities to increase their students' learning.

Based on the thoughtful feedback we received from the teachers and the researcher, we revised our initial embedded assessments, greatly reducing their numbers and focusing in on the overarching goal of explaining "why things sink and float." Afterward, when talking with teachers, we no longer spoke of embedded assessments, which we thought would trigger their stereotypes about assessments. Instead, we started calling them "Reflective Lessons" to emphasize their function as a component of the teaching and learning process.

The New Generation of Formative Embedded Assessments: The Reflective Lessons

A second look at the FAST unit and the information collected during the pilot study led us to a developmental progression of student ideas, which then became the basis for redesigning the original embedded assessment suites into Reflective Lessons (Figure 2.2, p. 30). This progression was aligned to the unit and based on different conceptions students have as they develop an understanding of sinking and floating. These conceptions develop from naive (e.g., "things with holes in them will sink") to scientifically justifiable conceptions (e.g., "sinking and floating depend on the relative densities of the object and the medium supporting the object").

Although Figure 2.2 may appear quite complicated, the ideas behind it are straightforward and consistent with students' different ideas about sinking and floating. Before instruction, students have all different kinds of ideas about sinking and floating, such as that heavy things sink, flat things float, things with air in them float. We would place these ideas at Level 1 or "Naive Conceptions." As students progress through the unit, they complete investigations that apply either mass or volume to sinking and floating; that is, a single uni-dimensional factor (Level 2), holding all else constant. Next, students simultaneously apply mass and volume, or multiple uni-dimensional factors, to explain sinking and floating (Level 3). Afterward, students integrate mass and volume into density, a single bi-dimensional factor, in their explanations (Level 4). Finally, students consider

Figure 2.2 Conceptual Development for Understanding Why Things Sink and Float

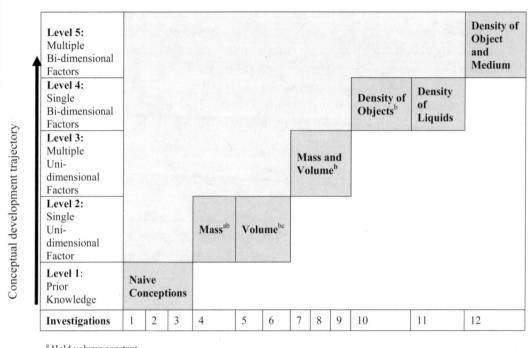

	Conceptual development trajectory										
Level 5: Multiple Bi-dimensional Factors											**Density of Object and Medium**
Level 4: Single Bi-dimensional Factors							**Density of Objects**[b]	**Density of Liquids**		
Level 3: Multiple Uni-dimensional Factors						**Mass and Volume**[b]				
Level 2: Single Uni-dimensional Factor			**Mass**[ab]	**Volume**[bc]						
Level 1: Prior Knowledge	**Naive Conceptions**									
Investigations	1	2	3	4	5	6	7	8	9	10	11	12

[a] Hold volume constant
[b] Hold liquid (water) constant
[c] Hold mass constant

the object's density and the liquid's density, or multiple bi-dimensional factors (Level 5), in their explanations (Yin 2005).

The final Reflective Lesson suites are shown at their critical junctures in Figure 2.3. Two types of Reflective Lessons were embedded in the unit. Each of the type one Reflective Lessons included a sequence of the following activities: (a) graphing and interpreting evidence and drawing conclusions about WTSF ("Why things sink or float"), (b) applying knowledge learned to predict and explain what would happen in a new situation (Predict, Observe, <u>E</u>xplain), (c) writing a brief explanation about why things sink and float, and (d) predicting and observing a surprise phenomenon to introduce the next set of lessons. The second type of Reflective Lesson was concept mapping, which encouraged students to make connections between the concepts they learned.

Figure 2.3 Reflective Lessons and Junctures at Which They Were Embedded in the Unit

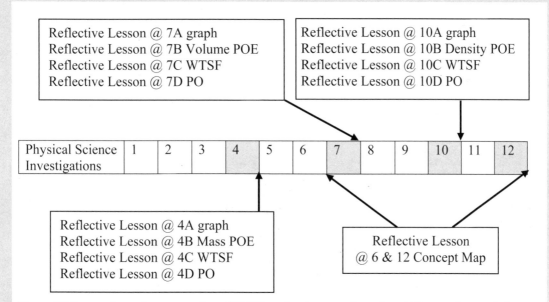

Reflective Lesson @ 7A graph
Reflective Lesson @ 7B Volume POE
Reflective Lesson @ 7C WTSF
Reflective Lesson @ 7D PO

Reflective Lesson @ 10A graph
Reflective Lesson @ 10B Density POE
Reflective Lesson @ 10C WTSF
Reflective Lesson @ 10D PO

Physical Science Investigations | 1 | 2 | 3 | 4 | 5 | 6 | 7 | 8 | 9 | 10 | 11 | 12

Reflective Lesson @ 4A graph
Reflective Lesson @ 4B Mass POE
Reflective Lesson @ 4C WTSF
Reflective Lesson @ 4D PO

Reflective Lesson
@ 6 & 12 Concept Map

Notes: POE = predict, observe, explain; WTSF = why things sink or float; PO = predict and observe

The Reflective Lessons were designed to enable teachers to (a) elicit students' conceptions, (b) encourage communication of ideas, (c) encourage argumentation (comparing, contrasting, and discussing students' conceptions), and (d) reflect with students about their conceptions. In this way, teachers could guide students along a developmental trajectory that they had in hand from naive conceptions of sinking and floating to more scientifically justifiable ones (Figure 2.2).

The Experimental Study

To test whether the final Reflective Lessons could help students improve learning, motivation, and conceptual change, we conducted a small experiment. We randomly assigned 12 teachers to teach either the regular inquiry curriculum (control group—6 teachers) or the curriculum with the Reflective Lessons included (experimental group—6 teachers). Teachers in the experimental group attended a training workshop with the researchers, curriculum developers, and one of the pilot teachers. During the training, teachers participated in the Reflective Lessons as students, talked about the process of the lesson, and then practiced teaching the Reflective Lessons themselves with lab school students. Teachers in the control group also attended a training workshop that oriented them to the study and invited them to share their assessment practices, among other things.

In the study, we gave pretests and posttests to the students in both groups. We examined the effect of the Reflective Lessons by comparing improvement made by the two groups, regarding students' motivation, achievement, and conceptions of sinking and floating (Figure 2.4) (Yin 2005).

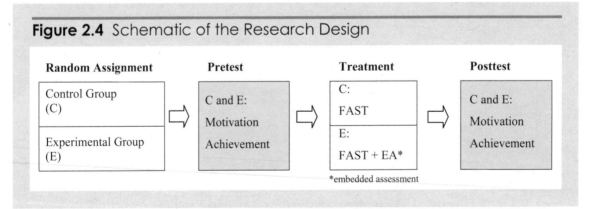

Figure 2.4 Schematic of the Research Design

Since the Reflective Lessons integrated formative assessment ideas, curriculum goals, and teachers' input, we expected that students in the experimental group would benefit from the Reflective Lessons and show higher learning gains than the control group. To our surprise, our findings did not support this conjecture. We found no statistically significant differences between average performance in the control and experimental groups. That is, students in the experimental group and control group did not differ, on average, on motivation, learning, or conceptual change. This finding persisted even after we accounted for differences among students' achievement and motivation before the study began.

Despite the fact that the study did not come out as expected, we learned a lot about how teachers actually used the Reflective Lessons in their classrooms. In each group, teachers varied substantially in producing differences in students' motivation, learning, and conceptual change. In viewing classroom videos we found that although the Reflective Lessons (embedded assessments) were implemented by teachers in the experimental group, not all the teachers used them effectively to give students feedback or modify teaching and learning (Ruiz-Primo and Furtak 2006, 2007). That is, among the teachers in the experimental group, those teachers whose students had higher learning gains relied more on the other two types of assessment techniques—on-the-fly and planned-for-interaction assessment—rather than on the Reflective Lessons.

To give an idea of the differences among teachers, let us consider two teachers in the experimental group, Gail and Ken.[1] Gail took an active role in using the Reflective Lessons with her students. She would build knowledge with students by challenging their ideas, asking them for empirical evidence to justify their ideas, and making clear how a model of sinking and floating was emerging. The Reflective Lessons created teachable moments for her, which she then took advantage of with informal assessment techniques. Ken, in contrast, relied on the Reflective Lessons themselves to help the students learn and looked at the activities as discovery learning; that is, he depended on the students to develop their own understandings with limited teacher intervention (Furtak 2006). He reasoned that it was not his role to act on the students' ideas about sinking and floating and to guide the students or tell them the answers; rather it was up to students to discover for themselves why things sink and float.

In Figure 2.5, page 34, we see the developmental trajectory for a typical student from Gail's class and another from Ken's. While Gail's student progressed along the trajectory, Ken's student held to her original explanation. The achievement test scores for the two students reflected the differences

---

[1]These names are pseudonyms. We use male and female names for writing ease (e.g., to avoid he/she, his/her). We did not find gender differences in teaching effects in our study.

Figure 2.5 Development of Understanding of Why Things Sink and Float in Two Experimental Teachers' (Gail's And Ken's) Students

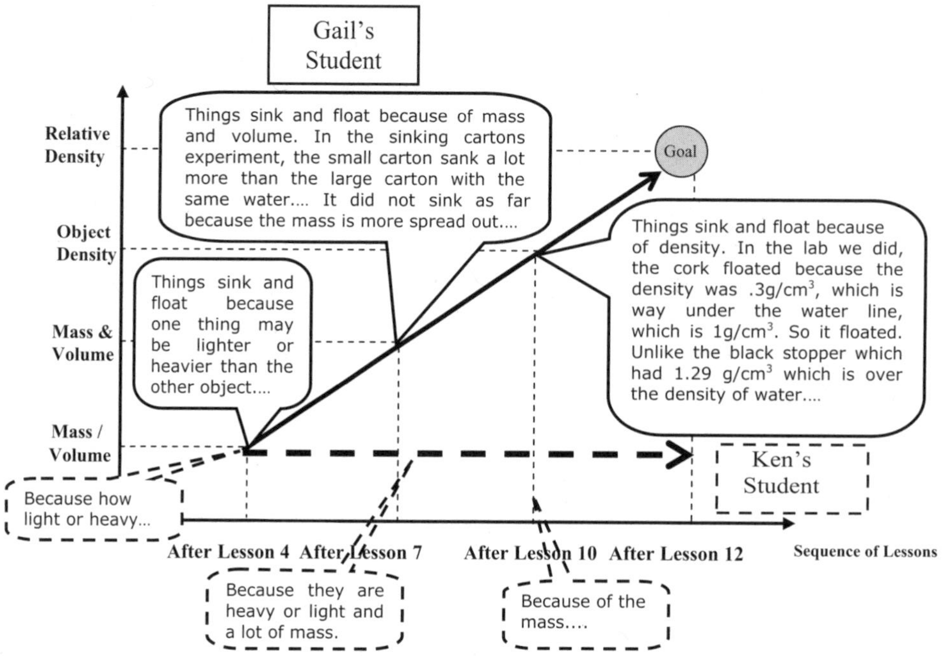

Gail's Student

Relative Density

Things sink and float because of mass and volume. In the sinking cartons experiment, the small carton sank a lot more than the large carton with the same water.... It did not sink as far because the mass is more spread out....

Goal

Object Density

Things sink and float because of density. In the lab we did, the cork floated because the density was .3g/cm^3, which is way under the water line, which is 1g/cm^3. So it floated. Unlike the black stopper which had 1.29 g/cm^3 which is over the density of water....

Mass & Volume

Things sink and float because one thing may be lighter or heavier than the other object....

Mass / Volume

Ken's Student

Because how light or heavy...

After Lesson 4 After Lesson 7   After Lesson 10 After Lesson 12   Sequence of Lessons

Because they are heavy or light and a lot of mass.

Because of the mass....

in learning (Gail's student: pretest 15 and posttest 36; Ken's student: 23 and 23, respectively) (Yin 2005).

Concluding Comments

As we know, when any new reform idea comes along, there is a lot of hype. Moreover, teachers are expected to pick up the new "tools" and implement the ideas perfectly on the first try, after they have been trained (briefly!) to do so. Even though we worked intensely with our experimental teachers to learn how to use Reflective Lessons and provided follow-up during the experiment, the kinds of knowledge, belief, and practice changes we wanted to bring about—conceptual changes—needed much more time. Those teachers who already believed in and had already incorporated some of the techniques in their practice that we sought to build in the experimental group performed largely as we had hoped. However, those teachers whose beliefs were somewhat different took even longer to acquire the habits of mind and teaching techniques required to use Reflective Lessons (formative assessment) effectively.

We continue to believe that formative assessment practices hold promise for improving science inquiry teaching, and for improving students' motivation, learning, and conceptual change. However, if we are to put formative assessment to the test fairly, we need time to work with teachers on their formative assessment knowledge, beliefs, and practices. Once a reasonable level of expertise has been reached, that is the time to try the experiment again (and again and again). If successful, we may have something that would help improve science education; if not, we know not to pursue this aspect of reform further. Perhaps not surprisingly, we are currently engaged in a replication (hopefully with appropriate improvements) of the experiment. Stay tuned!

References

Black, P. J., and D. Wiliam. 1998. Assessment and classroom learning. *Assessment in Education* 5(1): 7–73.

Furtak, E. M. 2006. *The dilemma of guidance in scientific inquiry teaching.* Doctoral diss., Stanford, CA: Stanford University.

Lucks, M. A. 2003. *Formative assessment and feedback practices in two middle school science classrooms.* Master's thesis. Stanford, CA: Stanford University.

Pottenger, F., and D. Young. 1992. *The local environment: FAST 1 Foundational Approaches in Science Teaching.* Manoa, HI: University of Hawaii, Curriculum Research & Development Group.

Ruiz-Primo, M. A., and E. M. Furtak. 2007. Exploring teachers' informal formative assessment practices and students' understanding in the context of scientific inquiry. *Journal of Research in Science Teaching* 44(1).

Ruiz-Primo, M. A., and E. M. Furtak. 2006. Informal formative assessment and scientific inquiry: Exploring teachers' practices and student learning. *Educational Assessment* 11(3/4): 237–263.

Shavelson, R. J. 2006. On the integration of formative assessment in teaching and learning: Implications for new pathways in teacher education. In F. Oser, F. Achtenhagen, and U. Renold (Eds.), *Competence-oriented teacher training: Old research demands and new pathways.* Utrecht, The Netherlands: Sense Publishers.

Yin, Y. 2005. *The influence of formative assessments on student motivation, achievement, and conceptual change*. Doctoral diss., Stanford University.

## CHAPTER 3

From Practice to Research and Back: Perspectives and Tools in Assessing for Learning

Jim Minstrell, Ruth Anderson, Pamela Kraus, and James E. Minstrell
FACET Innovations

There are certain things about really listening to kids, really planning based on what you hear—really making it so students [are] doing the thinking and the work and you're helping to facilitate … that can happen in all disciplines ….
—High school science and mathematics teacher

I was first drawn to research in my classroom by a keen desire to better understand how well my students were learning and to tailor my teaching to their learning needs. During the last 20 of my[1] 30 years as a teacher of high school science and mathematics, I regularly conducted research in my classroom—at first to try and confirm some of the research claims I read about and later to find answers to my own questions. Curiosity led me to experiment with the way I went about designing lessons and assessing what my students learned from those lessons. Once I started paying closer attention to how my students were learning (not just what they were learning), I

---

[1]Although we all collaborated on authoring this chapter, it is largely a history of how one of the authors (Jim Minstrell) developed his practice of formative assessment while working at the intersection of research and classroom practice in the teaching and learning of science.

found myself scrutinizing how I taught (not just what I taught) and making substantial changes based on my findings.

In this chapter I draw on research and teacher practice to share the story of my own professional inquiry that led me to an exciting marginal space between the worlds of classroom teachers and education researchers. In this space, I worked to build a bridge between my practice and research. It is a story that reiterates the importance of firsthand experience for adult learners—not just children—and highlights the need for greater alignment between the perspectives and practices of teachers and the research-based tools they are asked to use in their classrooms. This chapter is intended for fellow teachers and researchers who are committed to better understanding how students think and to finding focused, evidence-based ways of supporting them in their learning.

Moving Beyond "Did They Get It?"

I had been teaching for nearly a decade and was relatively comfortable in my practice. I knew the curriculum I was required to teach. My colleagues and administrators considered me competent and successful in the classroom, and I was popular with my students, who generally performed well on standardized tests. Nevertheless, at that 10-year point I felt a little restless, suspecting that there was something missing from my practice. For one thing, I noticed that while my students generally responded well to items on which I had "trained" them, they were not so successful when I "snuck up on them" and asked a deeper question postinstruction (at the end of the unit and/or end of the year). Also, there were some ideas that they never seemed to get, regardless of how many times we went over the material.

It was about this time that I first became aware of early research on students' preconceptions and misconceptions (Champagne, Klopfer, and Anderson 1980; McCloskey, Caramazza, and Green 1980; Caramazza, McCloskey, and Green 1981; Gunstone and White 1981; Posner et al. 1982). I also learned that students tended to leave their science classes exhibiting those same problematic conceptions. I became curious and decided to test these research findings in my own classroom. Sometimes I would administer a few questions from the research to see if my students had the same preconceptions. They did.

"No problem," I thought. I was confident that I could teach in a way that they would leave my classes thinking "correctly." At first, I would simply *tell* my students how they might be tempted to think and warn them about what was problematic in that thinking. Then I would move on with my usual teaching—mission accomplished, I thought. But at the end of the unit I found my students (even some of the strongest students) still exhibited those problematic conceptions on the deeper questions.

Clearly I needed to do something differently. My students needed more than my word that their thinking was problematic. I needed to redesign instruction so that I actually challenged the problematic aspects of their initial ideas and initiated or supported the learning goals. How would I know if I was successful? I knew I wouldn't be satisfied with a quiz or question that would tell me simply if they "got it" or not; I wanted to know what sense they were making of their class experiences, what they were learning, and the ways in which their ideas had changed. I also wanted to know early enough in the unit so I could adjust my next lesson activities and hopefully foster even more learning and practice with the ideas before I moved on to the next unit.

Toward an Assessment for Learning Cycle

As I explored approaches to finding out and better understanding what my students knew, I found myself posing the same questions and engaging in a cycle of activities that allowed me to really examine their ideas and do something with the information I was getting. The activities and questions were as follows:

1. **Gathering information** about my students' thinking.
   What is the learning goal or standard that we're really trying to achieve (in this unit, or even in this lesson)? What expressions would I expect to see in my students' work or responses if they really understood or could do that learning goal? What are my students thinking with respect to the learning goals? How will I know? Will it be through their ideas shared during a discussion or will I need to set a task (elicitation or probe) to collect information on their thinking?

2. **Interpreting the information** to hone in on essential issues to address.
   What problematic responses (misconceptions?) might I expect to see or hear? How might I interpret those responses? What is the range of ideas expressed by the class? What might the responses mean with respect

to the students' thinking? What are the strengths as well as the weaknesses in that thinking? What needs (cognitive or experiential) do the responses suggest?

3. **Acting with purpose** based on what I have learned.
   What specific action might I plan and implement to address their needs? What experience, next question, or feedback might I give students to build on the strengths of their thinking and to challenge problematic facets of their thinking? What action can I implement to get my students to reflect on and revise their prior ideas in light of their observations within the new experience?[2]

I would repeat the steps in the cycle (gathering, interpreting, acting) until I had evidence that students had achieved the learning goal—including knowing the reasons why they believed the new ideas and why the earlier conceptions no longer made sense. The processes and learning and assessment became so intertwined that I came to think of them as constituting an assessment/learning cycle, or an "assessment for learning cycle."

The length of the cycles I was implementing varied but generally became shorter and more frequent. For example, early on I elicited initial understandings from my students only at the start of an extensive unit; later, I did so a few times in a unit. Eventually, I was moving through the cycle formally or informally in nearly every lesson. Some lessons would include a series of "mini cycles" as my students and I refined the ideas that emerged with each new experience.

Like the variably ordered steps in the scientific method, the cycle, when fully engaged, became less like a march and more like a dance—but always moving forward to achieve a deep understanding of the learning goal. And although the steps in the cycle may seem simple and memorable, the details and subtasks for each step add complexity.

---

[2]Suppose I wanted students to understand Newton's third law about action and reaction forces. An extended example in the Appendix (p. 58) shows how teachers can address Forces as Interactions. Topics covered include relevant learning goals, identifying misconceptions, interpretive information, activities and questions to stimulate thinking about the critical ideas, and assessment questions with feedback to students and to teachers that clarifies learning targets and offers lessons to promote further reflection.

Gathering Information
- *Determine the learning goal.*
  This might initially be thought of as something like the statements in the National Science Education Standards (NRC 1996) or Benchmarks for Science Literacy (AAAS 1993) or the statements that exist in many of the state standards documents. But, those statements are typically very general and need to be unpacked into component pieces. What is the specific learning target?
- *Choose and implement an appropriate information-gathering strategy.*
  I needed to choose or determine a learning situation or task from which I could get information on student thinking. Ideally the task would allow me to learn about any problematic thinking as well as useful thinking. I might be able to get the information from the students' responses to a homework question or even their utterances during a class, small-group, or individual discussion. A better task will give me their rationale for their answer as well as the answer. Typically it is when students have to explain a situation or actually have to complete a procedural task that I can more easily get access to their thinking. Finally, I had to implement the task in a way that would give me the information I needed. For example, if students are not clear about the task, I may need to clarify it so that I get their thinking relevant to the learning goal. Are they answering the right question?
- *Anticipate student responses.*
  This was a surprisingly important step, allowing me to check the appropriateness of the strategy I had chosen (e.g., What responses do I expect to get? Will this activity really give me responses that represent a range of student thinking?) and to move my own thinking ahead regarding the needs I might have to address next.

Interpreting the Information
- *Identify problems and strengths in student thinking.*
  Ideally, I am prepared to sort responses (together with their rationale) into groups of like responses. Then I can attempt to interpret the meaning behind each of the groups. I assume students are trying to make sense of their experiences. I think about how that response might seem reasonable and I identify the strengths as well as the problematic aspects of that idea or approach.

- *Determine needs to move learning forward.*
  As a transition to the next step in the cycle, I try to determine the need of a student or students who seem to hold a particular kind of idea. This will help me to streamline my instruction rather than simply re-teaching and hoping they "get it" the second time around. In one case, students may simply need to make (or repeat) the correct observation and remember it in the process of trying to make sense of the situation. In another case, students may need to clarify and know the relationship between two variables. In yet another case, students may need to see multiple examples of similar phenomena from which to infer a pattern or see that their thinking will lead to a prediction that is not consistent with what is observed.

Acting With Purpose
- *Plan based on findings.*
  I needed to plan and implement a specific action that would address the students' thinking—the thinking I had just uncovered. The purpose of the action was to reinforce useful aspects of the students' thinking and challenge problematic thinking. The activity I chose couldn't simply be topically aligned; it had to give the students an opportunity to test (challenge) their initial ideas.
- *Target needs—not just topics.*
  Sometimes I simply created a specific situation in which students could observe what does happen. In many cases the activity initiated an inquiry on the part of the students who were then motivated to test their ideas. Occasionally I chose or designed specific feedback for a student, suggesting specifically what was problematic about her think-ing and reminding her of a previous experience that was inconsistent with her thinking. Other times I might come up with a specific real-world situation from which they could see that their collectively held idea didn't work, and then ask the students to make new sense of this specific related situation.

Mastering this "dance" did not happen overnight. As you might ex-pect, such a cycle depends on a teacher's content knowledge and question-ing skills. For me, it required creating a classroom culture that was "safe" enough for my students to feel comfortable expressing and respectfully critiquing ideas. They also had to learn to live with a certain amount of ambiguity rather than expecting to be told they were "right" or "wrong."

Initially, there was a fair amount of resistance when I asked my students to give their best answers before having them explore. They were not used to having to give their answers and rationale *before* exploring with materials. Some of my strongest students were particularly resistant because their success in science had been based largely on their ability to memorize information, internalize it, and give it back on the test. Asking them to chance being "wrong" up front made them uncomfortable.

On the other hand, those students who had been turned off by having to memorize material that didn't make sense to their thinking were much more open to this process. I found them actively engaged as we acknowledged their ideas and past experiences and used new experiences to challenge problematic aspects of those ideas. Eventually, however, most everyone came to appreciate the power of the process. As I passed out the elicitation activity at the start of a unit, I would encounter more than a few groans, but once when I tried to apologize ("Sorry, but you know it's this research project I'm doing ...") the students wouldn't let me. "I hate these things," one boy admitted, "because it seems like I'm always wrong. But they really help me see what I need to know." Many students seemed to agree and no one suggested we discontinue them.

It took time to develop both my skills and the classroom culture, but the payoff with my students was significant. I saw higher quality engagement in activities and gains in their learning that were substantially better than I had achieved through traditional instruction. Typically, posttest performance improved by 10% to 80% over what had been achieved in prior years (Minstrell 1982a; Minstrell 1982ba; Minstrell 1984; Minstrell 1989).

Connecting Practice to Research

I was accumulating long lists of problematic student thinking around concepts I covered in class. Some of these came from what I found in the research and confirmed in my class, but some came from what I discovered through my own interactions with students and their work.

Drilling deeper into my students' understandings, I found both correct and problematic pieces—not just "misconceptions." Since some of the problematic thinking involved both positive and negative pieces, I became bothered by the negative connotation of the term *misconception*. For example, more than 20% of my students seemed to believe that gravity was caused by air pressing down. At least they believed that gravity was due to an interaction between the object and something, rather than simply believing that gravity was a property of the object itself. Later, in my instruction, I could help them realize that gravity was an interaction between the object and Earth.

As I adapted my instruction to target specific pieces of problematic thinking I found myself becoming much more efficient. Instead of simply re-teaching a lesson or part of a lesson, hoping that more students "got it" the second time around, I could hone in on some specific problematic pieces of thinking and create experiences (e.g., questions, discussions, instruction, experiments) that enabled my students to confront those pieces. I didn't realize it at the time, but what I was engaging in was what is now commonly referred to as "formative assessment" or "assessment for learning." Teachers and researchers don't always agree among themselves—let alone with each other—on the definition of formative assessment. Like so many educational terms (*inquiry* and *engagement* to name a couple), *formative assessment* seems to take on different meanings with different individuals. As I write here, I have the following working definition in the back of my mind:

Purposeful, ongoing classroom practices and activities to make student thinking visible. Assessment practices that
- *provide teachers with useful information about their students' learning and inform instruction-related decisions and*
- *provide students with useful information about their own learning and build their capacity to improve their learning.*

This definition, however, only has meaning for me due to living the process in my classroom with my students. By the same token, findings of research around formative assessment ring true with me as well. Over the past 20 years for example, studies large and small (Black and Wiliam 1998a; Black and Wiliam 1998b) have shown that

- teachers who regularly use formative assessment have seen significant gains in student learning—especially among lower-achieving students;

- regular engagement in formative assessment prompts teachers to be more reflective about their practices and more in tune with their students' learning; it also prompts students to become more reflective in their learning and consequently more independent learners; and
- formative assessment (when effectively implemented) can do more to improve student achievement than any of the most powerful interventions, such as one-on-one tutoring, reduced class size, or cooperative learning.

Research has also provided several examples of a formative assessment "cycle," essentially a series of steps: (1) gathering useful specific information about student learning (the how and the why, not just the what), (2) effectively interpreting that information, and (3) acting on that information by providing feedback and/or designing the next piece of instruction based on that new knowledge (Sadler 1989; Black and Wiliam 1998a; Black and Wiliam 1998ba; Sadler 1998; Cowie and Bell 1999; Bell and Cowie 2001).

As the reader can see, my own assessment for learning cycle was very similar to the step-by-step illustrations found in the research—with two important exceptions: (1) mine lacks the strong emphasis on high-quality feedback[3] recommended by researchers (although I now agree that is an essential piece) and (2) I truly owned my cycle. The lack of a strong feedback step might have been more a function of how I structured my courses— lots of discussion and exploring student explanations through questioning. (Feedback largely came in the form of additional questions during large- and small-group investigations or whole-class discussions.)

The second point—that I "owned" the process—is probably the more important. If I had simply read "the steps" in some articles (as you are doing now), I would probably have been skeptical about its success and I am not sure I would have been motivated to develop the skills and classroom culture necessary to reap the positive effects. What I was doing in the classroom was not something I had seen at a workshop or been told by researchers that I should do. Instead, I *discovered* a process over time that worked for me through my own professional inquiry—motivated by my own curiosity about my students' learning and a sincere wish to improve my craft. Seeing

---

[3]High-quality feedback as defined by Black and Wiliam and others uses nonevaluative language and promotes learner independence among students.

positive results from those efforts over time with my own students is what kept me going.

Research has found that while most teachers strongly believe in (and practice) monitoring student understanding, very few effectively act on what they learn from their assessments. Nor do they prompt students to be more aware of and involved in their learning processes. In their milestone review study (1998a), Black and Wiliam suggest that insufficient training and skills are to blame. However, we know that there are other reasons that teachers don't implement formative assessment in the ways that research has found to be most effective—reasons that include the lack of a clear understanding of what formative assessment is, its purposes, and how it intersects with what they are currently doing in the classroom.

The work of a teacher is arguably a combination of art and science. Part of the art includes adopting strategies and tools as they fit personal styles and the needs of their students. Teachers rarely adopt teaching tools "as is." Why should an approach to assessment be any exception? As an experienced teacher, concerned with "keeping current," I would not have readily incorporated a complex system of assessment into my everyday practice. It would have seemed too resource intensive in terms of my time and preparation; not having "real data"—information from my students rather than from some research study subjects—I don't think I would have seen clearly the "return on my investment" in benefit to my students. I wouldn't see that benefit until I began exercising the "science" aspect of my teaching practice in an effort to find out what my students understood and revise what I was doing based on that evidence.

From Teacher Curiosity to Funded Research

I did not conduct all my classroom research in isolation and without help. Earning my master's and participating in a science curriculum development project encouraged me to reflect seriously on my practice. I then pursued and secured a grant that bought out a portion of my time to do research in my classroom, enabling me to explore what I had started on my own.[4] Subsequent grants either initiated by me or in collaboration with colleagues

---

[4] A reviewer of my first self-initiated proposal to the National Science Foundation later shared that the review committee thought the proposal was not clearly written and reflected lack of research experience BUT that it offered refreshing ideas, came out of the classroom, and was very inexpensive (my school district had a 3.5% overhead on federal grants). So NSF took the risk.

allowed me to refine the research and publish the results. My research projects and subsequent graduate work put me in contact with university researchers and psychologists who influenced my thinking and prompted me to further refine my thinking.

Personal Style or Shareable Practice?

The funding agency that had supported some of my early classroom research wanted to know to what extent my students' gains were the result of my approach to formative assessment or were tied to my personality, personal teaching style, or the positive relationship I had with my students. In short, could other teachers achieve similar success by applying the same learning-assessment cycle to focus on addressing students' problematic thinking?

To test this small step in generalizing the approach, I "drafted" two mathematics teacher colleagues who worked at my school. The three of us then proceeded to teach our physics and physical science classes in similar ways, applying the cycle. Both my colleagues were experienced classroom teachers and had taken physics courses as undergraduates, but neither had been specifically prepared to teach physics. I coached them in applying the cycle and in teaching physics. In the end we were actually coaching one another to set clearer and more specific learning goals and to determine more accurately what our students were thinking. More important, we were figuring out together how we could adapt our lessons to better address problematic ideas while building on useful ideas that we identified in students' work, responses, and comments.

Our skills of interpretation—the ability to "diagnose" problematic thinking—greatly improved through this collaborative process. My colleagues improved their understanding of physics and became skilled in anticipating student thinking around key concepts. Meanwhile, I was forced to be more explicit about what I did in the classroom and why. Together we also became much more adept at designing experiences to prompt our students to confront and revise their problematic ideas. Given the non-physics backgrounds of my colleagues and the fact that they were new to this sort of an assessment cycle, I had not anticipated any significant results from our collaboration until possibly the end of the two-year study. But, by the end of the first school year, when we compared gains from preinstruction to postinstruction on similar items, the results we were getting were virtually indistinguishable from one another (Hunt and Minstrell 1996).

Also, comparing posttest scores with the scores of comparable groups of students in other schools revealed that these two teachers were achieving better results.

Moving From "Misconceptions" to "Facets of Student Thinking"

As I mentioned earlier, I began to be bothered by the term *misconception* to describe students' problematic thinking because many of the ideas students apply have positive aspects as well as problematic aspects. My colleagues and I (teachers and university researchers[5]) coined the term *facets of thinking* to avoid the negative connotation of misconception. For us, a facet is a learner construction of one or more pieces of knowledge or reasoning the student uses to answer a question, solve a problem, or explain an event. One example of a facet about average speed is "the student determines the average speed by dividing the final position by the final clock reading." Although this procedure will work if the initial position and clock reading are both zero, often this is not the case. Thus, it is an idea that will give the correct answer in some situations, but not in others. The actual definition for average speed is the total distance traveled divided by the total elapsed time to travel that distance. Typically in textbooks, initial conditions are a position of zero and a clock reading of zero, and thus this problematic idea is not challenged and may later be misapplied in a real-world situation.

During the six years we worked together, my teacher colleagues and I created a Teacher Guide that included learning goals, typical problematic facets of thinking, assessment tasks to monitor learning, and lesson experiences adapted to address specific learning goals and related problematic thinking.

Extending the Practice to a Virtual Community of Colleagues

In a next round of research, we pushed the question of generalizability beyond close coaching within our school. Could we support distant teachers in obtaining similar positive results? We took this challenge to some of our physics teacher colleagues across the state of Washington and invited them to collaborate on the project. Twelve teachers from Port Angeles (on the Olympic Peninsula) to Spokane (in far Eastern Washington) and including the greater Seattle area volunteered to participate. We initiated this

---

[5]The university researchers were John Clement, Andy diSessa, Earl Hunt, and Emily van Zee. The teachers were Dottie Simpson and Virginia Stimpson.

research using a common end-of-year physics assessment. This was used as a baseline for each of the collaborating teachers. Because these colleagues taught in schools with student populations that varied greatly from each other, we used a research design that involved comparing end-of-year assessment results for subsequent cohorts with the results from that teacher's prior (baseline) cohort. Thus, each teacher was trying to improve his or her results from the previous year.

The summer after collecting the baseline results, 11 of the 12 teachers participated in a 4.5-day summer workshop led by me and the two teacher colleagues who had been close coached. (The 12th teacher had already been a participant in an earlier six-week workshop in which I had taught.) The first day teachers experienced an introduction to the assessment for learning cycle within three short topics: Obtaining the Best Value and Uncertainty in Measurement; Distinguishing Between Perimeter, Area, and Volume; and Determining Density. They were also presented with lists (facet clusters) of problematic facets associated with each of those topics. Then, they participated as students for two days in working through an entire unit on forces by fluids, not typically part of their curriculum but that related to their curriculum on forces. The latter was done to get each of the teachers to experience this formative assessment learning environment as learners. Throughout that unit the formative assessment cycle was continuously repeated by the instructors. Periodically the group stopped to reflect on what they had been thinking and experiencing both as learners and as teachers. We discussed what had happened during each cycle, what our students might have thought, what lesson or aspect of the instruction was intended to address which problematic idea, and to what extent the experience of addressing learner thinking had actually produced any reflection and improved understanding.

By the end of the workshop, every teacher (including one who had nearly a master's in physics) was able to say he or she had learned some physics as well as how to use an assessment cycle to address "misconceptions" (problematic facets of thinking). During the remaining days of the workshop we studied and discussed Teacher Guide materials, including facet clusters and assessment tasks and lesson strategies to address learning goals and related problematic facets. Several of these teachers had experienced a similar assessment cycle embedded in a professional development curriculum at the University of Washington (*Physics by Inquiry*, [McDermott, Shaffer et al.

1996]) but had not had the opportunity to focus on how to implement such a cycle in their own classrooms with their own curricula.

During the subsequent academic year, the teachers met face-to-face four times for about two hours each time and also corresponded by e-mail about once or twice a month. Teachers discussed difficulties and successes in using the assessment for learning cycle, the assessment tasks, and lesson materials. In one instance, a teacher said he was having little success with his students learning a particular idea. A second teacher asked the first whether he had had the students do the related experiences (from the Teacher Guide) designed to address that particular problematic facet. The first teacher said he hadn't, because he didn't think his students would need it. The second teacher responded with "That's why. Your students had that [problematic] facet, but you didn't lay out the experiences that addressed that idea, so your students didn't change their thinking." (This second teacher had used the Teacher Guide experiences that his colleague had skipped over and had had good results with his students.)

At the end of that first year, the teachers gave a posttest similar to the baseline posttest given the previous spring to their previous cohort of students. Across teachers, their next cohort of students performed an average of 15% better than their previous cohort, that is before the teachers incorporated the assessment for learning cycle (Hunt and Minstrell 1996; Minstrell and Matteson 1993). Improvement from the baseline cohorts to next cohorts ranged from 5% to 35%. The difference in gain scores seemed to be roughly correlated with the physics content knowledge and experience of the teacher. The lowest gains, for example, had been achieved by a biology teacher who had been "drafted" to teach physics, while the highest gains belonged to the teacher who had nearly completed a master's in physics.

Implementing a Strategy or Adopting a Practice?

At the end of the second year that the teachers were applying the approach, the student cohorts performed an average of 19% better than the baseline cohorts. But there was still a range among teachers' results that suggested more than differences in their physics content knowledge. What were teachers doing differently? Through interviews and classroom observations, we discovered that some teachers were only mechanically applying the cycle. That is, they were doing the suggested activities: assessing students' initial facets, assigning activities that we had suggested would address the problematic and goal facets, and then reassessing to see how students did. However, when they came to lessons in topic areas that were not included in the Teacher Guide materials, these teachers did not generally try to apply the cycle. In some cases, they would return to a "stand and deliver" kind of approach. In other words, they had not internalized the cycle—it wasn't part of their thinking and practice.

On the other hand, some teachers had clearly internalized the cycle and embedded it into their whole approach to planning and implementing instruction—not only in physics, but also in the other science courses they were teaching. One said that in all his interactions with students, "First I have to figure out what the students are thinking, then I have to figure out what [lesson] I'm going to do about that. This approach to teaching is exciting. I go home feeling great but totally exhausted." He and others were actively cultivating a classroom culture conducive to the approach so that students felt more comfortable sharing their thinking and challenging one another's ideas (including the teacher's) based on evidence.

Another collaborating teacher recently reflected on his experience with that teacher research group. He told me it had been a sort of mid-career "rebirth" for him, prompting a substantial shift in his focus from what he was teaching to what his students were saying:

> When I was [a younger teacher] I won a teaching award and all this stuff. But really it was a "stand and deliver" kind of model. And I was pretty good at the stand and deliver kind of model…the kids liked me and we had good interactions and good relationships…we had fun and they learned…. But to shift to more of a student-centered model and more student-responsive model has come as a direct result of my work in that workshop…and I found that when I shifted—and I have always taught other things in addition to physics—the big challenge has always been, How do I do these things that I hold so deeply as truth in content areas where there is not the research done on students' thinking? And so I really have to listen to the kids, and the things that the kids say become the facets in my head that inform what I do tomorrow.

As with any innovation, the degree to which the cycle "worked" in the classroom would depend largely on the degree to which the teacher came to "own" the approach. In retrospect, we may not have paid as close attention to this issue as we should have. We were pleased to know that with some professional development, teachers who adopted the approach could achieve significant gains with their students. We were eager now to find a way to scale up to create opportunities for many teachers across the country to improve student learning. We ultimately developed a suite of tools we call Diagnoser Tools. Before discussing the tools, however, I need to dip back again into a little personal history.

Development of Web-Delivered Tools to Support the Assessment for Learning Cycle

In 1985, long before the ubiquity of internet communication or the World Wide Web, I was invited to participate in a conference titled "Technology in Education: Looking Toward 2020." I had been conducting research in my classroom with virtually no information technology. Nevertheless, the conference organizers encouraged me to look into the future and envision how technology might assist me in the work that I did as a teacher. The result was a chapter that included the story of a futuristic teacher, Matt, and his electronic "teaching assistant" TACFU (teaching assistant computer for understanding). The following excerpts from the fictional account of Matt and his class reveal my own dream of tools to support the work that I was, at the time, painstakingly enacting with paper-and-pencil instruments:

> *The fall of 2020… It's time to finish preparations for today's teaching and learning. Matt leaves one computer and goes into the next room to engage TACFU.*
>
> *TACFU, connected to the phone line, has been monitoring student progress wherever students log on with their electronic notebooks (a personal hand-carried computer that can communicate with large-frame computers) to accomplish their assigned work.… This allows TACFU and Matt to collect large amounts of data on students and to construct student profiles they can use to guide individual and group learning.…*
>
> *Matt requests individual responses for nearly every student activity. That takes an incredible amount of reading, analysis, synthesis, and response. Once again TACFU proves useful.*
>
> *For problem situations, TACFU has kept a record of each solution given by every student when he or she was logged onto a large frame machine. The collections of answers have been synthesized into a Framework of Knowledge Profile for each student for each concept area. These frameworks were con-*

structed out of past research on the conceptual understanding of novices and the development of those conceptions toward expert thinking….

Matt sees his role as one of fostering general intellectual growth rather than merely passing on traditional content. The research on the integration of general thinking skills with content understanding helped form a base for the programs that keep TACFU probing for development.

TACFU describes students' thinking, diagnoses possible difficulties, and suggests experiences that would foster development for each student. Matt compares his own assessments of the status of thinking with assessments done by TACFU…. There is a list of students who are just beginning to distinguish between action on a body and property of a body…. Matt considers how much easier it will be for those students to understand the formal notions of net force and resulting acceleration….

Because Matt has grown to trust the research-based decisions of TACFU, in most instances he goes with the machine's suggestions. In some cases, however, Matt chooses to override the machine because he feels he has more recent or pertinent information…. Between TACFU's input and his own, Matt decides what activities to recommend to [each] student. From these several suggestions, the student will choose her or his assignment…[so that] Matt can be more certain that Melinda…Michelle…and James…will attack the assignment in a way that will be appealing to each of them…. (Minstrell 1988)

Around the year 2000, together with Earl Hunt and colleagues at the University of Washington, we began to realize our own TACFU. Based on the Teacher Guide we had used in our peer coaching project, we built an online suite of tools we call the Diagnoser.[6] (The Appendix, p. 58, describes the Diagnoser and its tools in detail.) The present version, developed by FACET Innovations, has been operational and available free of charge to teachers since fall of 2004 at *www.diagnoser.com*. New content is added as developed.

We have found that teachers use Diagnoser Tools in a variety of ways, including improvement of their own science content knowledge. While we are glad to know the tools meet teachers' needs in several ways, we also are finding that some uses are more effective than others.

---

[6]Diagnoser Tools support several topics in physical science and some in human body systems. Other science content area materials are currently under development.

Diagnoser Tools are designed to support a classroom practice of assessment for learning, and they are most effective in the hands of teachers who have such a practice. This was exemplified most recently by a Seattle-area teacher who was using Diagnoser Tools extensively. In an interview she described her thinking and use of Diagnoser: "I'm curious about what students are thinking. My [teaching/learning] lab is right here in the classroom…. When I think I have taught something, [sometimes] I see that the students haven't fully made all the connections that I think I taught. So it forces me to go back and say 'wow' [and] to think, What other kinds of experiences can I do?"

Interviewing this teacher, we found that her perspective and approach to learning and teaching was consistent with the assessment for learning cycle embedded within the Diagnoser Tools. We also noticed that her students were outperforming similar populations of students in other classes in her school and other schools in her district. We do not claim that the tools are responsible for her student gains, but recognize (as she does) that they effectively support her in the work she is doing.

This teacher, and others who are using the tools to their potential, assign sets of questions several times during a unit and assess students early and formatively by using some version of Elicitation Questions. They use the data to tune their lessons to address students' problematic thinking by prompting students to reflect on initial ideas and recent experiences, looking for inconsistencies and ways to resolve them. These teachers assess continuously, and reflect on and interpret the results for possible implications for next instruction. Students of these teachers have achieved positive results on later summative district assessments (Hunt E., T. Madhyastha, P. Kraus, and J. Minstrell [unpublished]). We think these teachers get good results not necessarily because of the tools but because they use the tools to carry out formative assessment in the classroom.

Students of these teachers receive learning benefits as well. Results of the Elicitation Questions and subsequent discussions suggest hypotheses that students are motivated to test. They want to see if "their" idea works. The sets of questions offer assessment of whether they have learned the key ideas from class activities. Students are encouraged to talk with each other about their answers and in a multiple-choice question to think together why each particular answer might make sense to a student as well as why the keyed answer might be most consistent with class experiences.

Many Diagnoser-registered teachers, however, are getting no effects (no harm but no improved results either) on state or district assessments. Preliminary results from ongoing research suggest the "no difference" may be because the teachers are not using the tools to their potential. Just as early on I needed to deepen my understanding of the content and I needed to learn more about formative assessment, these teachers are also in the early stages of their content learning and/or of the pedagogy related to effective use of formative assessment. They are not yet able to use the tools in a formative way as designed. For example, some use the sets of questions in a summative way, administering one Diagnoser Question Set as a pretest and a second as a posttest as I did early in my experience with "misconceptions." Their view of assessment may not include using Diagnoser formatively to support "next-day" learning or instruction. Still other teachers have assigned several sets as review for the state test or a big district test, which amounts to practice in answering questions and accomplishes little in the way of monitoring and promoting learning.

Some teachers do assign Diagnoser Question Sets while students are in the early stages of their learning, but these teachers may not access the student reports and take the time to interpret and act on the information. In this way the Question Sets are reduced to one more activity for students to get out of the way. Other teachers do access the reports, but just as I did early on, they re-teach or go over right answers rather than changing their instruction to address new learning. While each of these uses sidesteps or shortcuts the idealized version of the assessment for learning cycle presented earlier, we recognize that the teachers' use of Diagnoser Tools is similar to my early development of effective practice, wherein I did not really think about and address my students' thinking in planning and implementing assessment and instruction. But, while impact on students appears minimal, there may be significant impact on the teacher. Effective practice of formative assessment is a process that is learned over time.

We encourage the reader to try adopting a perspective of genuine interest in what students are thinking. Try implementing the assessment for learning cycle. Then try using the Diagnoser Tools to support those efforts. We look forward to hearing your stories of success.

Conclusion

In the past 30 years, we have learned a lot about how people learn and what we can do to support improved learning in all classrooms (Bransford, Brown, and Cocking 1999). Much of the research has highlighted the importance of formative assessment in helping teachers and students to focus their attention on what still needs to be learned in order to move on.

We started this chapter by examining the complexity of an effective cycle of assessment for learning and ended by presenting an example of tools designed to support such a cycle. More than a set of strategies, questions, or tools, however, assessment for learning requires a particular teacher perspective in order to be effective. At the core is a genuine interest in what students are thinking and their rationale for that thinking. From there, the teacher must have the desire and the skill to act so that students' needs are "diagnosed" and classroom experiences address those needs. As the personal story in this chapter suggests, it is not a perspective or practice that is developed overnight and may take years to refine.

As teachers develop their skills in assessing for learning, they will experience a growing need for tools and resources to support their efforts. Hopefully, this need will create yet another opportunity and common ground for teachers and researchers to come together in an effort to bridge research and practice in the interest of improved student learning.

Acknowledgments

The authors would like to acknowledge the collaboration of the many teachers who have shared their experiences with us. Special thanks to teachers Angie DiLoreto, Eric Magi, Jim Slavicek, and Sherm Williamson, who have contributed extensively to the research and development of Diagnoser Tools. Special thanks also to Earl Hunt both for his leadership in the technological development of an early version of the tools and for the related psychometric research that he, Tara Madhyastha, and other colleagues have contributed over time. We are grateful to the National Science Foundation for funding the development of Diagnoser Tools (NSF Grant 129406), the effects of use of the tools (NSF Grant 0435727), and research and development on teachers' understanding and practice in formative assessment (NSF Grant 0535818). Any opinions, findings, and conclusions or recommendations expressed in this chapter are those of the authors and do not necessarily reflect the views of the National Science Foundation.

References

American Association for the Advancement of Science (AAAS). 1993. *Benchmarks for science literacy*. New York: Oxford University Press.

Bell, B., and B. Cowie. 2001. The characteristics of formative assessment in science education. *Science Education* 85(5): 536–553.

Black, P., and D. Wiliam. 1998a. Assessment and classroom learning. *Assessment in Education* 5(1): 7–73.

Black, P., and D. Wiliam. 1998b. Inside the black box: Raising standards through formative assessment. *Phi Delta Kappan* 80(2): 139–148.

Bransford, J. D., A. L. Brown, R. R. Cocking, eds. 1999. *How people learn: Brain, mind, experience, and school*. Washington, DC: National Academy Press.

Caramazza, A., M. McCloskey, and B. Green. 1981. Naive beliefs in "sophisticated" subjects: Misconceptions about trajectories of objects. *Cognition* 9(2): 117–123.

Champagne, A. B., L. E. Klopfer, and J. Anderson. 1980. Factors influencing the learning of classical mechanics. *American Journal of Physics* 48: 1074–1079.

Cowie, B., and B. Bell. 1999. A model of formative assessment in science education. *Assessment in Education* 6(1): 101–116.

Gunstone, R. F., and R. T. White. 1981. Understanding of gravity. *Science Education* 65: 291–299.

Hunt, E., and J. Minstrell. 1996. Effective instruction in science and mathematics: Psychological principles and social constraints. *Issues in Education* 2(2): 123–162.

Hunt, E., T. Madhyastha, P. Kraus, and J. Minstrell. Unpublished. The relation between conceptually based instruction and nature of assessment.

McCloskey, M., A. Caramazza, and B. Green. 1980. Curvilinear motion in the absence of external forces: Naive beliefs about the motion of objects. *Science* 210(5): 1139–1141.

McDermott, L. C., P. Shaffer et al. 1996. *Physics by Inquiry*. New York: John Wiley and Sons.

Minstrell, J. 1982a. Conceptual development research in the natural setting of the classroom. *Education for the 80's: Science*. Washington, DC: National Education Association.

Minstrell, J. 1982b. Explaining the "at rest" condition of an object. *The Physics Teacher* 20: 10–14.

Minstrell, J. 1984. Teaching for the development of understanding of ideas: Forces on moving objects. *AETS yearbook: Observing science classrooms*.

Minstrell, J. 1988. Teachers' assistants: What could technology make feasible? In R. S. Nickerson and P. P. Zodhiates (Eds.), *Technology in education: Looking toward 2020*. Hillsdale, NJ: Lawrence Erlbaum.

Minstrell, J. 1989. Teaching science for understanding. In L. Resnick and L. Klopfer (Eds.), *ASCD 1989 yearbook: Toward the thinking curriculum: Current cognitive research*. Alexandria, VA: Association for Supervision and Curriculum Development.

Minstrell, J., and R. Matteson. 1993. Adopting a different view of learners: Effects on curriculum, teachers and students. Unpublished video report to the James S. McDonnell Foundation.

National Research Council (NRC). 1996. *National science education standards*. Washington, DC: National Academy Press.

Posner, G. J., K. A. Strike, P. W. Hewson, and W. A. Gertzog. 1982. Accommodation of a scientific conception: Toward a theory of conceptual change. *Science Education* 66(2): 211–227.

Sadler, D. R. 1989. Formative assessment and the design of instructional systems. *Instructional Science* 18: 119–144.

Sadler, D. R. 1998. Formative assessment: Revisiting the territory. *Assessment in Education* 5(1): 77–84.

Appendix

Diagnoser Tools

Below we briefly describe the Diagnoser Tools, their relationship to the research on formative assessment, and how they build on what our collaborating teachers learned about support for the assessment for learning cycle. The tools and associated services are presently available to teachers free of charge at *www.Diagnoser.com*. Teachers can register online.[7]

After registering for the site, the teacher has access to the content resources for planning and embedding an assessment for learning cycle into instruction. To explain the Diagnoser Tools here we will use a physical science strand called Nature of Forces. More specifically, we will look at Forces as Interactions. Suppose we have as our learning goal to help students learn about action and reaction forces, how they compare, and that each force in the pair acts on a separate object. This is essentially Newton's third law, a subtle idea in most introductory physical science or physics classes. Clicking on Forces as Interactions, we gain access to the tools to support formative assessment in this mini-unit.

Diagnoser Tools

1. Learning Goals: Clear Learning Targets

———————

[7]To begin using the tools, click on "Teacher Login." If you are not in one of the districts currently working more closely with Diagnoser, click "Continue" and complete the required fields on the Teacher Registration Page. At this writing there are over 1,200 teachers, researchers, developers, and teacher educators registered. The teachers have registered nearly 50,000 students who have completed nearly 75,000 question sets and submitted over 700,000 responses to items.

Learning Goals were drawn from National Science Education Standards (NRC 1996) and Benchmarks for Science Literacy (AAAS 1993). In Diagnoser, these resources provide the teacher with the learning targets aligned with each of the other tools in the subtopic. *Clicking on Learning Goals under Forces as Interactions will give the teacher access to the learning goals for that subtopic.*

2. Elicitation Questions: Assessment of Students' Initial Understanding

Elicitation Questions (EQs) include some questions that were designed to open up a few of the critical issues, both the learning goals and some of the common misconceptions, in the subtopic. The questions have a "printer friendly" page so teachers can copy them and have students respond with their initial ideas. The results give the teacher a reading of what the students are thinking, but the teacher should not tell students whether their thinking and answers are right or wrong. The teacher should grade students only on the apparent honest effort they put into describing their thinking at this point. The EQ is also intended to open up the learning issues for students. It can help students know the sorts of questions they need to be able to answer by the end of the instruction, and a subsequent class discussion of the EQ can expose a student to the thinking of fellow students. Student questions and hypotheses about the subtopic get generated from a well-run EQ discussion and can be tested and answered in the next activities. Thus, the EQ and discussion can initiate subsequent inquiry.

Clicking on Elicitation Questions under Forces as Interactions will present the EQ, as shown in Figure 3.1. Student answers to the Skaters and Sam and Shirley questions typically raise issues of whether properties (like size or strength) of interacting objects or the resulting effects (like who moves more or the direction of movement) tell us anything about the relative forces two interacting objects exert on each other. These hypothesized factors that might affect the relative magnitudes of the forces exerted by each participant in the interaction become the focus of inquiry in the next lessons to test initial ideas and develop ideas more consistent with science.

3. Facet Cluster: A Framework for Interpreting Student Responses

A Facet Cluster is a list of the various goal facets (specific learning goal statements) and problematic facets of students' thinking related to the subtopic. When the teacher reads the responses to Elicitation Questions or listens during a class discussion of the various responses, the teacher can use the Facet Cluster as a guide to interpreting the student comments and identifying major misconceptions (problematic facets) to address in subsequent lessons. The facets are a middle language between research and teacher practice. They are slight abstractions of what students actually say or do and so provide teachers with a scheme for interpreting and coding apparent thinking. The facets are roughly ranked in the cluster from the more problematic (codes 9X or 8X) to less problematic (3X or 2X) to goal facets (1X or 0X). The rough ranking of the facets allows the teacher to track possible development or at least changes in students' thinking during lessons related to that subtopic. Many teachers print the Facet Clusters and put them in a notebook for reference during subsequent lessons.

Figure 3.1 Example of an Elicitation Activity in Forces as Interactions

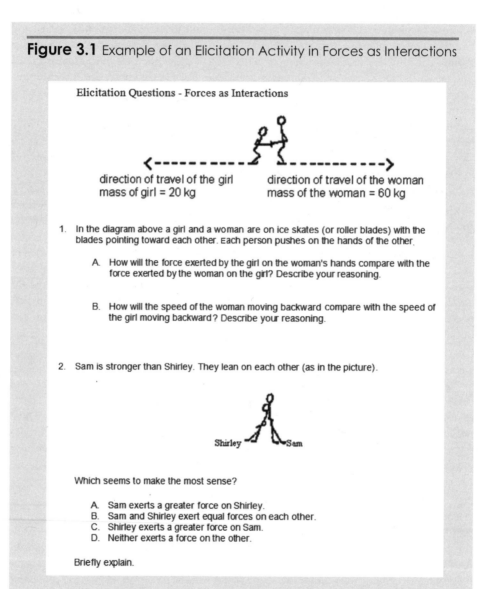

Elicitation Questions - Forces as Interactions

direction of travel of the girl
mass of girl = 20 kg

direction of travel of the woman
mass of the woman = 60 kg

1. In the diagram above a girl and a woman are on ice skates (or roller blades) with the blades pointing toward each other. Each person pushes on the hands of the other.

    A. How will the force exerted by the girl on the woman's hands compare with the force exerted by the woman on the girl? Describe your reasoning.

    B. How will the speed of the woman moving backward compare with the speed of the girl moving backward? Describe your reasoning.

2. Sam is stronger than Shirley. They lean on each other (as in the picture).

Shirley   Sam

Which seems to make the most sense?

    A. Sam exerts a greater force on Shirley.
    B. Sam and Shirley exert equal forces on each other.
    C. Shirley exerts a greater force on Sam.
    D. Neither exerts a force on the other.

Briefly explain.

Clicking on Facet Cluster under Forces as Interactions will produce the screen shown in Figure 3.2. Notice that facets 60 through 64 identify problematic ideas related to students focusing on relative properties of the interacting objects, and 50 through 54 identify ideas related to the relative effects produced. While identifying properties and effects are important to note in science, it turns out they are not helpful when thinking about the action and reaction forces exerted by each of two objects that are interacting with each other. See goal facets 00 through 03 for the key related ideas we hope students learn.

4. Developmental Lessons: Lesson Guidelines to Address Preconceptions
Another important set of tools in Diagnoser are the Developmental Lessons. They give general guidelines and/or specific lesson activities and other experiences that will address the pre-instruction ideas that come out of the discussion of the Elicitation Questions. The lessons are also presenting students with information from which they can arrive at the goal facets. Many teachers use the students' explanations given during the discussion of Elicitation Questions as hypotheses for students to test in an inquiry activity like those

Figure 3.2 Facet Cluster for Forces as Interactions

Facet Cluster - Forces as Interactions

Facets and facet clusters are a framework for organizing the research on student conceptions so that it is understandable to both discipline experts and teachers. Facet clusters include the explicit learning goals in addition to various sorts of reasoning, conceptual, and procedural difficulties. Each cluster contains the intuitive ideas students have as they move toward scientifically accurate learning targets.

Facets are arranged with the Goal Facets at the top of the page followed by the more problematic facets. Each facet has a two-digit number. The 0X and 1X facets are the learning targets. The facets that begin with the numbers 2X through 9X indicate ideas that have more problematic aspects. In general, the higher facet numbers (e.g., 9X, 8X, 7X) are the more problematic facets. The X0's indicate more general statements of student ideas. Often these are followed by more specific examples, which are coded X1 through X9.

Forces as Interactions Facet Cluster

00 The student understands that all forces arise out of an interaction between two objects and that these forces are equal in magnitude and opposite in direction.

    01 All forces arise out of an interaction between two objects.

    02 The force pairs are equal in magnitude.

    03 The force pairs are opposite in direction.

40 The student identifies equal force pairs, but indicates that both forces act on the same object. (For the example of a book at rest on a table, the gravitational force down on the book and the normal force up by the table on the book are identified as an action-reaction pair.)

50 The student uses the effects of a force as an indication of the relative magnitudes of the forces in an interaction.

    51 More damage indicates one of the interacting objects exerted a larger force.

    52 If an object is at rest, the interaction forces must be balanced.

    53 If an object moves, the interaction forces must be unbalanced.

    54 If an object accelerates, the interaction forces must be unbalanced.

60 The student indicates that the forces in a force pair do not have equal magnitude because the objects are dissimilar in some property (e.g., bigger, stronger, faster).

    61 The 'stronger' object exerts a greater force.

    62 The moving object or a faster moving object exerts a greater force.

    63 The more active or energetic object exerts more force.

    64 The bigger or heavier object exerts more force.

90 The student believes that inanimate/passive objects cannot exert a force.

suggested in the Developmental Lessons. In these lessons designed to address critical issues in the subtopic, we expect that the teacher will need to help keep students focused with questions like "How will this experiment tell you whether your hypothesis/prediction seems correct? How will you know if it is not correct?"

Clicking on Developmental Lessons under Forces as Interactions gives suggestions for activities or experiments for students to test their tentative hypotheses/predictions such as "The stronger person will exert the greater force" (facet 61). One Developmental Lesson in this case suggests using two students of obvious unequal strength each pulling with an identical spring scale on the other. Teachers will need to guide students in the design of their experiments so the focus is on their driving question. Guiding questions can help, e.g., "How will your experiment tell you whether the relative strengths of interacting objects are different or if the force that each exerts on the other turns out to be the same?" In this subtopic, from students' summaries of results and conclusions, students should eventually come to the tentative conclusion that in all their experiments the forces exerted by the two objects were of equal magnitude: "For every action there is an equal and opposite reaction [force]" (facets 00 to 03).

5. Question Sets: Formal Formative Assessment and Feedback for Students
Each Diagnoser Question Set typically includes 6 to 10 questions for checking on students' understanding of the subtopic associated with the learning goals of that facet cluster. After a few lessons the students and teacher may want to check whether students have the key idea or still are using one or more problematic ideas. Question Sets can be assigned by the teacher and are delivered online to each student. For multiple-choice and numerical-response questions, each answer is coded with one or more of the facets in that facet cluster. Thus, the system is making a tentative diagnosis of what thinking that student seems to be applying at that time for that situation. Since Diagnoser is a formative assessment tool, the focus should not be on grading the student but on guiding the student in further learning. Students get feedback after responding to each question or pair of questions. At the end of the set, students are also asked a metacognitive question about their level of confidence with that subtopic. After submitting, students are presented with short reports telling them the fraction of questions they got correct (consistent with a goal facet) and identifying up to two aspects of the subtopic on which they seem to need more work (as evidenced by problematic facets diagnosed.) The students' answers are then automatically sent to a Teacher Report tool in Diagnoser.

If as a teacher you clicked on the tab titled "View/Assign Question Sets" and assigned Forces as Interaction Set 1 to a class, the first question a student would see is shown in Figure 3.3. Notice that this question revisits a "leaning people" situation students would have seen in an Elicitation Question for this same subtopic. We want to see if students can apply what has been learned through the Developmental Lessons to a situation seen before. Other questions will focus on a more distant transfer of the idea. For this question, if the student clicked on the fourth answer, because she seemed to be focusing on the effects of whether the people moved one way or the other, she would get a feed-

back screen associated with facet 53 shown in Figure 3.4. The fact that the student chose an answer to question 1 that is consistent with facet 53 is sent to a Teacher Report.

6. Teacher Report: Data From Which to Make Instructional Decisions

Facet codes associated with each response by all students in the class are automatically accumulated in a Teacher Report. The first part of the report is displayed as a matrix with student identifiers down one side and question numbers along the top. Each cell in the matrix represents the facet code associated with the answer that that student gave to that question. Summary Statistics, the second part of the Teacher Report, quickly summarizes information for which problematic facets seem to show up with the most students and which facets seem to be more resilient. The purpose of the report is to provide student learning data that teachers can use to make decisions about potential focus for next lessons.

Figure 3.3 An Example Multiple-Choice Question From Diagnoser Set 1 for Forces as Interactions

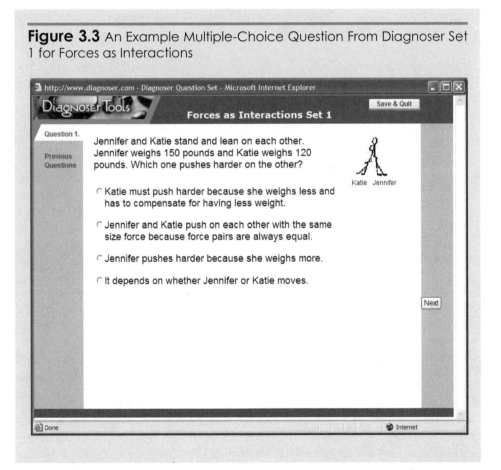

Figure 3.4 An Example of Feedback for Student Whose Diagnosis Was Facet 53 for a Diagnoser Question in Forces as Interactions

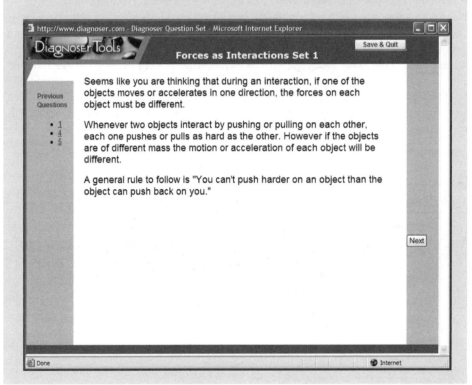

By clicking on the tab for Teacher Reports and choosing Forces as Interactions for your classes you can get the coded data from students' responses to the question set. See Figure 3.5. By reading the codes in the matrix associated with each question for each student, the teacher can identify individuals who most need help (e.g., student 1239-0-1) and students who seem to have the idea so far (e.g., student 1239-0-12). Responses to the metacognitive self-rating question can also be read where "1" would represent extreme confidence and "6" would indicate extreme lack of confidence. Sometimes if a student (e.g., student 1239-0-5) got a particular facet on first answering a particular question (e.g., facet 53 for question #6), he was given feedback and asked to try that question again. In that case both answers are coded in the cell for that item for that student.

The quick Summary Statistics at the bottom of Figure 3.5 show that problematic facets in the 50s and 60s seem to be showing up most (at least once) and that those are most resistant (more than once). This conclusion can also be reached by doing a quick scan of the matrix. Data are used to inform decisions about the next steps in instruction.

7. Prescriptive Activities: Specific Lessons to Address Identified Facets
When particular problematic facets have been reported in the Teacher Report, Prescriptive Activities tools provide lessons from research or best practice that have been found effective in addressing particular facets. One of the places where teachers new to formative

Figure 3.5 An Example Teacher Report for a Particular Class Being Assessed by Set 1 in Forces as Interactions

Teacher Reports

Refresh | Print

Note: Student Reports show the data for the first time through the assignment. If students are currently taking the sets being reported on, there may be updates. Click here to refresh.

Forces as Interactions Set 1			Questions							
Student ID	Date Completed	Self Rating	1	2 All	3	4	5	6	7	8 All
1239-0-1	2005-10-22	5	50	text		40	54	90 , 60	62 , 51	text
1239-0-2	not yet	none	63			00	02	02	01	
1239-0-3	2005-10-22	3	02			53	02	02	01	
1239-0-4	2005-10-22	3	50	text		00	53	02	01	
1239-0-5	2005-10-22	2	63			00	02	53 , 02	53 , 01	
1239-0-6	2005-10-22	2	64			00	02	02	01	
1239-0-7	2005-10-22	2	02			00	02	53 , 02	53 , 01	
1239-0-8	2005-10-22	5	64			00	53	02	01	
1239-0-12	2005-10-22	2	02			00	02	02	01	
1239-0-13	2005-10-22	3	02			63	53	02	01	
1239-0-14	2005-10-22	6	02			63	02	02	01	
1239-0-15	2005-10-22	2	50	text		00	02	53 , 02	53 , 01	
1239-0-16	2005-10-22	4	02			63	64	Unk , 02	62 , 01	

Summary Statistics for Forces as Interactions Set 1

Facet	At least once	More than Once
00	92.3%	92.3%
40	7.7%	0%
50	61.5%	38.5%
60	61.5%	15.4%
90	7.7%	0%
Unk	7.7%	0%

Explanation for Facets and notes, Text of questions, Prescriptive activities assessment break the cycle is by not using the data gathered to inform their instructional decisions. Instead of choosing an activity that specifically addresses problematic understanding, many teachers just choose another activity that asserts or confirms the goal idea. But unless misconceptions are specifically addressed, there is little change in understanding (Bransford, Brown, and Cocking 1999.) Thus, unless Prescriptive Activities, or some lesson like them, is conducted to address major misconceptions, the act of formative assessment has not been completed and students have not had sufficient opportunity to learn.

Prescriptive Activities can be accessed by clicking on the resource at the bottom of the Teacher Report page or by clicking on Prescriptive Activities on the list of tools under Forces as Interaction on the Content Resources page. Figure 3.6 shows one of the activities that research and best practice have shown can be used to address problematic facets in the 50s and 60s, which were prominent in the Teacher Report shown in Figure 3.5 (C. Camp, J. Clement et al. [1989] Preconceptions in Mechanics: Lessons Dealing With Students' Conceptual Difficulties. Dubuque, Iowa: Kendall/Hunt). After students complete these activities, the teacher might assign an additional Diagnoser Question Set to check students' understanding formally. Since this is still formative to learning, we recommend students not be graded on their performance on these question sets. Let that happen on a later summative quiz or test where questions from Diagnoser Question Sets could be used or adapted and used for grading. But, for now, the focus is still on assessment to promote learning.

For questions or comments about Diagnoser Tools, contact us at *Info@ FACETInnovations.com* or visit the website at *www.FACETInnovations*.com.

Figure 3.6 Prescriptive Activity to Address Facet 50 in Forces as Interactions

Activity to address Facet 50:
If you have several students whose diagnosis is 60 and several with 50, you may want to set up several of the situations described for each and allow students to investigate. Follow up with a discussion to arrive at consensus as to the relative magnitude of forces during the interaction.

Although numbers 2 and 3 below are easier to set up, students are likely to need the first-hand experience they can gain from number 1 below. Consider doing the three activities in the order described below.

1. Consider the results of the following set up:
Materials needed:
• Two identical force scales, e.g. two big laboratory spring scales for pulling (100n scales work well) OR two bathroom scales for pushing adjusted to zero when held oriented up and down rather than horizontally. (In all situations when spring scales are used, calibrate the scales first to make sure they give identical readings when pulled against each other or when lifting the same object.)
• One cart, wagon, or skateboard that is sturdy and stable enough for a person to ride safely on it.
Procedure: Use common sense to do this experiment safely.
One person sits on the wagon, holding one scale. The other person connects the hook of his scale onto the hook of the other person's scale. Predict which scale will read the larger value and why. Then the standing person (and his spring scale) pulls on the seated person (and her spring scale).

Compare the two scale readings when the cart is not moving, when it is moving at a constant velocity, and when the cart is accelerating.

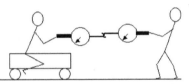

For pushing, the two hold their bathroom type scales against each other and the standing person pushes while each reads his or her scale. Sometimes the scales are more easily read by someone looking over the shoulder of the two people who are pushing the scales.

Note to Teachers: The dials of the two scales should read the same, within experimental error, indicating the people are exerting equal forces on each other even though one is making the other move.

Probing Students' Understanding Through Classroom-Based Assessment

Section 2 examines several classroom-based assessment formats, the information they provide, and their impact on students' learning.

Discussion Questions

- How might traditional student work and classroom routines contribute to formative assessment?
- What models exist to help teachers become skilled in using formative assessment techniques?
- What have researchers learned about the impact of formative assessment on student learning?
- What are some of the most effective formative assessment practices?
- What are the characteristics of a good formative assessment task?

Chapter Summaries

Jacqueline Jones of the Division of Early Childhood Education, New Jersey State Department of Education, and Rosalea Courtney of Education Testing Service are researchers who study the development of young students' ideas about the natural world. They have written extensively about the value of classroom artifacts, such as drawings, transcriptions of group discussions, and teachers' anecdotal notes, to document and assess children's ideas and track their change over time. Using samples of children's drawings and conversations about a class pet, Jones and Courtney describe the sense-making processes they use in their work and model a process teachers can apply in their own classrooms.

Alicia Alonzo at the University of Iowa employs students' entries in their science notebooks, particularly their drawings, as a formative assessment tool. She presents a series of drawings collected over several months during her research in third- and fourth-grade classes in California schools. Acknowledging the variety in both format and purpose of science notebooks, Alonzo presents a set of decisions teachers should make to enhance the value of their students' notebooks as assessment tools. This chapter includes a series of notebook entries illustrating the progress of one young student and Alonzo's analysis process. Insightful commentary accompanies each entry, modeling how teachers can use this assessment tool to interpret subtle aspects of student work.

Katherine McNeill of Boston College and Joseph Krajcik from the University of Michigan are researchers and curriculum developers. They have conducted extensive research with middle school science students, studying their abilities to produce written scientific explanations to demonstrate their understanding and reasoning skills. These scientific explanations, according to the authors, are powerful assessment tools that offer teachers insights into the depth of students' understanding. McNeill and Krajcik describe the framework they use to help students develop these writing and reasoning skills and create explanations that clearly distinguish between claim, evidence, and reasoning. The authors also provide the rubrics they use to assess student work and discuss the types of feedback they give to students.

Olga Amaral, from San Diego State University, and her colleague Michael Klentschy, former superintendent of the El Centro, California, school district, describe their ongoing studies of the roles of science notebooks in assessing the content and literacy development of English language learners. Their chapter explores the use of students' notebooks as tools for learning, assessment, feedback, and language acquisition. The authors emphasize the value of using student notebooks to help teachers compare what they teach to what students are learning and how closely they are achieving the standards. The authors describe the professional development they use with teachers and acknowledge that demanding professional development is essential if notebook assessment is to achieve the multiple goals they advocate.

Curriculum developer Arthur Eisenkraft from the University of Massachusetts and teacher Matthew Anthes-Washburn of Boston International High School discuss the assessment of laboratory investigations from two perspectives: first, the quality of the investigation itself, and second, the student performance and achievement that result from the investigation. The authors frame their discussion in a broad context beginning with exploring the attributes of a "quality" laboratory investigation, one that addresses standards-based student learning goals and demonstrates essential features of inquiry. Emphasizing the importance of students' self-assessment, the authors present rubrics that clarify both the requirements and the criteria for success. These rubrics are intended for use by both teacher and student, ideally with some level of agreement.

Kathy Long, Larry Malone, and Linda De Lucchi from the Lawrence Hall of Science at University of California at Berkeley are researchers and developers of the FOSS elementary science curriculum. They have developed a comprehensive embedded assessment system entitled Assessing Student Knowledge (ASK) that provides teachers with multiple probes to monitor several aspects of student learning during the course of a multi-week science unit. Working closely with teacher collaborators, Long, Malone, and De Lucchi developed, tested, and refined the ASK system's multiple items, which are intended for use early, midway, and at the end of an extended unit. Collectively they reveal how student learning is progressing over time, helping teachers to evaluate and pace instruction and enabling students to monitor their own learning

The chapter on the Calipers Project discusses the assessment potential of computer-based simulations that allow students to investigate and manipulate multivariable systems, many of which are too complex or dangerous to actually create in a laboratory or classroom. These simulations invite active manipulation, testing of variables, and problem solving, and, moreover, they ask students to answer assessment questions and provide them with instant feedback and their teachers with detailed records of each student's responses and progress over time. The author-developers are Edys Quellmalz (WestEd), Angela DeBarger, Geneva Haertel, and Patricia Schank (SRI International), Barbara Buckley, Janice Gobert, and Paul Horwitz (Concord Consortium), and Carlos Ayala (Sonoma State University in California). The authors describe the design of their assessments, which address standards-based content, are anchored to the simulation scenario, are aligned with common curricula, and take advantage of the innovative technology.

Page Keeley and Francis Eberle of the Maine Mathematics and Science Alliance discuss their development of a bank of formative assessment probes based on cognitive research and standards-based learning goals. The authors describe the design process and its professional development component that engages teachers in the construction and critical analysis of formative assessments. Readers will find examples of assessment questions followed by detailed analyses of what they reveal about students' understanding and misconceptions. The chapter concludes with reflections by a middle school teacher about how the responses he anticipated from his students contrasted with what he learned from their actual responses.

## CHAPTER 4

Documenting Early Science Learning

Jacqueline Jones
New Jersey State Department of Education

Rosalea Courtney
Educational Testing Service

Young children are fascinated by the natural world. They think about how things work, what is alive, and why some things change their shape and form. Science explorations such as planting, animal studies, and cooking are a natural part of early childhood classrooms and can provide the settings for understanding how children are making sense of the world around them. The real evidence of children's early science understanding comes directly from these everyday classroom settings. Records of children's conversations, anecdotal notes and photographs of their actions (e.g., Figure 4.1, p. 74), and samples of their drawings and constructions form the classroom-based data that help teachers see how children are thinking about the natural world. Early childhood educators can provide more appropriate learning environments by engaging in the documentation/assessment process of collecting, describing, and interpreting evidence of young children's emerging science understandings.

Guiding Principles

An ongoing collaboration between education researchers and preschool and early elementary teachers (see Chittenden and Jones 1998) resulted

This chapter originally appeared, in a slightly different form, as Jones, J., and R. Courtney. 2003. Documenting early science learning. In D. Doralek and L. J. Colker (Eds.), *Spotlight on Young Children and Science*, pp. 27–32. Washington, DC: National Association for the Education of Young Children. Reprinted with permission from the National Association for the Education of Young Children.

Figure 4.1 First-Grade Children Observing Soil

© Rosalea Courtney, ETS

in a set of principles to guide the classroom-based documentation process. The three principles are

1. Collect a variety of forms of evidence.
2. Collect the forms of evidence over a period of time.
3. Collect evidence that shows the understanding of groups of children as well as the understanding of individuals.

1. Collect a variety of forms of evidence.
Young children express their ideas through their conversations, drawings, play, and constructions. We can learn a great deal about their thinking by looking carefully at records of children's language and samples of their work. Samples of various forms of evidence that are a part of most early childhood classrooms are described below. Most of the evidence of young children's learning should be a part of the everyday life of the classroom.
A *drawing*: A preschool child's observations of the spots on the class rabbit, the shape of its ears, and its bushy tail can be seen in her drawing (Figure 4.2).

Figure 4.2 Drawing of the Class Rabbit

Drawing and dictation: Drawings alone may not always reflect a young child's ideas and perceptions. The child's thinking about the rabbit, Baby, is more visible when the child's dictated comments are added to the drawing (Figure 4.3).

Figure 4.3 Drawing and Dictated Comments About the Rabbit

BABY

Kevin   7/3/9?
The Baby is
cute. The baby
sometimes feels like
walking if it does not
want to hop.

Photographs: Images of children at play can reveal their emerging science thinking. After a preschool class had finished shucking corn, the teacher took a photograph of a child's spontaneous construction of a "corn garden" in the block area (Figure 4.4). She pretends to water the garden.

Figure 4.4 A Child's Construction of a Corn Garden.

© Rick Toone, ETS

2. Collect the forms of evidence over a period of time.

Children's understandings of big ideas such as living things and changes in matter are not established firmly with one experience. Children need time to return to these ideas and concepts, to ask new questions, and to fit new learning into established ideas. The evidence of young children's learning is most useful when it is viewed over a period of weeks or months. For example, some children in a first-grade classroom observed the life cycle of silkworms from eggs to adult moths over several weeks (Figure 4.5). Some of their observations were recorded in their science journals (Figure 4.6). In one child's journal, a number of entries describe the silkworms' size and color: "They are brown when they are little and they are green and brown when they are big." She also notes differences between the newly hatched caterpillars and human babies: "The little ones that just hatched is not moving like us when we were babies. We did not know how to do anything like them." In a later entry, she writes about how the silkworm caterpillars are dependent on humans to find food for them: "They don't live in trees. They don't live in the ground like the other worms. They don't live by themselves. You have to take care of them."

Figure 4.5 Moths Emerging From Cocoons

© Rick Toone, ETS

Figure 4.6 Life Cycle of Silkworms: Science Journal Entries

4/10: They are brown when they are little and they are green and brown when they are big.

4/18: Their whole body looks like a elephant's trunk.

4/25: The little ones that just hatched is not moving like us when we were babies.

4/26: Their heads is fatter than their body. And their heads are white and their body is brown.

5/1: Now they are growing and changing colors. When they are babies, they were brown, but now they are white.

5/2: You have to take care of them. You have to give them some leaves. One is eating and Tyler is right. They do start by eating from the edge.

5/4: Now they are growing and they are learning how to eat fast. But when they were babies, they eat slow, but they are not little and more so now they eat faster. They were very hungry and my big fat one is eating fast just the way I said …

5/10: I have a question. What's the black stuff on top of their face?

5/11: One of my silkworms were making silk and I thought it was shedding skin, but it was not shedding skin. It was making silk.

5/19: When I cleaned out my container and gave my silkworms some leaves, … some of them have to share and they was fighting together and then was eating and one silkworm eat a whole leaf. And he still eating another leaf. He was really hungry.

Figure 4.7 What's Happening to These Caterpillars?

Jason: They don't look like caterpillars. They look like celery 'cause they are green.

Gabrielle: Two of them are starting to change into cocoons. They look like celery. One of the caterpillars is moving.

Morris: One is already in a cocoon. Four are still caterpillars. One is not moving so it's a cocoon. Cocoons can't move.

Sam: I see one moving

Jaella: One has something sticking out.

Kiri: Two look like trees. Some stuff is sticking out of the body.

Tom: One looks like a leaf. One is all squished up.

Zack: Two don't look the same. One looks like it has bumps on the side.

Jason: They don't look like themselves so they must be cocoons. They're not long like the others.

Figure 4.8 Preschool Children's Drawings of a Rabbit

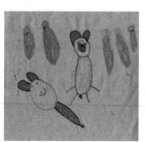

3. Collect evidence that shows the understanding of groups of children as well as the understanding of individuals.
Science is an inherently social activity, and children should be encouraged to discuss their ideas with other children.
Record of a class discussion: A kindergarten teacher has recorded the class discussion of what the children think is happening to a collection of caterpillars that the class has been observing (Figure 4.7). The children are engaged in close observation of the changes in the life cycle of a butterfly. They use metaphors to describe their observations (e.g., "They look like celery").

Preschool children made the collection of drawings in Figure 4.8 when a rabbit was brought to class. Although no carrots were present in the classroom during this visit, each of these drawings shows the class rabbit with the food. The association between rabbits and carrots appeared to be shared by many of the children.

The Documentation/Assessment Process
With the three guiding principles as a foundation, documentation and assessment of young children's emerging science understanding consists of a five-stage cycle: identifying, collecting, describing, interpreting, and applying the classroom-based evidence in order to plan more appropriate experiences and environments.

1. Identify appropriate
- science-related goals and concepts,
- activities and experiences, and
- classroom settings.

It is important to have some agreed upon notion of what we want children to experience, explore, and understand. In addition to specific curriculum goals, teachers who participated in this process often used the Benchmarks for Science Literacy (AAAS 1993) or the National Science Education Standards (NRC 1996) to guide their expectation for young children. The benchmarks and standards were especially useful in providing a focus for collecting those samples of children's work that highlight specific science goals.

2. Collect evidence of children's learning, including
- records of children's conversations and
- children's work samples.

Records of children's conversations and their work samples can take a variety of forms, including whole-class discussions, individual interviews, drawings, constructions, and diagrams. Consider which forms of evidence will give the best indication of how children are coming to understand the selected science goals and concepts.

3. Describe evidence of children's learning
- without judgment and
- with colleagues.

The first step in coming to understand what children are learning is to take a close look at what is actually in the language records, drawings, and constructions *prior to making a judgment* (Himley and Carini 2000). Description is a skill that takes some practice. It is easier to talk about what is missing or what is incorrect in children's statements or work sample than it is to focus on the knowledge that is represented. In addition, working with colleagues on a careful description of children's work samples and records of their language can provide new and useful insights into children's learning. Another teacher, or a parent, can bring a new perspective, often seeing things in the work that the child's teacher may miss.

4. Interpret evidence of individual and group understanding by
- connecting to learning goals and
- identifying patterns of learning.

At this stage the children's work should be compared to the standards and goals that were identified by the teacher during an earlier part of the cycle (Stearns and Courtney 2000). Does the work demonstrate the intended goals, such as observation or prediction? Are some additional types of work samples needed to demonstrate understanding? Are patterns of understanding emerging for the whole class? If, for example, the teacher wants the children to observe living things in the classroom (e.g., note changes that take place over time) and to ask questions about their observations, then journal entries can be used as evidence that these goals have been met.

5. *Apply* new information and understanding to the improvement of
- instruction and curriculum and
- future assessment.

The major purpose of assessment is to inform instructional practice. Therefore, the information from the documentation/assessment process must be tied directly to new planning. The process begins anew as the teacher uses information and insights gained from the process to identify the next set of the science-related goals and experiences. The process is an ongoing cycle in which children's emerging science understanding can be nurtured and documented in the everyday life of an early childhood classroom. Teachers have found this process valuable for understanding the learning of individuals and groups, for guiding instruction, and for reporting to parents.

References

American Association for the Advancement of Science (AAAS). 1993. *Benchmarks for science literacy.* New York: Oxford University Press.

Chittenden, E., and J. Jones. 1998. Science assessment in early childhood programs. In *Dialogue on early childhood science, mathematics, and technology education.* Washington, DC: American Association for the Advancement of Science.

Himley, M., and P. F. Carini, eds. 2000. *From another angle: Children's strengths and school standards.* New York: Teachers College Press.

National Research Council (NRC). 1996. *National science education standards.* Washington, DC: National Academy Press.

Stearns, C., and R. Courtney. 2000. Designing assessment with the standards. *Science and Children* 37: 51–55, 65.

## CHAPTER 5

Using Science Notebooks as an Informal Assessment Tool

Alicia C. Alonzo
University of Iowa

S cience notebooks have been promoted as a means to enhance students' scientific and literacy skills and as formative assessment tools for teachers. However, as with so much in educational practice, there are no magic bullets, and actually implementing science notebooks in the classroom and making them work for teachers and students can be challenging. This chapter won't provide easy answers or tricks. However, I hope it will help you to find ways to make science notebooks more effective tools for you and those you work with (whether teachers or students).

Research Basis

The ideas presented in this chapter draw on research with ethnically and linguistically diverse students in grades 3–5. Study #1 was conducted with 34 third-grade teachers in 15 schools in a large school district in Southern California. The district had a long history of using both hands-on kits and notebooks as part of its elementary science curriculum. The students and teachers in this study were conducting investigations of floating and sinking, using the Elementary Science Study (ESS) (1986) Clay Boats unit. In the examples presented here, students were trying to shape pieces of clay to float and carry cargo. Study #2 involved 19 teachers in four districts in California. Depending on the district curriculum, the unit being studied—Educational Development Center (EDC)'s (1990) Circuits and Pathways—was taught at either fourth or fifth grade. In the notebook entries considered here, students were trying to light a small lightbulb with a battery and 1–2 wires. Study #3 was conducted with 10 teachers in the Bay

Area of California and involved two Full Option Science System (FOSS) kits: Mixtures and Solutions (FOSS 1993a) and Variables (FOSS 1993b).

Scope

A wide variety of formats—ranging from collections of worksheets from kit-based curricula to structured records of hands-on work to informal records of students' scientific thinking—have all been described as "science notebooks." For example, Ruiz-Primo and colleagues (Ruiz-Primo and Li 2004; Ruiz-Primo et al. 2004) examined science notebooks that often included pages from FOSS curriculum materials. Others (e.g., Baxter, Bass, and Glaser 2000; Campbell and Fulton 2003; Klentschy 2005) have focused on science notebooks as collections of blank pages to be filled with records of students' hands-on scientific work. (This type of science notebook would typically include components such as those shown in Figure 5.1. However, there is a wide variety in the degree of structure and format of these records as well.) Finally, some notebooks serve more like "journals," places for students to record their ideas. As we will explore later in this chapter, each has its strengths and weaknesses.

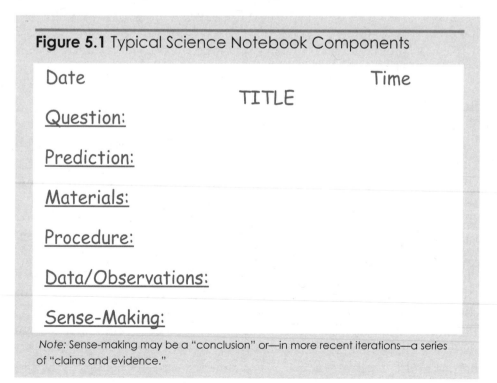

Figure 5.1 Typical Science Notebook Components

Date   TITLE   Time

Question:

Prediction:

Materials:

Procedure:

Data/Observations:

Sense-Making:

Note: Sense-making may be a "conclusion" or—in more recent iterations—a series of "claims and evidence."

CHAPTER 5

As its title implies, this chapter focuses primarily on the use of science notebooks for a specific type of assessment. While notebooks may be collected and assessed in formal ways (both formatively and summatively), this chapter considers more *informal* uses of notebooks as sources of assessment information. Informal assessment may be defined in terms of its "spontaneity" (e.g., Oosterhof 1994)—that is, the information we collect is what we happen to notice, in this case about students' notebook writing as we circulate through the classroom. However, the spontaneity of capitalizing on moment-to-moment classroom interactions is definitely not the same as unplanned assessment. This chapter examines the decision-making processes that prepare us to take advantage of the informal assessment opportunities offered by having students write in science notebooks. Because students' notebooks may contain a record of what they are doing during an investigation, we can get more information about their work than we might if we just saw what they happen to be doing at the time we look their way.

I will explore three key decision points in the use of science notebooks. To illustrate each point, I raise a question (or questions) to ask yourself about the use of science notebooks in your own classroom. I use student work to illustrate the importance of those questions, present related research results, and conclude with a brief discussion of the decision point.

Decision Point #1
Consider the following two questions about your use of science notebooks:

- Why am I having students write in science notebooks?
- When I look at students' notebooks, what do I want to learn?

Before reading any further, take a moment to examine the student notebook sample in Figure 5.2, page 86. Make a list for yourself of the things you notice about this student's work.

Figure 5.2 A Third Grader's Science Notebook Entry

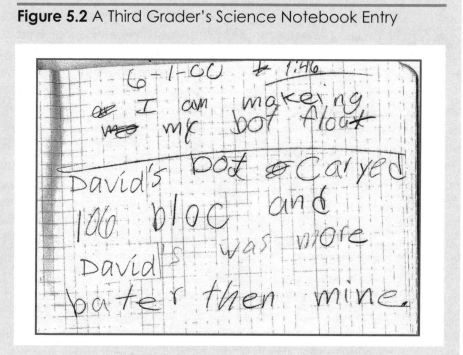

If you're like most elementary school teachers, one of the first things you noticed was the student's spelling. Although your experience reading text like this may have meant that you had no trouble figuring out that the student was making his boat float and that David's boat was better because it carried 106 blocks, the first thing that jumped out at you may have been the unconventional spelling and grammar. This is understandable, given the heavy emphasis on literacy in elementary school—and the expertise of many elementary school teachers. However, this focus can mean that we are less likely to notice scientific aspects of the student's work. For example, a more scientific lens may lead us to notice that the student has recorded data (an exact number of blocks that David's boat held) and a comparison (which boat held more). But we don't know how many blocks his boat held or anything about differences between the two boats that may account for the different number of blocks.

In Study #1, I found that the majority of written feedback provided to third graders focused on language issues—spelling, grammar, and neatness of writing (Alonzo 2001). You may wonder if teachers' *oral* comments to students in class focused on more substantive scientific issues, but classroom observations revealed that most of the talk between teachers and students about the notebooks focused on missing elements. (For example, this student wrote the date and time at the top of his paper, most likely as a result of teacher prompting.)

In Study #2, in which teachers were engaged in professional development to support their use of science notebooks, 0–27% of the notebook feedback addressed language issues (Alonzo and Aschbacher 2005). This professional development focused, in part, on what to look for in students' notebooks and how to provide effective feedback to students. However, even with this explicit focus, a full quarter of some teachers' written comments focused on writing. It's a hard bias to overcome!

Because informal assessment is so dependent on what we happen to notice about a piece of student work, it's especially important to recognize this tendency. It makes sense to notice students' writing if that's the purpose of our assessment, but in science notebooks, we are often interested in students' understanding of scientific content and processes. Being clear about the purpose in our own minds helps to ensure that we take note of students' scientific understandings.

If your purpose in having students write in science notebooks *is* for literacy development, then it makes sense to focus your assessment strategies on literacy. However, many other purposes for using science notebooks—for example, helping students to think and write scientifically (e.g., Stokes, Hirabayashi, and Ramage 2003) or providing a record of student thinking for reflection and formative assessment (e.g., Aschbacher and Alonzo 2006; Shepardson and Britsch 2000)—require noticing different aspects of students' work. If you start with the purpose and tie what you notice to that purpose, notebooks are more likely to give you the information you are looking for.

Decision Point #2

We now turn to the second decision point in the use of science notebooks. Given your purposes for having students write in science notebooks and looking at what they have written, this section provides another set of questions to consider:

- What do I need to ask students to write in their notebooks?
- What type of notebook structure will best meet my goals?

Before reading any further, examine the student notebook sample in Figure 5.3. Make a list for yourself of the things you notice about this student's work. In particular, consider what you think this student understands about the day's investigation.

Figure 5.3 A Third Grader's Science Notebook Entry

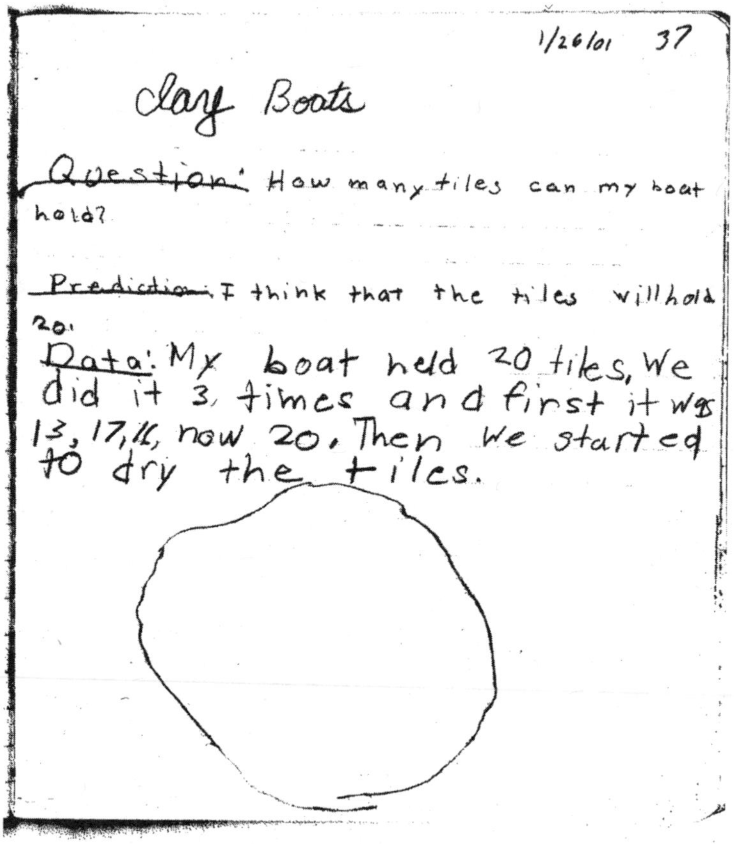

There is a lot to see in the student's entry in Figure 5.3. Based on her written work, we might conclude that she understands something about repeated measures. We might question her understanding of a prediction, noticing that her boat "magically" held the exact number of tiles she had predicted. We might notice that she has provided some information about the shape and size of her boat (by tracing its outline), but that other information (such as the height of its sides) is missing. But what do we know about what the student learned from this investigation? Returning to our list of what components might be found in a student's notebook (Figure 5.1, p. 84), we notice that the student's entry stops with her data. There is no sense-making or record of the student having used her investigation to understand something about the way the world works.

And, in fact, this student's entry is not at all atypical. As we move down the list in Figure 5.1, we see that the components are of *increasing* importance in terms of students' scientific understandings. Research has shown, however, that student notebooks tend to have a *decreasing* presence of those components. Sense-making writing is rare in students' notebooks. Findings from Study #1 (Alonzo 2002b) are shown in Table 5.1, page 90. The third-grade teachers in this study were more than twice as likely to ask their students to record their data as they were to ask them to write anything about what that data meant. And students complied with requests for data more than twice as often as they did requests for sense-making writing. While this is not surprising, given the often rushed and chaotic nature of the time at the end of a hands-on investigation, the result is that only 11% of the students' notebooks contained any sense-making writing. Similar results were found in Study #3 (Ruiz-Primo and Li 2004), in which notebooks were 5 and 13 times more likely to contain an entry that reported results than they were to contain an entry that made sense of those results in the Variables unit and the Mixtures and Solutions unit, respectively.

Table 5.1 Opportunities for and Occurrence of Data and Sense-Making Writing in Students' Notebooks

	Notebook Component	
	Data	**Sense-Making**
% of lessons in which teachers asked students to write the notebook component	84%	37%
% of notebooks that included the notebook component when requested	72%	29%

Note: This data result from an in-depth analysis of observations and notebooks from a subset (n = 6) of teachers in the larger Study #1 during two consecutive academic years. These teachers represent a range of facility with the use of science notebooks and the teaching of Clay Boats (ESS 1986). Classroom observations (6–10 per teacher) and class sets of notebooks (15–20 per teacher per year) were analyzed.

But, it is the sense-making components that have the greatest potential for advancing student understanding. Literature on writing-to-learn tells us that the act of writing to make sense of difficult content may actually improve students' understanding of that content (e.g., Bereiter and Scardamelia 1987; Rivard 1994). This is supported by the results of Study #3 (Ruiz-Primo et al. 2004). Students were given pretests (at the beginning of each unit) and posttests (at the end of each unit), and classrooms were classified as low, medium, or high according to the gains in student achievement. In the Variables unit (FOSS 1993b), notebooks in the high-gain classrooms were four times more likely to include sense-making components than were those in the medium- and low-performing classrooms. (While we cannot conclude from these results that writing sense-making components *caused* the higher performance, this possibility is certainly worth exploring through research in more controlled studies.)

In addition, it is the sense-making components, in which students are writing about (and thus revealing to the teacher) their thinking about a concept, that offer the richest opportunities for formative assessment. Through these sense-making writings, we can learn what students are thinking about the content and then adapt instruction to better serve their developing understanding.

Here, I advocate a "less is more" approach. Why spend a lot of class time and energy nagging students to write notebook components that do not meet your goals for their learning? Although there may be times when you want students to produce a "complete" notebook entry, with the date, time, materials, and so forth, most of the time you will get more productive writing from your students by focusing attention on the notebook entries that will help students to reach your goals *and* allow you to learn what you need to from their notebooks.

More structured notebook entries—for example, in which a question is provided, along with a sentence-starter for students' predictions—may lend themselves to focusing on students' data and resulting claims and evidence. On the other hand, if a major focus of your instruction is teaching students how to write their own questions and make predictions, a more open-ended format would be more appropriate. It all comes down to your purpose for having students write a particular science notebook entry.

Decision Point #3

We now turn to the third and final decision point in the use of science notebooks. Given what you have decided to have students record in their notebooks and your goals for those entries (both for your students and yourself), ask yourself:

- What support will I need to offer my students so that their writing satisfies my purposes?

Before reading any further, examine the student's notebook sample in Figure 5.4, page 92. What do you notice about this student's work? What did he learn from the day's investigation?

Figure 5.4 Part of a Third Grader's Science Notebook Entry

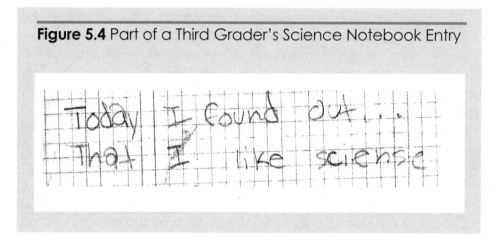

Using science notebooks requires a delicate balance between providing enough support so that notebook entries are useful to both students and teacher and providing so much support that the entry no longer represents the student's thinking. The example in Figure 5.4 is quite common. At the end of a lesson, teachers often ask students to write what they learned or "found out" through their hands-on investigations. However, without indications of what an acceptable "conclusion" might look like, students will write about whatever is most salient to them. In addition to conclusions like that in Figure 5.4, I have seen many notebook entries in which students wrote that what they are taking away from a lesson is that they don't like working with a particular classmate. This may be the idea about the lesson that is foremost in the student's mind, and maybe the difficulty with working cooperatively was such that this is *all* the student learned; however, it is more likely that the child wasn't sure what else to write. This student wasn't adequately scaffolded in his writing so that he could reveal what he had learned.

In this case, an important form of scaffolding could come through the words chosen to prompt student writing. Instead of asking students what they "found out," support students by asking a more specific question, focused on the targeted content for the day's lesson. This helps to ensure that notebook writing represents what students understand about the content and not just what they (mis)understand about what they are supposed to be writing. As students become more familiar with different notebook components, they may require less of this type of support. For example, through-out the year, a teacher might help students to see their sense-making writing as responses to the questions posed to frame their investigations so that, by the end of the year, she doesn't have to specify what students should be writing about. Her students have come to understand what sort of writing is expected in this section of their notebooks.

Entries like the one in Figure 5.4 or that in Figure 5.5 make it difficult to distinguish between a child who doesn't understand the genre and one who doesn't understand the content. This is where teachers have a *tremendous* advantage over outside observers. You can use notebook entries like these to consider understanding of both content and process because you know what supports were offered for each. Looking at the notebook entry in Figure 5.5, you would know whether the idea of providing evidence for a claim is new to this student and, thus, what is reasonable to expect in terms of the quality of evidence provided. Ideally, the student would be able to supply evidence from the hands-on investigation with batteries and bulbs to support his claim, rather than appealing to your authority. This evidence is important as part of "doing science"; it also helps to show student thinking about the science content. If this student could provide evidence for his claim from his own investigation, you might be more confident that he understands "the critical point" and hasn't just written down something he heard during a class discussion.

Figure 5.5 Part of a Fifth Grader's Science Notebook Entry

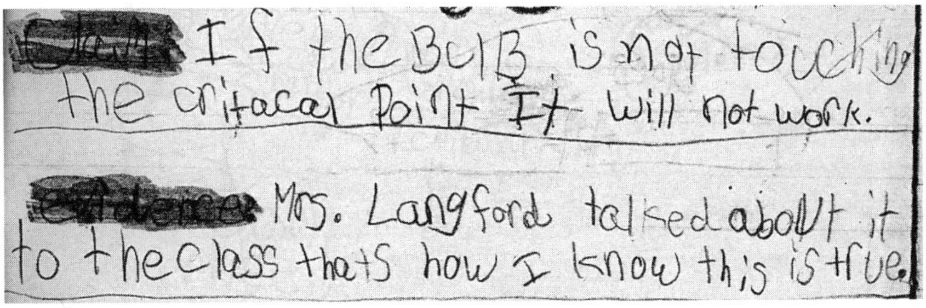

An important, and often overlooked, piece of support for science note-books is in the recording of data. Consider the drawing in Figure 5.6. Does this student understand how to light a bulb? Because her bulb is not drawn with enough detail to show the "critical contact points" (the base and side of the bulb, which must be touched in order for it to light), we can't tell whether her circuit would light and thus whether she is developing an understanding of electric circuits. It is easy to confuse sloppiness with conceptual difficulties. In Study #2 (Alonzo and Aschbacher 2005), teachers would often dismiss students' notebook entries such as these by saying that the students were simply being sloppy. They assumed that students understood how to light the bulb but were just not recording neatly what they knew. These teachers were *shocked* to discover that their students could not complete a fairly simple predictions page—placed toward the middle of the Circuits and Pathways unit (EDC 1990)—in which students were shown a series of circuits and asked whether each would light the bulb. Showing students how to record meaningful data helps them to develop important scientific skills (observ-ing and recording data carefully) *and* allows you to determine whether their notebook entries represent confusion or understanding.

Figure 5.6 Does This Fourth Grader Understand How to Light a Bulb?

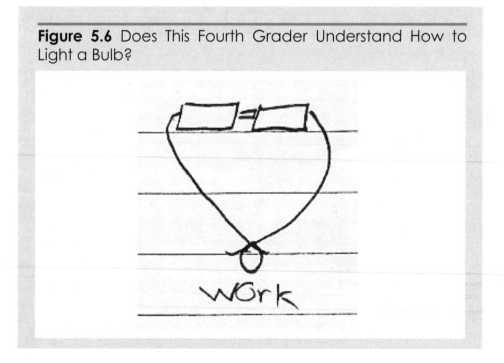

In Study #1 (Alonzo 2002a), several examples of how to scaffold students' developing understanding of recording data were documented. In one class, students learned to accurately depict the clay boats they were building. The data table shown in Figure 5.7 was the result of several cycles of modeling and practicing. The teacher started by holding up a student's clay boat and showing how she looked at the boat from the top and from the side in order to draw different views that together would represent a 3-D object on paper. She asked student volunteers to come to the front of the class and guided them through the process of depicting their boats on a class chart. And finally, she gave students feedback and guidance as they recorded their own shapes in their notebooks.

Figure 5.7 A Third Grader's Science Notebook Entry Produced After Significant Modeling and Practice

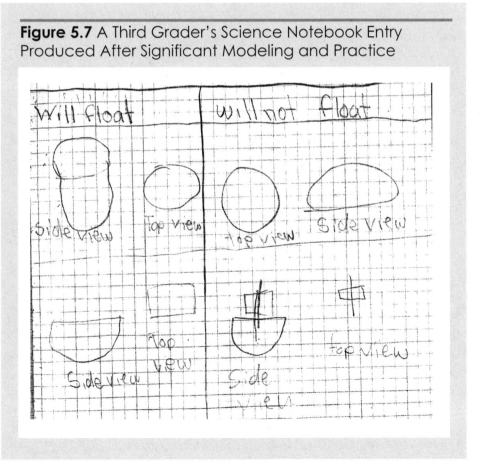

Another teacher had her students critique a sample notebook entry (Figure 5.8) from a previous class. She asked her students what they could tell from the student's entry and what they thought the student should do differently to help them better understand what he did and found out. This helped her students to internalize criteria for high-quality notebook entries to improve their own entries.

Figure 5.8 A Third Grader's Science Notebook Entry Used in a Notebook-Critiquing Discussion

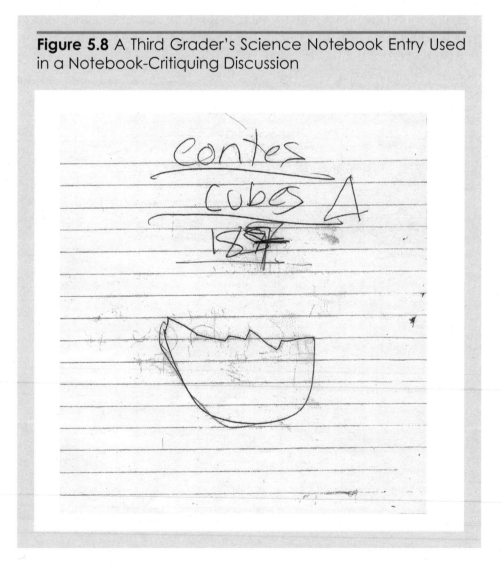

CHAPTER 5

Summary

The use of science notebooks requires a set of decisions. The first involves the purpose. What do you want students to gain from writing in their science notebooks? (Common purposes include science content knowledge, science-process skills, and literacy skills, but you may have others as well.) Given this purpose, what information do you hope to gain from students' science notebooks for your own formative assessment purposes?

The second decision involves the format and content of the science notebook. Choices range from collections of curriculum worksheets to informal records of students' scientific thinking but most commonly take a "lab report" format—a structured record of students' hands-on work, including some variation of the components shown in Figure 5.1. What format and components will best support your purposes for having students write in science notebooks?

The final decision involves scaffolding. What supports do students need in order to record information in their notebooks that helps both them and you to meet your purposes? This scaffolding includes both the prompts (what you actually ask them to write), as well as the support provided to help them understand the different components of a science notebook and what a high-quality entry might look like.

The bottom line is this: There is a lot that students can write in their notebooks, but writing too much can mean that none of it is done well. What is your purpose for having students write in science notebooks? How does your use of science notebooks support this purpose?

Acknowledgments

Study #1 was funded by a National Science Foundation (NSF) Postdoctoral Fellowship in Science, Mathematics, Engineering and Technology Education (PFSMETE), Grant # DGE-9906498. Study #2 was conducted in collaboration with Pamela R. Aschbacher (PI) and colleagues in the research division of the Caltech Pre-College Science Initiative, under NSF Grant # REC-0106994. Study #3 was conducted by Maria Araceli Ruiz-Primo, Min Li, Carlos Ayala, and Richard Shavelson, at the Stanford Education Assessment Laboratory. Their work was supported by NSF Grants # SPA-8751551 and TEP-9055443. Any opinions, findings, conclusions, or recommendations expressed in this chapter are those of the author. They do not necessarily represent the official views or policy of the National Science Foundation or those of the PIs and others involved in Study #2 and Study #3.

References

Alonzo, A. C. 2001. Using student notebooks to assess the quality of inquiry science instruction. In P. R. Aschbacher (Chair), *Challenges in assessing evidence of learning and teaching in elementary science.* Symposium conducted at the annual meeting of the American Educational Research Association, Seattle (April).

Alonzo, A. C. 2002a. Making science notebooks work in elementary school classrooms. Presentation at the annual meeting of the National Science Teachers Association, San Diego, CA (March).

Alonzo, A. C. 2002b. Validity issues in using students' science notebooks as a window into classroom practice. In A. C. Alonzo (Chair), *Challenges of validity and value in using students' science notebooks to improve science teaching and learning.* Symposium conducted at the annual meeting of the American Educational Research Association, New Orleans, LA (April).

Alonzo, A. C., and P. A. Aschbacher. 2005. Factors which contribute to teachers' adoption of science notebooks as a formative assessment tool. In A. C. Alonzo (Chair), *Teachers' use of formative assessment in science classrooms: Factors in the development of expertise.* Symposium conducted at the annual meeting of the American Educational Research Association, Montreal, Canada (April).

Aschbacher, P. R., and A. C. Alonzo. 2006. Examining the utility of elementary science notebooks for formative assessment purposes. *Educational Assessment* 11: 179–203.

Baxter, G.P., K. M. Bass, and R. Glaser. 2000. *An analysis of notebook writing in elementary science classrooms.* (Center for the Study of Evaluation Report 533). Los Angeles: National Center for Research on Evaluation, Standards, and Student Testing.

Bereiter, C., and M. Scardamelia. 1987. *The psychology of written composition.* Hillsdale, NJ: Lawrence Erlbaum.

Campbell, B., and L. Fulton. 2003. *Science notebooks: Writing about inquiry.* Portsmouth, NH: Heinemann.

Educational Development Center (EDC). 1990. *Insights: Circuits & pathways.* Newton, MA: Educational Development Center.

Elementary Science Study (ESS). 1986. *Clay boats teacher's guide.* Hudson, NH: Delta Education.

Full Option Science System (FOSS). 1993a. *Mixtures and solutions.* Chicago, IL: Encyclopedia Britannica Educational Corporation.

Full Option Science System (FOSS). 1993b. *Variables.* Chicago, IL: Encyclopedia Britannica Educational Corporation.

Klentschy, M. 2005. Science notebook essentials. *Science and children* 43(3): 24–27.

Oosterhof, A. 1994. *Classroom applications of educational measurement* (2nd ed.). New York: Merrill.

Rivard, L. P. 1994. A review of writing to learn in science: Implications for practice and research. *Journal of Research in Science Teaching* 31: 969–983.

Ruiz-Primo, M. A., and M. Li. 2004. On the use of students' science notebooks as an assessment tool. *Studies in educational evaluation* 30: 61–85.

Ruiz-Primo, M. A., M. Li, C. Ayala, and R. J. Shavelson. 2004. Evaluating students' science notebooks as an assessment tool. *International Journal of Science Education* 26: 1477–1506.

Shepardson, D. P., and S. J. Britsch. 2000. Analyzing children's science journals. *Science and Children* 38(3): 29–33.

Stokes, L., J. Hirabayashi, and K. Ramage. 2003. *Writing for science and science for writing: The Seattle elementary Expository Writing and Science Notebooks program as a model for classrooms and districts.* Inverness, CA: Inverness Research Associates. Retrieved May 3, 2005 from *www.inverness-research.org/reports/seanotebks_nov03/SeanotebksReport03.pdf*

Assessing Middle School Students' Content Knowledge and Reasoning Through Written Scientific Explanations

Katherine L. McNeill
Boston College

Joseph S. Krajcik
University of Michigan

I think that probably one of my favorite components is writing the scientific explanations, because the kids actually have to think and they actually have to be scientific.... They have to look at a data chart and pull the evidence out of there.... You gotta know what you are doing in order to be able to complete it. It touches on the science, it touches on the writing skills, and it touches on critical thinking.
—Ms. Hill, practicing middle school teacher in a large urban area

Science is fundamentally about explaining phenomena in the world. Explaining involves making claims and justifying those claims with appropriate evidence and reasoning. The National Research Council's *National Science Education Standards* (NRC 1996) states that "scientists evaluate the explanations proposed by other scientists by examining evidence, comparing evidence, identifying faulty reasoning, pointing out statements that go beyond the evidence, and suggesting alternative explanations for the same observations" (p. 148).

Moreover, the NRC argues that it is important to engage students in a similar process of constructing and critiquing scientific explanations in science classrooms. Science education research has shown that when students construct scientific explanations it may change their image of science (Bell and Linn 2000), foster deeper understanding of important science concepts (Zohar and Nemet 2002), and help them write stronger explanations in which they justify their claims (McNeill et al. 2006).

Besides the numerous benefits for student learning, students' written scientific explanations can serve as an important assessment tool for teachers. Having students engage in this complex inquiry practice makes students' scientific thinking and reasoning visible to teachers, allowing them opportunities for assessment (Osborne, Erduran, and Simon 2004). By examining students' written scientific explanations, teachers can develop an understanding of how well their students apply and use scientific concepts to explain phenomena. In this chapter, we discuss an instructional framework for scientific explanations, provide a set of rubrics to assess student work, illustrate types of feedback to give students, and describe how to develop assessment tasks to assess students' written scientific explanations.

Instructional Framework for Scientific Explanation

Although scientific explanations are an important learning goal, students often have difficulty appropriately justifying their claims (Sadler 2004). Consequently, we developed an instructional framework for scientific explanation based on our experiences working with middle school teachers in Detroit and on the existing science education research literature (McNeill et al. 2006; Moje et al. 2004). In other work, we discuss in more detail different instructional strategies teachers can use to introduce and support students in scientific explanation (McNeill and Krajcik 2008) and the connections between scientific explanations and literacy (Sutherland et al. 2006). In this chapter, we focus on issues relating to assessment.

Our instructional framework breaks down scientific explanation into three components: claim, evidence, and reasoning. The *claim* is a statement or conclusion that answers the original problem or question that the students are trying to answer. We found that this is the easiest component for students to construct. The *evidence* is scientific data that support the claim. The evidence can come either from investigations that the students complete or from secondhand sources such as newspapers, books, or the internet. The evidence should be both appropriate and sufficient. By appropriate, we mean it is directly relevant to the current problem and allows the student to figure out his or her claim. Sufficiency stresses the idea that students should not rely on just one piece of data; rather, they should consider and use multiple pieces of evidence to support their claim. The *reasoning* is the justification for why their data count as evidence to support their claim, which often requires the use of scientific principles. Scientific principles can help students determine what data do or do not count as evidence for a particular claim. We will illustrate this framework with student examples.

Assessing Students' Written Explanations

We developed a general or base explanation rubric (Appendix A, p. 115) for scoring scientific explanations across different content and learning tasks (Harris et al. 2006; McNeill et al. 2006). It includes the three components of a scientific explanation and offers guidance to think about different levels of student achievement for each component. The base rubric needs to be adapted to create a specific rubric for a particular task. The specific rubric combines both the general structure of a scientific explanation with the appropriate science content for the particular task. Appendix B, page 115, provides the specific rubric for the chemistry assessment task in Figure 6.1, page 104, which is part of a middle school science curriculum unit that focuses on properties, substances, chemical reactions, and conservation of mass (McNeill et al. 2004). This assessment task requires that students apply the scientific principles that different substances have different properties and that a property (such as density, color, and melting point) is a characteristic of a substance that does not change even if the amount of the substance changes. Figure 6.2, page 104, gives an example from a seventh-grade student who wrote a strong scientific explanation for this problem.

Figure 6.1 Scientific Explanation Assessment Task for Substance and Properties

Examine the following data table:

	Density	Color	Mass	Melting Point
Liquid 1	0.93 g/cm³	no color	38 g	-98 °C
Liquid 2	0.79 g/cm³	no color	38 g	26 °C
Liquid 3	13.6 g/cm³	silver	21 g	-39 °C
Liquid 4	0.93 g/cm³	no color	16 g	-98 °C

Write a **scientific explanation** that states whether any of the liquids are the same substance.

Figure 6.2 Student Example of a Strong Scientific Explanation

Write a **scientific explanation** that states whether any of the liquids are the same substance. Liquid 1 and 4 are the same substance. They both have a density of .93g/cm³, have no color, and start to melt at -98°C. For substances to be the same, they must have the same properties. Since Liquids 1 and 4 have the same properties, they are the same substance. The other 2 liquids are different substances because they have different properties.

We can use the specific scientific explanation rubric for this task (Appendix B, p. 115) in order to assess the student's response in Figure 6.2. The student would receive the highest score of a 2 for claim, because she provided an accurate and complete claim that "Liquid 1 and 4 are the same substance."

She would also receive a 2 for evidence, because she provided three appropriate pieces of data as evidence by including density, color, and melting point. Finally, she would receive a 2 for reasoning, because she links her evidence to her claim by noting that same substances have the same properties while different substances have different properties. She states, "For substances to be the same, they must have the same properties. Since Liquids 1 and 4 have the same properties, they are the same substance. The other 2 liquids are different substances because they have different properties."

The next example (Figure 6.3) illustrates a seventh-grade student who had more difficulty writing a scientific explanation for this task. Again, this student would receive a 2 for his claim, because he provides an accurate and complete claim. Yet, he has a more difficult time justifying or supporting his claim with appropriate evidence and reasoning. He provides two pieces of appropriate evidence when he writes "Liquid 1 and liquid 4 have the same density and melting point." Because he does not also mention that they have the same color, he would not receive a 2 for his evidence. Consequently, we would give the evidence a score of 1, because it is not sufficient. Finally, we scored the reasoning a 0, because the student did not provide a link for why having the same density and melting point count as evidence to support the claim that Liquid 1 and 4 are the same substance. To receive points for reasoning, the student needs to discuss that samples of the same substances have the same properties.

Figure 6.3 Student Example of a Weaker Scientific Explanation

Two of these substances are the same because Liquid 1 and liquid 4 have the same density and melting point. But not the mass, but their could be more water in Liquid 1 than Liquid 4 or a biger container.

In Figure 6.3, the student did provide a bit of a discussion of why having different mass might not be important in that he wrote "their could be more water in Liquid 1 than Liquid 4 or a larger container." Unfortunately, the first statement about there being more water in Liquid 1 than Liquid 4 is not a scientifically accurate reason for why the masses would be different, even if the two liquids were the same substance. If there were more water in one compared to the other, then the compositions would not be the same and the densities would also be different. The student's response provides the teacher with some insight into his thinking, suggesting confusion about how to differentiate between substances and about the concept of mass. Teachers can also use the knowledge they obtain of their students' understanding to make decisions about future instruction.

We provide these rubrics (Appendixes A and B, p. 115) as flexible tools for other teachers, administrators, and researchers to adapt for their needs. In evaluating students' evidence for this particular task, another individual could decide that having three levels (0, 1, and 2) is not enough to completely capture the students' understandings. We have also scored this task using four levels for evidence: a level 3 included three pieces of appropriate evidence, a level 2 included two pieces of appropriate evidence, a level 1 included one piece of appropriate evidence, and a level 0 included no appropriate evidence. There are tradeoffs between detail and depth of rubrics for assessments and how much time it takes to examine student work. The individuals using the tool should determine the right balance to meet their needs. But we have found the rubrics to be a helpful resource for ourselves and the teachers we work with in assessing students' abilities to apply science concepts to explain phenomena.

Providing Feedback

Teachers can use students' scientific explanations to obtain information about their students' understanding. They can also use the explanations as the basis for the feedback they give their students, which in turn will help students improve their explanations. Feedback that focuses on what needs to be done—with the specific goal of helping students learn (rather than being simply a rating of achievement)—can improve student performance (Black 2003).

Teachers can provide feedback on explanations that students create as well as on other examples of student work. Teachers can use anonymous written student examples from other classes as a focus of classroom discussion to critique strengths and weaknesses and create a class consensus of a satisfactory response (Duschl 2003). Teachers can also provide feedback on spoken explanations offered during class discussion.

Feedback on scientific explanations can address a variety of different features. For instance, teachers can use the rubric to guide their comments on the different components of a scientific explanation (i.e., claim, evidence, and reasoning). In other cases, a teacher may find that a student's scientific explanation illustrates a particular difficulty a student is having with the science concepts. That teacher may choose to focus on the science content instead of the structure of the scientific explanation. Finally, it is important to realize, although we use the three components (claim, evidence, and reasoning) to guide students' explanation construction, our ultimate goal is that students write a coherent explanation and not three separate, unrelated components. Consequently, teachers may comment on the holistic quality of a scientific explanation and how it hangs together as a whole.

Comments on students' explanations should be explicit and clear. Simply providing feedback such as "good work" or "needs more detail" is rarely useful for a student. The student might not understand what was good about his or her explanation or what aspects need more detail. Specifically pointing out strengths and weaknesses can help students understand what to include in their writing. For example, for the second example (Figure 6.3) a teacher might say, "The explanation did a great job talking about density and melting point, but it did not include a discussion of color." For a particular weakness, the teacher might also provide suggestions on how to improve that aspect of the scientific explanation. In terms of color, a teacher might comment, "You should talk about whether the color of liquid 1 and 4 is the same or different and whether or not that piece of data is important for your claim." Finally, it can also help to ask students questions to promote deeper thought. With the student's reasoning in the second example, it might be beneficial to ask him questions to get a better sense of what he was thinking when he was writing the explanation. For example, a teacher could ask, "Why did you use density and melting point as evidence for your claim? Why are these pieces of data important? Do you think mass is important to determine if they are the same substance? Why or why not?" These types of questions can help the teacher better understand the student's thinking as well push the student's own understanding about both the science content and scientific explanation.

Creating Explanation Assessment Tasks

So far we have used an example scientific explanation task that we developed to assess students' understanding of substances and properties. Developing good assessment tasks that align with learning goals or standards can be difficult. In this section, we describe the six steps that we go through in order to develop scientific explanation assessment tasks. Following these steps helps ensure that the task will call on students to combine their understanding of the content and their ability to write scientific explanations. The steps also help align the task with the desired national science content standards and make certain that the task actually measures the desired content. Previously, we discussed a chemistry example. In this section, we focus on a middle school life science standard to illustrate how students can engage in scientific explanations across different content areas.

Step #1: Identify and unpack the content standard.
The first step involved in writing an assessment task is to identify the content standard you wish to assess. Unfortunately, there are often far more ideas in one content standard than you might initially think. It is important to closely examine and unpack the content standard to determine the science ideas that you are trying to help students learn and assess. When interpreting standards it can be helpful to (1) break down the standard into related concepts, (2) clarify the different concepts, (3) consider what other concepts are needed, and (4) make links if needed to other standards. For example, one of the American Association for the Advancement of Science (AAAS 1993) benchmarks for life science is given in Figure 6.4. We will use this benchmark to develop an explanation assessment task.

Figure 6.4 The Interdependence of Life Benchmark

Two types of organisms may interact with one another in several ways: They may be in a producer/consumer, predator/prey, or parasite/host relationship. Or one organism may scavenge or decompose another. Relationships may be competitive or mutually beneficial. Some species have become so adapted to each other that neither could survive without the other. (p. 117; 5D, Grades 6 through 8)

This benchmark includes a variety of different ideas. One important idea is predator and prey relationships, which we decided to focus on for this assessment task. Addressing this idea meant that we needed to clarify the meaning behind the phrases "organisms may interact" and "predator/prey relationship." We clarified their meaning in the following manner: A biological interaction is a relationship between two species in the natural world. One type of relationship that two organisms can have is a predator and prey relationship. A predator is an animal that captures and consumes other organisms for food; the prey is the organism that is captured and consumed for food. This relationship can result in stable populations for both species. However, if a predator enters a new environment it can cause the population size of a prey to drop, even possibly resulting in no more prey in that environment.

It is also important to consider what prior knowledge students will need to understand this concept as well as what nonnormative ideas or misconceptions they may have about the content. You may decide to include some of the misconceptions in your design of the assessment task. For example, our clarification in the previous paragraph assumes that students are already familiar with the concepts of organism, food, and populations. Some students, however, may not understand the link between the death of one organism and the change in a population size over time.

Step #2: Unpack the scientific inquiry practice.
It is important to consider what you are looking for in terms of the scientific inquiry practice (e.g., design, models, scientific explanation). In this chapter we specifically focus on scientific explanation, which we broke down into three components: claim, evidence, and reasoning. These components correspond to the base rubric (Appendix A, p. 115) and clearly articulate what the teacher is looking for in terms of the assessment task. Unpacking the scientific inquiry practices specifies what it is that you will want students to do with their understanding of the science content. If you create an assessment task for a different scientific inquiry practice (e.g., design of investigation), you would need to unpack that inquiry practice and create a corresponding base rubric.

Step #3: Create learning performance.
The scientific explanation assessment task measures both students' understanding of the science content and the scientific inquiry practice. To help us think

about what it means to combine the content and inquiry practice, we create "learning performances." Learning performances specify what students should be able to do with the content knowledge. We do not want students to simply memorize the content standard and recite it back to us. Rather, we want them to apply the content as they engage in scientific inquiry. A learning performance can be thought of as a cross between the content standard and an inquiry practice. The learning performance clarifies how the content knowledge is used in reasoning about a scientific phenomenon. Figure 6.5 is an example learning performance. It combines the content learning goal (i.e., predator/prey) with the scientific inquiry learning goal (i.e., scientific explanation).

Figure 6.5 Example Learning Performances

Content Standard	x Scientific Inquiry Standard	= Learning Performance
Two types of organisms may interact with one another in…a predator/prey relationship. (AAAS 1993, 5D: M2 , p. 117)	• Develop descriptions, explanations, predictions, and models using evidence. (NRC 1996, p. 145) • Think critically and logically to make the relationships between evidence and explanation. (NRC 1996, p. 145)	Students construct a scientific explanation that includes a *claim* about whether two organisms have a predator/prey relationship, *evidence* in the form of the characteristics of the organisms and their populations, and *reasoning* that predators consume prey and that when predators enter a new environment they can cause a drastic decrease in the prey population size.

Step #4: Write the assessment task.
Next we design an assessment task that would result in students applying both their content and scientific explanation understandings to create the

desired product. We do this by aligning the assessment task to the learning performance that we previously created. Figure 6.6 shows the assessment task that we created for the predator and prey scientific explanation learning performance.

Figure 6.6 Scientific Explanation Assessment Task for Predator and Prey

Animal A lives in a grassland area where it eats grasses, roots, and weeds. Animal A lives in small tunnels it creates under the soil. At time point 7, Animal B moves into that environment. Animal B is a much larger animal and eats small animals, insects, and reptiles. Write a scientific explanation that tells what type of relationship these two animals have with each other.

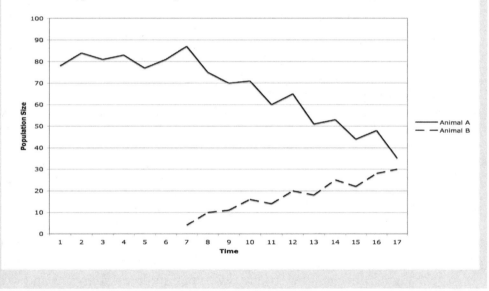

Step #5: Review the assessment task.
After creating the assessment task, we use three questions adapted from the Project 2061's assessment framework (DeBoer 2005) to review our task:

1. Is the knowledge needed to correctly respond to the task?
2. Is the knowledge enough by itself to correctly respond to the task or is additional knowledge needed?
3. Is the assessment task and context likely to be comprehensible to students?

These questions help us reflect on whether our assessment task aligns with the desired learning goal and whether or not it will be accessible to the students.

Step #6: Develop specific rubrics.
We next use the base scientific explanation rubric to create a specific rubric for the particular assessment task by determining what would count as appropriate and sufficient claim, evidence, and reasoning. As discussed above, the specific rubric differs from the base rubric in that it is specific to a task and shows clearly what content knowledge the students should apply. Appendix C, p. 116, provides the specific rubric for the predator and prey task.

Concluding Comments

Developing scientific explanation tasks and specific rubrics is challenging. Yet, once developed, the tasks and rubrics can help you better assess students' understanding of the science content as well as students' reasoning behind their understanding of phenomena. Moreover, engaging your students in constructing scientific explanations helps them develop an understanding of science and the scientific practice of explanation. As discussed in the National Science Education Standards (1996), constructing explanations is one of the essential features of doing inquiry. Giving students specific feedback is critical to help them develop this ability.

Using rubrics can allow you to determine students' strengths and weaknesses, which can inform your future instructional design. For example, if you find that your students are having difficulty differentiating between evidence and opinion or that they are having a hard time with a particular science concept, like understanding that mass is not a property, this can help focus your future lesson plans. Using the same base rubric across different content areas can also help you develop a richer sense of students' ability to engage in particular scientific inquiry practices. You can use the base rubric across content areas and time to determine how students' ability to engage in scientific explanations is developing. Furthermore, providing your students feedback on their scientific explanations can push their thinking and help them develop a richer understanding than they would without the feedback. As the quote from Ms. Hill at the beginning of the chapter illustrates, engaging in scientific explanation with your students has multiple benefits in terms of both learning and assessment because "It touches on the science, it touches on the writing skills, and it touches on critical thinking."

Acknowledgments

This research was conducted as part of the Investigating and Questioning Our World through Science and Technology (IQWST) project and the Center for Curriculum Materials in Science (CCMS), supported in part by the National Science Foundation grants ESI 0101780 and ESI 0227557, respectively. Any opinions expressed in this work are those of the authors and do not necessarily represent those of the funding agency, the University of Michigan, or Boston College.

References

American Association for the Advancement of Science (AAAS). 1993. *Benchmarks for science literacy*. New York: Oxford University Press.

Bell, P., and M. Linn. 2000. Scientific arguments as learning artifacts: Designing for learning from the web with KIE. *International Journal of Science Education* 22(8): 797–817.

Black, P. 2003. The importance of everyday assessment. In J. M. Atkin and J. E. Coffey (Eds.), *Everyday assessment in the science classroom* (pp. 1–11). Arlington, VA: NSTA Press.

DeBoer, G. E. 2005. Standard-izing test items. *Science Scope* 28(4): 10–11.

Duschl, R. A. 2003. Assessment of inquiry. In J. M. Atkin and J. E. Coffey (Eds.), *Everyday assessment in the science classroom* (pp. 41–59). Arlington, VA: NSTA Press.

Harris, C. J., K. L. McNeill, D. L. Lizotte, R. W. Marx, and J. Krajcik. 2006. Usable assessments for teaching science content and inquiry standards. In M. McMahon, P. Simmons, R. Sommers, D. DeBaets, and F. Crowley. (Eds.), *Assessment in science: Practical experiences and education research* (pp. 67–88). Arlington, VA: National Science Teachers Association Press.

McNeill, K. L., C. J. Harris, M. Heitzman, D. J. Lizotte, L. M. Sutherland, and J. Krajcik. 2004. How can I make new stuff from old stuff? In J. Krajcik and B. J. Reiser (Eds.), *IQWST:*

Investigating and questioning our world through science and technology. Ann Arbor, MI: University of Michigan.

McNeill, K. L., and J. Krajcik. 2008. Teacher instructional practices to support students writing scientific explanations. In J. Luft, R. Bell, and J. Gess-Newsome (Eds.). *Science as inquiry in the secondary setting.* Arlington, VA: NSTA Press.

McNeill, K. L., D. J. Lizotte, J. Krajcik, and R. W. Marx. 2006. Supporting students' construction of scientific explanations by fading scaffolds in instructional materials. *The Journal of the Learning Sciences* 15(2): 153–191.

Moje, E. B., D. Peek-Brown, L. M. Sutherland, R.W. Marx, P. Blumenfeld, and J. Krajcik. 2004. Explaining explanations: Developing scientific literacy in middle-school project-based science reforms. In D. Strickland and D. E. Alvermann (Eds.), *Bridging the gap: Improving literacy learning for preadolescent and adolescent learners in grades 4–12* (pp. 227–251). New York: Carnegie Corporation.

National Research Council (NRC). 1996. *National science education standards.* Washington, DC: National Academy Press.

Osborne, J., S. Erduran, and S. Simon. 2004. Enhancing the quality of argumentation in school science. *Journal of Research in Science Teaching* 41(10): 994–1020.

Sadler, T. D. 2004. Informal reasoning regarding socioscientific issues: A critical review of research. *Journal of Research in Science Teaching* 41(5): 513–536.

Sutherland, L., M., K. L. McNeill, J. Krajcik, and K. Colson. 2006. Supporting students in developing scientific explanations. In R. Douglas, M. P. Klentschy, K. Worth, and W. Binder (Eds.), *Linking science and literacy in the K–8 classroom.* (pp. 163–181). Arlington, VA: National Science Teachers Association Press.

Zohar, A., and F. Nemet. 2002. Fostering students' knowledge and argumentation skills through dilemmas in human genetics. *Journal of Research in Science Teaching* 39(1): 35–62.

Appendix A

Base Explanation Rubric

Component	Level		
	0	**1**	**2**
Claim— A conclusion that answers the original question.	Does not make a claim, or makes an inaccurate claim.	Makes an accurate but incomplete claim.	Makes an accurate and complete claim.
Evidence—Scientific data that support the claim. The data need to be appropriate and sufficient to support the claim.	Does not provide evidence, or only provides inappropriate evidence (evidence that does not support claim).	Provides appropriate but insufficient evidence to support claim. May include some inappropriate evidence.	Provides appropriate and sufficient evidence to support claim.
Reasoning—A justification that links the claim and evidence. It shows why the data count as evidence by using appropriate and sufficient scientific principles.	Does not provide reasoning, or only provides reasoning that does not link evidence to claim.	Provides reasoning that links the claim and evidence. Repeats the evidence and/or includes some—but not sufficient— scientific principles.	Provides reasoning that links evidence to claim. Includes appropriate and sufficient scientific principles.

Appendix B

Scientific Explanation: Substance and Properties

Component	Level		
	0	**1**	**2**
Claim— A conclusion that answers the original question.	Does not make a claim, or makes an inaccurate claim.	Makes an accurate but incomplete claim.	Makes an accurate and complete claim.
	States that none of the liquids are the same or specifies the wrong solids.	Vague statement, like "some of the liquids are the same."	Explicitly states "Liquids 1 and 4 are the same substance."
Evidence— Scientific data that support the claim. The data need to be appropriate and sufficient to support the claim.	Does not provide evidence, or only provides inappropriate evidence (evidence that does not support claim).	Provides appropriate but insufficient evidence to support claim. May include some inappropriate evidence.	Provides appropriate and sufficient evidence to support claim.
	Provides inappropriate data, like "the mass is the same" or provides vague evidence, like "the data table is my evidence."	Provides one or two of the following pieces of evidence: density, melting point, and color of liquids 1 and 4 are the same. May also include inappropriate evidence, like mass.	Provides all three of the following pieces of evidence: density, melting point, and color of liquids 1 and 4 are the same.
Reasoning– A justification that links the claim and evidence. It shows why the data count as evidence by using appropriate and sufficient scientific principles.	Does not provide reasoning, or only provides reasoning that does not link evidence to claim.	Repeats evidence and links it to the claim. May include some—but not sufficient scientific principles.	Provides accurate and complete reasoning that links evidence to claim. Includes appropriate and sufficient scientific principles.
	Provides an inappropriate reasoning statement, like "they are like the fat and soap we used in class" or does not provide any reasoning.	Repeats that the density, melting point, and color are the same and states that this shows they are the same substance. Or provides an incomplete generalization about properties, like "mass is not a property so it does not count."	Includes a complete generalization that density, melting point, and color are all properties. Same substances have the same properties. Since liquids 1 and 4 have the same properties, they are the same substance.

Appendix C

Scientific Explanation: Predator and Prey

Component	Level		
	0	1	2
Claim — A conclusion that answers the original question.	Does not make a claim, or makes an inaccurate claim.	Makes an accurate but incomplete claim.	Makes an accurate and complete claim.
	States that they do not have a relationship or that they have the wrong relationship, like a parasite and host relationship.	Less explicit statement like "Animal B caused the population size of Animal A to decrease."	Explicitly states "Animal A and B have a predator and prey relationship."
Evidence— Scientific data that support the claim. The data need to be appropriate and sufficient to support the claim.	Does not provide evidence, or only provides inappropriate evidence (evidence that does not support claim).	Provides appropriate, but insufficient evidence to support claim. May include some inappropriate evidence.	Provides appropriate and sufficient evidence to support claim.
	Provides inappropriate data, like "both lines go up and down a lot," or provides vague evidence, like "the graph is my evidence."	Provides one or two of the following pieces of evidence: (1) When Animal B entered the environment, the population of Animal A decreased. (2) The two animals eat different food so they are not competing for food. (3) Animal B eats organisms like Animal A so it could be eating Animal A.	Provides all three of the following pieces of evidence: (1) When Animal B entered the environment, the population of Animal A decreased. (2) The two animals eat different food so they are not competing for food. (3) Animal B eats organisms like Animal A so it could be eating Animal A.
Reasoning— A justification that links the claim and evidence. It shows why the data count as evidence by using appropriate and sufficient scientific principles.	Does not provide reasoning, or only provides reasoning that does not link evidence to claim.	Repeats evidence and links it to the claim. May include some scientific principles, but not sufficient.	Provides accurate and complete reasoning that links evidence to claim. Includes appropriate and sufficient scientific principles.
	Provides an inappropriate reasoning statement, like "they are different animals," or does not provide any reasoning.	Repeats the evidence or provides an incomplete generalization about predator and prey relationships, like "predators eat other animals."	Includes a complete generalization that predators consume prey and that when predators enter a new environment they can cause a drastic decrease in the prey population size.

Making Meaning: The Use of Science Notebooks as an Effective Assessment Tool

Olga Amaral and Michael Klentschy
San Diego State University–Imperial Valley Campus

In an era of standards, assessment, and accountability, increased demands are placed on students to demonstrate an understanding of science content and on teachers to assess and determine the depth of student understanding. One of the major goals of elementary science instruction in this context is the development of scientific literacy in students. The American Association for the Advancement of Science (AAAS 1993) recommends that science literacy include students' ability to determine relevant from irrelevant information, to explain and predict scientific events, and to link claims to evidence to make scientific arguments. The National Research Council (NRC 2000) also recommends that both student knowledge and content understanding should be included in the assessment of science outcomes. These recommendations place increased demands on teachers to find ways to effectively assess students' knowledge and content understanding.

Furthermore, the NRC (Bransford, Brown, and Cocking 1999; Donovan and Bransford 2005) recommends that instructional planning and classroom instruction should focus on maximizing student opportunity to learn. Teachers can maximize student opportunity to learn when instructional planning, classroom instruction, and student assessment focus on three principles:

1. Engaging students to activate prior knowledge
2. Developing competence in an area of inquiry, including

- attaining a deep foundation of factual knowledge,
- understanding facts in the context of conceptual frameworks, and
- organizing knowledge in ways that facilitate retrieval and application

3. Recognizing that metacognitive approaches to instruction can help students take control of their own learning by defining goals and monitoring their progress

In addressing these three principles in instructional planning, teachers must also determine appropriate methods to assess student progress. The three principles align to a model of metacognition developed by Glynn and Muth (1994). In this model students develop metacognitive ability through learning science when they access prior content knowledge; use science-process skills; and apply reading, writing, listening, and speaking skills to learn content. In this model, student science notebooks and class discussion may be effective tools to assess student knowledge and content understanding.

Songer and Ho (2005) identify three challenges of science instructional programs that focus on the development of reasoning skills closely aligned to science literacy. They are as follows:

1. Formulation of scientific explanations from evidence (and the actual linking of claims to evidence)
2. Analysis of various types of scientific data (evidence)
3. Formulation of conclusions based upon relevant evidence

Student science notebooks may be one of the most effective strategies for teachers to use to address these three challenges, to plan effective instruction, and to assess student understanding. The application of language arts is essential for students not only to develop a deep understanding of science content but also to attain science literacy.

Student science notebooks have proved to be the best record of what science content is actually taught and learned by students; they provide an excellent ongoing assessment and feedback tool for teachers (Ruiz-Primo,

Li, and Shavelson 2002). Student thinking and making meaning from science instruction is enhanced by providing an opportunity for students to express a "voice" in their science notebooks and has led to increased student achievement not only in science, but in reading and writing as well (Amaral, Garrison, and Klentschy 2002; Jorgenson and Vanosdall 2002; Klentschy and Molina De La Torre 2004; Ruiz-Primo, Li, and Shavelson 2002; Saul et al. 2002; Vitale, Romance, and Klentschy 2005).

The science notebook is used during students' science experiences, in social interactions, as a tool for reflection, and as a tool for constructing meaning. In this context, the science notebook is an effective assessment tool. Ruiz-Primo, Li, and Shavelson (2002) also suggest that science notebooks are an excellent assessment tool to measure the quality of communication and conceptual understanding attained by students within a unit of study.

How students write about science is dependent on the student's familiarity with the phenomena, equipment, and length of exposure to the program of instruction (Amaral, Garrison, and Klentschy 2002). In unfamiliar situations, students' entries reflect the immediately observed science investigation; in familiar situations, the children's entries are based on their experiences with the phenomena—that is, their ability to place the science investigation in a real-world context (Shepardson 1997). An analysis of the initial use of science notebooks in Imperial County, California, showed that the students' first science notebooks took on the form of a narrative or procedural recount (Amaral, Garrison, and Klentschy 2002). Eventually, teachers must guide students to be more reflective about their work. Science notebooks have the potential to move students beyond simply completing the task to making sense of the task.

Scaffolding Writing in the Inquiry Process

By giving students writing prompts, teachers help students learn to use science notebooks as a permanent record of classroom science investigations. Carefully developed writing prompts enhance student science content understanding and provide a rich opportunity for formative assessment. Klentschy (2008) suggests several writing prompts or scaffolds that serve this purpose, as depicted in Figure 7.1, page 120.

Figure 7.1 Science Notebook Components

Component	Purpose	Writing Prompt (Scaffold)
Question	Questions and question-formulating strategies are central to science. The formulation of a question forms the basis for high-quality instruction in science.	*How does…?* *What does…?*
Prediction	The prediction provides a reasonable explanation by the learner as to the result of the investigation. Using "because" also activates prior knowledge.	*I think…will happen because…*
Planning -General -Operational	The general plan determines which variable will be changed and which will be kept constant and what will be observed or measured. The operational plan describes the sequence of events and the materials that will be used to conduct the investigation.	General *…will be changed.* *…will be kept the same.* *…will be observed or measured.* Operational *First…* *Second…* *Next…* *Finally…*
Data—Observations/ Measurements	Data and the collection and display of data become the evidence for the investigation.	Data charts, tables, graphs and labeled diagrams and illustrations.
Claims/Evidence	Scientific explanations help frame the goal of inquiry as understanding natural phenomena and articulating and convincing others of that understanding.	*I claim…I know this because…*
Conclusion	A conclusion is the justification that links the claim and evidence together.	*Today I learned…*
Reflection	Written reflection is essential to promote students' explorations of their own thinking and learning processes.	*Questions that I have now are…* *I wonder if…*

Source: Klentschy, M. 2008. *Using science notebooks in the elementary classroom.* Arlington, VA: NSTA Press.

These prompts are based on the lessons learned from the Valle Imperial Project in Science from 1996 to 2004 (Klentschy and Molina De La Torre 2004; Klentschy 2005; Amaral, Garrison, and Duron-Flores 2006). Some researchers believe that writing forces the integration of new ideas and relationships with prior knowledge and encourages personal involvement with new information (Kleinsasser, Paradis, and Stewart 1992). Science notebooks then become an extended opportunity for students to explain, describe, predict, and integrate new information, and they allow students to make conceptual shifts and facilitate retention (Fellows 1994). Students benefit from strong scaffolding with respect to building explanations from evidence (Songer 2003). Questioning, predicting, clarifying, and summarizing are strengthened through scaffolding. Clarifying promotes comprehension monitoring. Students benefit from scaffolding when analyzing data and building explanations from evidence (Hug, Krajcik, and Marx 2005). A process of scaffolded inquiry, reflection, and generalization develops students' metacognitive knowledge (White and Fredrickson 1998). Writing is an important tool for transforming claims and evidence into knowledge that is coherent and structured; writing also appears to enhance the retention of science learning over time (Rivard and Straw 2000).

Science Notebooks as an Assessment Tool for Content and Literacy

Science notebooks play an important role in the triad of what should be taught (standards), what is actually taught (classroom instruction), and what content students learn. In this context, the science notebook is the means by which students communicate their understanding. Ruiz-Primo, Li, and Shavelson (2002) indicate that the student science notebook provides insights into the depth of student content understanding and the quality of students' communication of that understanding. Klentschy and Molina De La Torre (2004) also indicate that the development of both the depth of student content understanding and quality of their communication are *developmental* and can be enhanced through scaffolding. Equally important, science notebooks can effectively assess students' ability to formulate scientific explanations from evidence. These factors make science notebooks an effective tool for teachers to use in assessing student science conceptual understanding and complex reasoning.

As a formative assessment tool, science notebooks provide evidence of student understanding. Entries in notebooks identify the gaps between where individual students are in their learning and where they need to be. Was the content standard actually understood by students in its relationship to a big idea in science? What was the quality of student communication and conceptual understanding? Similarly, class-wide results inform teachers about the gaps and strengths in the lesson design.

Classroom teachers can begin the process of using science notebooks by asking themselves the following questions:

- What content skills and/or process skills should all students learn in this unit?
- How do the science notebooks reflect student learning?
- What evidence should support their understandings? (What are my criteria?)
- What are the implications for further instruction?

In answering these questions, teachers can use an assessment template similar to the one depicted in Figure 7.2 as a means of getting started. The template can be used for examining a science notebook entry for an individual lesson or for the entire unit of study. The template is a means of learning how well the student has used the notebook to understand the objective of the lesson or the big idea of the unit.

Science instruction has the greatest impact on student learning when students apply evidence from instructional experiences to their prior knowledge and to the world around them. When students have to explain, as is the case with science notebooks, their ideas about their learning and perhaps the way in which they are thinking about scientific processes, they must focus on what they have learned, especially what they have internalized. Teachers should give students the opportunity to understand the criteria for scoring, to review their own work for completeness or revision, and to comment on what they have learned. This is extremely important in an era of standards, assessment, and accountability. Figure 7.3, page 124, is an example of a scoring guide with a specific set of criteria. This type of scoring guide is distributed to the student prior to the lesson. In a classroom discussion, the teacher explains the criteria or standard to the students. The target that should be expected for all students is at least "proficient."

Figure 7.2 Science Notebook Assessment Template

Elements and Criteria	NA	Present	Lacking	Meets	Exceeds
Big Idea					
Question Purpose					
Student generated: in own words Relates to purpose/"Big Idea" Clear and concise Investigable	COMMENTS:				
Prediction					
Connects to prior experience Is clear and reasonable Relates to question Gives an explanation/reason	COMMENTS:				
Planning					
Relates to investigable question Has clear sequence/direction Identifies variables/control Includes data organizer States materials needed	COMMENTS:				
Data/Observations					
Relates to question and plan Includes student-generated drawings, charts, graphs, narrative Organized Accurate	COMMENTS:				
What Have You Learned?					
Student generated: in own words Clear statement of what was learned Based on question/planning/evidence Reflective Shows rigor in thinking	COMMENTS:				
Next Steps/New Questions					
Student generated: in own words Extension/new application of original question Researchable or investigable WOW factor Can be recorded throughout	COMMENTS:				
Remarks/Considerations					
Creativity in evidence Growth over time (process and content)	COMMENTS:				

Source: Klentschy, M. 2008. *Using science notebooks in the elementary classroom.*
Arlington, VA: NSTA Press.

After students have completed the investigation, they use the scoring guide to review their science notebook entries for completeness or revision.

Figure 7.3 Scoring Guide With Specific Criteria

Measuring Time – Lessons 7-9
El Centro California
May 29, 2003

SCORING GUIDE

Student Self-Assessment	Teacher Assessment	
		ADVANCED (Expert)
✓		All items listed in proficient
✓		Plus: <u>diagrams of each water clock trial</u>
✓		Plus: diagrams and charts are <u>completely</u> labeled
✓		Plus: appropriate/advanced use of scientific language
		PROFICIENT
		4 of the following 5:
✓		Focus question relates to main idea of lesson
✓		A prediction that relates to the question
✓		A plan that relates to the question
✓		Data: Diagrams are clear and accurate
✓		Data: Diagrams of trials that worked/did not work
		All of the following –Claims and Evidence
✓		All claims are supported by evidence
✓		Descriptions/diagrams include correct labeling
✓		A chart with data from each trial
✓		What you learned
		PROGRESSING (Basic)
		5-7 Proficient points
		DOES NOT MEET STANDARD
		(Far Below Basic)
		4 or fewer Proficient Points

Teacher Feedback

When viewed as a formative assessment tool, teacher feedback can play an important role in the development of student conceptual understanding and complex reasoning. Furthermore, as Marzano, Pickering, and Pollock (2001) reported in a review of nine research studies, feedback that guides students, rather than tells them what is right or wrong on a test, can attain an effect size of .90 or higher. The most appropriate forms of feedback are asking guiding questions face-to-face with the student or writing guiding questions in his or her science notebook. The feedback can also grow into a personal, written conversation between the teacher and the student in the student's science notebook. Teachers considering the use of science notebooks as an assessment tool should stress this conversational form of feedback, which can best be summarized as "issues, evidence, and you." For example:

- *What evidence do you have to support your claims?*
- *What claims can you make from your evidence?*
- *Is there another explanation for what happened?*

The timing of feedback also appears to be critical to its effectiveness. In general, the more delay that occurs in giving feedback, the longer it takes students to clear up misconceptions (Marzano, Pickering, and Pollack 2001; Shepardson and Britsch 2001). Feedback should also address specific, developmentally appropriate levels of skill or knowledge expected of students. Research has consistently indicated that this form of feedback has a more powerful effect on student learning than simply reporting to students their standing in relation to their peers (Marzano, Pickering, and Pollack 2001).

Developmental "storylines," which show the sequence of lessons in a unit and how they lead to big ideas in science (Amaral, Garrison, and Klentschy 2002), form the basis for rubrics that teachers can use to give students written feedback in their science notebooks. Students who are not given written corrective feedback usually do not show improvement over time and a significant number actually decline in quality of communication and conceptual understanding (Ruiz-Primo, Li, and Shavelson 2002).

Research on feedback also indicates that students can effectively monitor their own progress (Marzano, Pickering, and Pollack 2001). Such self-evaluation is important in the construction of meaning by the students.

Challenges Teachers Face in Assessing Student Knowledge and Understanding

Assessment practices in science in the elementary classroom have most often been relegated to end-of-unit tests provided by commercial textbooks. These are usually in the form of multiple-choice questions with answers that can be easily found, and sometimes even highlighted, in the unit itself. They are most often factual questions and rarely ones that require students to go beyond recall or engage in any type of critical thinking. Procedural knowledge may also be assessed, but items intended to assess other aspects of learning may be limited or not available at all.

Questions intended to help students review and prepare for tests reveal the process that most students follow when attempting to learn the material. The savvy student is able to read the question, identify the section in the text that addresses it, and select the correct response from the three or four choices provided. If the correct answer is selected, this information hopefully will remain in the student's short-term memory and the student will be able to repeat the selection for the actual test.

What these assessments do not provide is a good representation of the students' thinking either about the concept studied or about how one student's thinking differs from that of his or her peers. They also do not necessarily give us an indication that students understand science for the long term as most questions relate directly back to what was just reviewed rather than what was experienced. De Fronzo (2006) points to conceptual models such as Glynn and Muth's Model of Student Cognitive Processes (1994) to describe the process students use to construct meaning. Many textbook assessment tasks do not allow teachers to examine how students do that nor do they show the extent to which students understand the information and how much information becomes part of the students' long-term memories. The National Research Council (NRC) probably says it best: "Understanding the contents of long-term memory is especially critical for determining what people know; how they know it; and how they are able to use that knowledge to answer questions, solve problems, and engage in additional learning" (2000, p. 102).

Traditional assessments rarely allow teachers to check for students' misconceptions because the selection of one answer on a multiple-choice test does not give the teacher any insight into why the student selected it. The National Research Council (NRC 2000, 2005) lists as one of its instructional principles the recognition that metacognitive approaches to instruction can help students to learn to define their own learning goals and monitor their progress. Harlen (2001) reports that science instruction has the greatest impact on student learning when students can apply evidence from instructional experiences to the knowledge they bring to the task and to the world surrounding them. When students have to explain their ideas about that learning and perhaps the way in which they are thinking about scientific processes, as is the case with notebook writing, students must focus on what they have learned—most likely what they have internalized from their learning experiences.

In the teaching and learning tradition of "the best way to learn something is to have to teach it," the student who must explain something in a notebook has an audience to "teach." So why not abandon those end-of-unit tests and simply use other forms of assessments such as science notebook writings? Teachers will be the first to describe the obstacles they face when they consider doing so.

The first challenge that teachers face in using science notebooks for assessment purposes is that of *time constraints*. Surveys (Amaral and Garrison 2005) have shown that practitioners would like to be more effective in assessing student knowledge by this means, but find that they barely have enough time to teach all that needs to be covered as the demands of standards and the pressures of preparing students for statewide tests loom larger each year. The idea of taking home a classroom set of notebooks to correct or score is frightening to many teachers. This is primarily due to their fear that they should correct everything that appears in the notebooks, every lesson, every day, for each student.

Another challenge that teachers often face is their inadequacy with their own *lack of content knowledge*. Scoring science notebooks requires a level of understanding on the part of the teacher that goes well beyond simply scoring an end-of-unit test based on the answer sheet provided by the textbook. Teachers can be the first to admit that they do not always understand what students might have meant by an entry in their science notebooks because not only does the teacher have to try to follow the student's reasoning but they also need to have a deep understanding of content knowledge if they are to determine the degree to which an answer or entry may be on target in its representation of the scientific concept. Furthermore, they need content knowledge to give students appropriate feedback and guide them appropriately to their next level of conceptual understanding.

Teachers also can have difficulty with interpreting assessment results for all students using an *equitable assessment process*. Teachers today are increasingly faced with groups of students with very wide ranges of ability in literacy and with students who lack full proficiency in the English language. Assessing student knowledge for some of these groups requires additional skills from teachers who may not have received appropriate training.

Yet another difficult task for teachers is to *interpret students' writing and decipher their thinking*. Assuming that the student is not present while the teacher is reviewing the notebook entries, the teacher may have to rely on what he or she thinks the student meant by recalling comparable situations in the classroom when the student has responded in similar fashion. If relying on this technique does not bring good results, the teacher may have to write a follow-up question for the student, which results in more work and delayed assessment. If the teacher uses techniques allowing for notebook scoring while still in class, then another problem might arise: the process to document those scores.

Perhaps the greatest challenge for the teacher is to *make effective use of assessments*. If the assessment is used solely to record student progress at a point in time, then a fairly simple process can be used. On the other hand, if the teacher uses it to track progress for individual students over time, perhaps over the weeks or months it takes to teach a variety of units, or to track the progress of an entire group of students on one concept in particular, then the interpretation and *application of the results* will need to be considered differently. Learning how to align assessment practices and link them to instruction and refinements of curriculum and implementation requires very special training.

Finally, the teacher may not have developed appropriate skills for providing students with the very best feedback on student writing, feedback that is reportedly one of the most powerful teaching tools when done correctly (Marzano, Pickering, and Pollock 2001). According to Klentschy and Molina De La Torre (2004), the "most appropriate form of feedback in a knowledge-transforming mode of instruction is one of asking guiding questions or writing guiding questions in students' science notebooks" (p.

350). For the teacher to give appropriate feedback, assessment must occur and a determination must be made by the teacher about the level of understanding that is reflected by the student writing. This is what will guide the level of prompting or questioning that the teacher uses. This task can be difficult for teachers because it is a task completed individually for each student, a demanding task that requires great attention from the teacher and, therefore, perhaps more time than correcting an average quiz or test.

Examples of Assessment Instruments for Classroom Use

The idea that assessments should be more closely linked to instructional practices is not a new one (Donovan and Bransford 2005; Marzano, Pickering, and Pollack 2001). The question of how to accomplish that for practical classroom use has been getting more attention recently as examples of assessments that include rubrics to evaluate student work are increasingly included in the literature. Katz and Olson (2006), for example, not only discuss principles for conducting appropriate in-classroom assessments but also describe steps to actively plan for assessment and ways to collect assessment information. They also provide sample rubrics that teachers can use with certain topics in science. One sample guides teachers in the evaluation of a grade 2 life-cycle poster project and involves descriptions rated into four categories: *excellent*, *very good*, *satisfactory*, and *improvements needed* (Figure 7.4).

Figure 7.4 Grade 2 Rubric for a Life-Cycle Poster Project

Excellent	Very Good	Satisfactory	Improvements Needed
All stages of life cycle shown in proper order, and clearly indicates that reproduction begins a new cycle	All stages of life cycle present, but conveys a linear fashion of one series of stages without showing how reproduction leads to a new cycle	Some stages of life cycle missing, or present but out of order	Life cycle stages missing and out of order; does not convey an understanding of a life cycle
Illustrates examples of two or more organisms, accurately labeling life cycle phases in each example	Illustrates one organism and accurately labels its life cycle stages	Life cycle stages are inaccurately labeled for one organism	Does not show life cycle stages for an organism

Source: Katz, A., and J. K. Olson. 2006. Strategies for assessing science and language learning. In A. Fathman and D. Crowther (eds.), *Science for English language learners: K–12 classroom strategies*. Arlington, VA: NSTA Press.

This rubric reflects a learning outcome that was important to the teacher who created it and is very specific to a concept that was introduced and studied—that is, this teacher wanted to record the degree to which students had understood the stages of a life cycle. Results usually range from the very least possible information provided by the student (the *Improvements Needed* category) to the most elaborate of information (found in the *Excellent* category).

Creating and using a rubric such as this one appears to be a simple enough task. However, it is not always as easy as it sounds. Teachers often find that there is student work that does not easily fit the rubric. In such cases, teachers can re-evaluate the rubric and make refinements to it. Again, this takes some skill and time. Teachers often have not had any training in the design of rubrics nor do they have much time to devote to these tasks. As a result, some companies selling science textbooks or other materials (e.g., *Scott Foresman Science* 2006) are now creating and providing rubrics that address specific tasks that teachers require of students.

Broader rubrics can be designed to be applied across science concepts. The rubric shown in Figure 7.5 is intended to record student progress relative to students' ability to record and interpret data as well as their ability to link this knowledge to the scientific concept studied. This type of rubric is used with the understanding that students have opportunities to carry out the desired tasks and that the teacher guides students to make the linkages necessary for students to see the bigger picture. It was created to address the problems that arise when lessons end without closure when the bell rings or end with teachers telling students to put things away but without any meaningful discussion about the process they have just undergone and what has been learned from it. Even when the teacher is skilled at bringing closure to the lesson, there can be limited or no attempts to have students think about the broader aspects and implications of the target lesson to the larger unit of study.

Rubrics also lend themselves well to recording students' performance and students' ways of thinking about science. Marzano, Pickering, and McTighe (1993) list standards in three categories that should concern educators in the assessment of scientific thinking:

Figure 7.5 Rubric for Use Across Science Concepts

	Level 1	Level 2	Level 3	Level 4	Level 5
Achieved Curriculum: (what students have learned): Student Notebook Tool	Students don't gather or report data.	Data is not clear or is inaccurate, incomplete, or not original.	Students clearly and accurately record data from the lesson.	Students record data clearly and accurately and interpret the findings.	Students record data clearly and accurately, interpret data, and relate findings to the key concepts.

Source: Amaral, O., and L. Garrison. 2006. Missing the forest for the trees. *Journal of Science Education and Technology.* New York: Springer Publications.

1. *Self-regulation.* The student
   - is aware of own thinking,
   - makes effective plans,
   - is aware of and uses resources,
   - is sensitive to feedback, and
   - evaluates the effectiveness of own actions.
2. *Critical thinking.* The student
   - is accurate and seeks accuracy,
   - is clear and seeks clarity,
   - is open-minded,
   - restrains impulsivity,
   - takes a position when the situation warrants it, and
   - is sensitive to the feelings and levels of knowledge of others.
3. *Creative thinking.* The student
   - engages intensely in tasks even when answers or solutions are not immediately apparent,
   - pushes the limits of own knowledge and abilities,
   - generates, trusts, and maintains own standards of evaluation, and
   - generates new ways of viewing a situation outside the boundaries of standard conventions. (p. 23)

These principles can guide us in the development of assessments that keep in sight the value of going beyond factual knowledge and include reasoning. We can see evidence of some of these standards in individual lessons. For example, "makes effective plans" can easily apply to an entry in a student notebook about an experiment that students will conduct. "Seeks clarity" can be demonstrated in the way a student describes the conclusion reached about an inquiry activity.

The performance tasks most likely to find favor with teachers are those that are embedded in instruction. The Science Teaching Action Research Project (Schilling et al. 1990) produced a series of tasks in which students were taken on imaginary journeys and participated in class projects. They then had to produce posters, complete activities, and ask and answer questions that were designed to elicit short written responses. The students engaged in activities that would lead them to gain sufficient knowledge about the topics to be able to both address the questions and ask new ones. One example of student work includes an activity on sundials that asks students to record data relative to the length of a shadow at various times during the day and then asks the question about a 4:00 p.m. entry that was blank because it had been cloudy. The question was "How did you decide where and how to draw the shadow?" (p. 120). Questions that begin with "how," "where," and "why" can be very powerful in eliciting student thinking.

The examples of student notebook entries in Figures 7.6–7.9 provide a window into the kind of writing students can do in elementary grades based on teacher guidance. Teachers can interpret these entries and determine their students' levels of understanding. Figure 7.6 comes to us from a third-grade student who is attempting to bring closure to the lesson and explain what has been learned.

Figure 7.6 A Third-Grade Student's Work: Which String Had the Highest/Lowest Pitch?

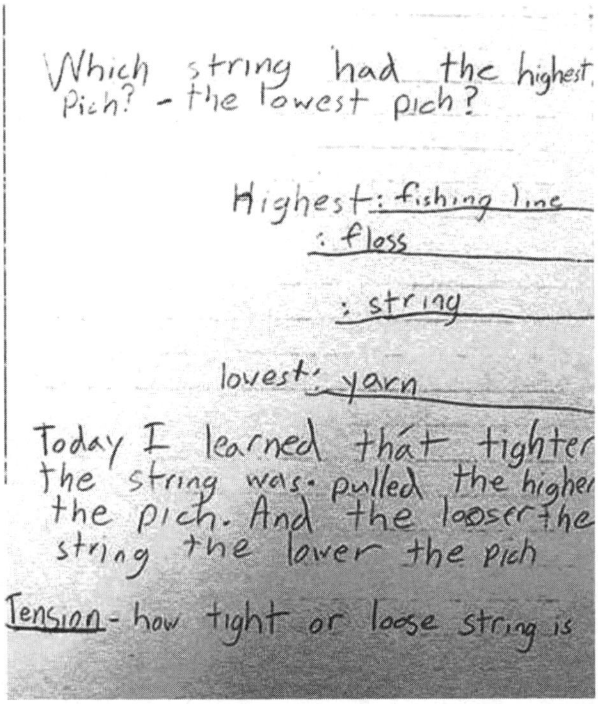

Which string had the highest pich? - the lowest pich?

Highest: fishing line
: floss

: string

lowest: yarn

Today I learned that tighter the string was pulled the higher the pich. And the looser the string the lower the pich

Tension - how tight or loose string is

Note: The spelling errors have been kept to reflect student's actual notebook entry.

A more sophisticated level of thinking may be reflected by questions that go beyond requiring a yes or no answer, such as the example in Figure 7.7, page 134, by a fifth grader.

Figure 7.7 A Fifth-Grade Student's Work: What's Inside a Seed?

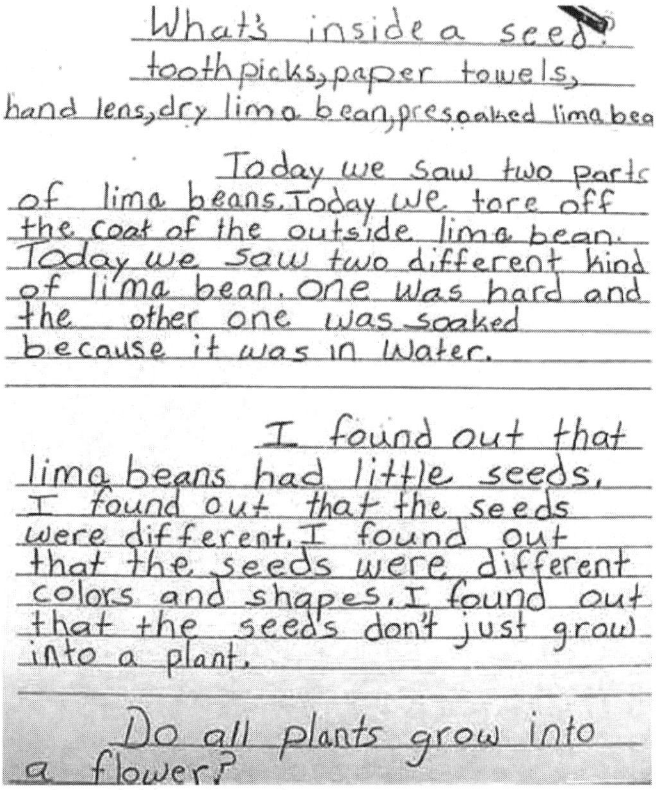

What's inside a seed?
toothpicks, paper towels,
hand lens, dry lima bean, presoaked lima bea

Today we saw two parts
of lima beans. Today we tore off
the coat of the outside lima bean.
Today we saw two different kind
of lima bean. One was hard and
the other one was soaked
because it was in water.

I found out that
lima beans had little seeds.
I found out that the seeds
were different. I found out
that the seeds were different
colors and shapes. I found out
that the seeds don't just grow
into a plant.

Do all plants grow into
a flower?

Figure 7.8 is an example from a fifth grader who comes up with a conclusion and links it to a prediction but does not show any conceptual understanding beyond the procedural tasks completed during the lesson.

Figure 7.8 A Fifth-Grade Student's Work: Conclusion

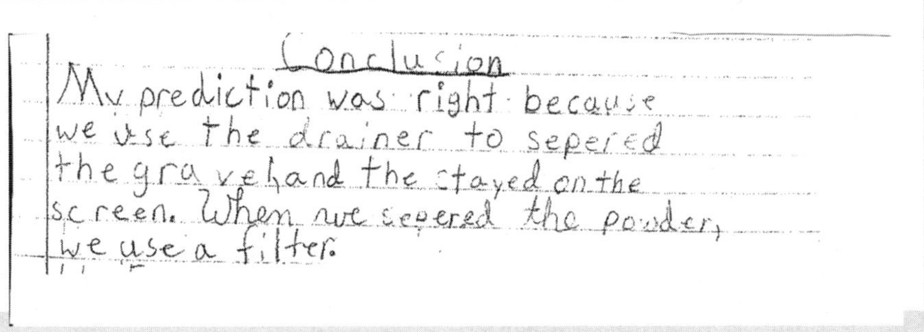

> Conclusion
> My prediction was right because
> we use the drainer to separed
> the gravel and the stayed on the
> screen. When we separed the powder,
> we use a filter.

The student conclusion in Figure 7.9 comes closer to reflecting some understanding but still focuses on procedures rather than on the "so what?" of the lesson.

Figure 7.9 Student Work: Conclusion

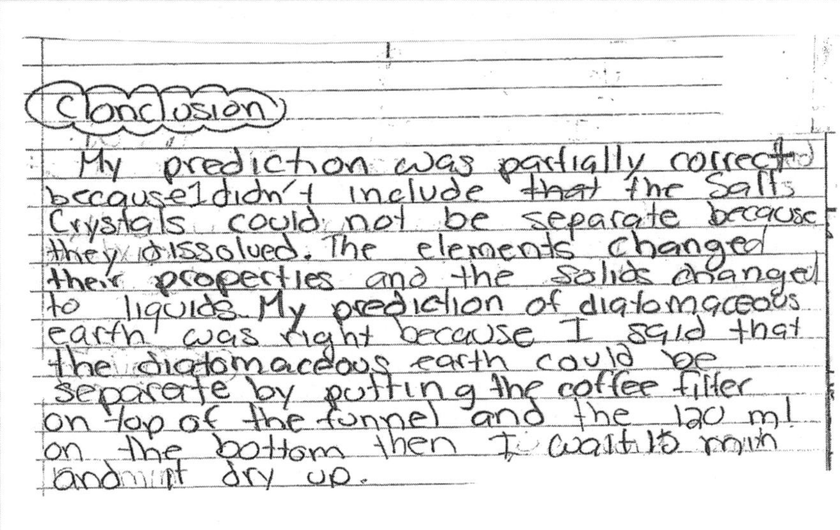

> Conclusion
> My prediction was partially correct
> because I didn't include that the salts
> crystals could not be separate because
> they dissolved. The elements changed
> their properties and the solids changed
> to liquids. My prediction of diatomaceous
> earth was right because I said that
> the diatomaceous earth could be
> separate by putting the coffee filter
> on top of the funnel and the 120 ml
> on the bottom then I wait 15 min
> and it dry up.

All of these notebook entries could have been assessed using a teacher-developed rubric. These, however, are but a very small sample of the representations of performance and skills that we want to see students accomplish and gain. Marzano, Pickering, and McTighe (1993) summarized rubrics that included the following groups of standards:

1. Content standards
    i.   declarative
    ii. procedural
2. Complex thinking standards
    i.   tasks with a clear purpose
    ii. uses of reasoning strategies
3. Information processing standards
    i.   interpreting and synthesizing information
    ii.   information-gathering techniques and resources
    iii. assessing the value of information
    iv.   recognizing how projects would benefit from additional information
4. Communication standards
    i.   clear expression of ideas
    ii.   communicating with diverse audiences
    iii. communicating in a variety of ways
    iv.   communicating for a variety of purposes
    v.   creating quality products
5. Collaboration and cooperation standards
    i.   working toward achievement of group goals
    ii.   demonstrating effective interpersonal skills
    iii. contributing to group maintenance
    iv.   performing variety of roles within group
6. Habits of mind standards
    i.   being aware of own thinking
    ii.   making effective plans
    iii.   being aware of and pursuing necessary resources
    iv.   being sensitive to feedback
    v.   evaluating effectiveness of own actions
    vi.   being accurate and seeking accuracy
    vii.   being clear and seeking clarity
    viii. being open-minded
    ix.   restraining impulsivity
    x.   taking a position when warranted

xi.   being sensitive to feelings and knowledge of others

xii.   pushing limits of own knowledge and ability

xiii.   generating, trusting, and maintaining own standards of evaluation

xiv.   generating new ways of viewing a situation outside the boundaries of standard conventions (p. 94)

Although this list may at first appear to be overwhelming, it is important to note that not every one of the items needs to be assessed during each lesson, or even each unit. What is important is for teachers to become increasingly more aware of the need to consider a variety of these tasks as part of the teaching and learning process *as well as* the assessment process. But how do we best prepare teachers to most effectively engage and evolve in this process of diversifying assessment practices and linking findings to instruction? That question brings us to the matter of professional development.

Professional Development for Enhancing Assessment Practices

The first step in preparing teachers to use student notebooks for assessment purposes is training teachers to understand the reasons for using notebooks in the first place. Second, they must buy into the premise that the benefits of using notebooks outweigh the challenges described earlier. Taking small steps toward the implementation of notebook writing may be the best approach in that it does not immediately scare teachers and result in a total refusal to cooperate. They must reach a level of comfort with the integration of writing activities during science instruction. Once familiarized with the process, teachers can begin to increase their levels of usage, and progress incrementally, increasing their sophistication about the meaning of what students write.

Training sessions are most effective when teachers can practice using rubrics to evaluate real student work. The sessions should begin with a presentation and discussion of the theoretical underpinnings of assessment. The training session leaders should share samples of student work to illustrate the practical aspects of implementation. This process becomes even more powerful if teachers are then asked to return to their classrooms, go through one cycle of using student notebooks for assessment (perhaps with just one task), and then return to discuss findings and implications. It is ideal if teachers work with a trainer who can serve as a coach or mentor to guide the discovery process that teachers encounter when they write feedback on student notebooks, make decisions about students' performance based on a predetermined rubric, or design a rubric based on a target task.

Training can also be effective if groups of teachers work together on assessment and evaluation tasks. If teachers are in the same grade level and use the same curriculum, for example, they can share their impressions of student notebook work and their perceptions of student learning. This sharing can enrich their conversations, result in their ability to use multiple ways of assessing, and lead to greater motivation to devote time to assessment issues.

Professional development must also consistently focus on the relationships among standards, curriculum, and assessments and the impact these have on instruction. Teacher training at both the preservice and the inservice levels needs to focus on the integration of cognition and measurement and its implications.

> *Children have rich intuitive knowledge of their world that undergoes significant change as they mature. Learning entails the transformation of naive understanding into more complete and accurate comprehension, and assessment can be used as a tool to facilitate this process. To this end, assessments, especially those conducted in the context of classroom instruction, should focus on making students' thinking visible to both their teachers and themselves so that instructional strategies can be selected to support an appropriate course for future learning.* (Pellegrino, Chudowsky, and Glaser 2001, p. 103)

Time constraints and all of the other challenges that teachers face can at times prevent us from advancing professional development in this area. In fact, it is not uncommon to hear teachers say that they don't have sufficient time to do all of this "classroom assessment" because they are too busy teaching and preparing students for "those" tests—the real tests, the state's tests. When the use of student notebooks is well-conceived and assessment practices are embedded, however, students gain experiences that take them beyond the short-term memorization of factual information; thinking and reasoning are incorporated as students write and explain their thinking. These experiences will likely serve them well in answering questions of other types, including those typically found in more traditional tests, such as state tests.

Some might say that the scoring of student notebooks is too subjective a process. On the other hand, if the notebooks are scored for the purpose of informing instruction, then a teacher can be a powerful observer of student work and can redirect lessons to address student weaknesses and misconceptions. The teacher will likely be the best judge of the work at that level. If the notebooks are to be scored for purposes of documenting progress beyond individual student competence, then processes have been established and protocols tested with research showing that they can be reliably scored (Ruiz-Primo et al. 1999).

Current Research

The Imperial Valley, California, Mathematics–Science Partnership is currently conducting research in collaboration with Tennessee State University researchers on the effects of scaffolded inquiry and the use of student science notebooks as the principal means of assessing student science content understanding from standards-based classroom science instruction.

In 2005, two studies were conducted with more than 1,200 fifth-grade students in Imperial County classrooms (Vanosdall et al. 2007). Forty teachers were grouped into matched pairs based on background factors including experience and science professional development hours attained. One teacher from each matched pair was placed in a treatment group and one placed in the control group. All teachers had previously received professional development on the use of the designated instructional materials aligned to the California Fifth-Grade Physical Science Standards. The experimental group received an additional eight hours of professional development on the use of enhanced lessons with scaffolded writing prompts embedded into the lesson plans. All teachers received similar amounts of classroom coaching from Imperial Valley Mathematics–Science Partnership science resource teachers. All students were pre- and posttested using a publisher-validated instrument. All student work was collected at the end of the instructional unit. All students were assessed using the California Standards Test–Science Subtest.

There were no significant differences between treatment and control groups on the pretest. The treatment group significantly outperformed the control group on the posttest. The treatment group significantly outperformed the control group on the Fifth-Grade Physical Science Standards Subtest of the California Standards Test. (See Vanosdall et al. 2007 for a complete discussion and analysis of the data from this study.)

Student science notebooks were scored using a validated scoring measure developed by Ruiz-Primo, Li, and Shavelson (2002). Students receiving instruction with writing scaffolds embedded into the instruction in both grades four and five produced student notebooks that were significantly different ($p < .001$) than those of the control group with respect to

- quality of communication,
- science conceptual understanding, and
- use of scientific vocabulary.

In all cases there were no significant differences for gender, language, and socioeconomic status with students in the treatment group. In fact, English learners in the treatment group made the greatest gains between pre- and posttest scores, significantly closing the achievement gap. The study was replicated in 2006 and 2007 in fourth- and fifth-grade Imperial County classrooms and in 2007 in fourth- and fifth-grade classrooms in the Wake County Public School System in Raleigh, North Carolina.

Conclusion

Communication has a central role in the process of inquiry. Words and language are used as ways of trying out a framework for understanding—students need to have space to reflect on ideas. Vygotsky (1978) states, "The three most important aspects of a teacher's role in elementary science are providing materials for students to observe and investigate, asking the right kinds of questions, and helping students to communicate their thinking and developing ideas" (p. 87). Student science notebooks are a valuable tool to accomplish these three important aspects.

References

Amaral, O., and L. Garrison. 2005. *Evaluation of CaMSP in the Imperial Valley.* Sacramento, CA: Department of Education.

Amaral, O., and L. Garrison. 2006. Missing the forest for the trees. *Journal of Science Education and Technology.* New York: Springer Publications.

Amaral, O., L. Garrison, and M. Duron-Flores. 2006. Taking inventory. *Science and Children* 43(4): 30–33.

Amaral, O., J. Garrison, and M. Klentschy. 2002. Helping English learners increase achievement through inquiry-based science instruction. *Bilingual Research Journal* 26 (Summer): 2, 213–239.

American Association for the Advancement of Science (AAAS). 1993. *Benchmarks for science literacy*. New York: Oxford University Press.

Bransford, J. D., A. Brown, and R. Cocking, eds. 1999. *How people learn: Brain, mind, experience, and school*. Washington, DC: National Academy Press.

De Fronzo, R. 2006. Comprehension strategies and the scientist's notebook: Keys to assessing student understanding. In R. Douglas, M. Klentschy, K. Worth, and W. Binder (Eds), *Linking science and literacy in the k-8 classroom*. Arlington, VA: NSTA Press.

Donovan, M. S., and J. D. Bransford, eds. 2005. *How students learn: History, mathematics and science in the classroom*. Washington, DC: National Academy Press.

Fellows, N. 1994. A window into thinking: Using student writing to understand conceptual change in science learning. *Journal of Research in Science Teaching* 31 (9): 985–1001.

Glynn, S., and D. Muth. 1994. Reading and writing to learn science: Achieving scientific literacy. *Journal of Research in Science Teaching* 31 (9): 1057–1073.

Harlen, W. 2001. *Primary science, taking the plunge*. (2nd ed.) Portsmouth, NH: Heinemann.

Hug, B., J. Krajcik, and R. Marx. 2005. Using innovative learning technologies to promote learning and engagement in urban science classrooms. *Urban Education* 40: 446–472.

Jorgenson, O., and R. Vanosdall. 2002. The death of science? What are we risking in our rush toward standardized testing and the three r's. *Phi Delta Kappan* 83 (8): 601–605.

Katz, A., and J. K. Olson. 2006. Strategies for assessing science and language learning. In A. Fathman and D. Crowther (Eds.), *Science for English language learners:K-12 classroom strategies*. Arlington, VA: NSTA Press.

Kleinsasser, A., E. Paradis, and R. Stewart. 1992. Perceptions of novices' conception of educational role models: An analysis of narrative writing. Paper presented at the annual meeting of the American Educational Research Association, San Francisco, CA (April).

Klentschy, M. 2005. Science notebook essentials. *Science and Children* 43 (3): 24–27.

Klentschy, M. 2008. *Using science notebooks in the elementary classroom*. Arlington, VA: NSTA Press.

Klentschy, M., and E. Molina De La Torre. 2004. Students' science notebooks and the inquiry process. In W. Saul (Ed.), *Crossing borders in literacy and science instruction: Perspectives on theory and practice*. Newark, DE: International Reading Association Press.

Marzano, R., D. Pickering, and J. Pollock. 2001. *Classroom instruction that works: Research-based strategies for increasing student achievement.* Alexandria, VA: Association for Supervision and Curriculum Development.

Marzano, R., D. Pickering, and J. McTighe. 1993. *Assessing student outcomes: Performance assessment using the dimensions of Learning Model.* Alexandria, VA: Association for Supervision and Curriculum Development.

National Research Council (NRC). 2000. *Inquiry and the national science education standards.* Washington, DC: National Academy Press.

Pellegrino, J. W., N. Chudowsky, and R. Glaser, eds. 2001. *Knowing what students know: The science and design of educational assessment.* Washington, DC: National Academy Press.

Rivard, L., and S. Straw. 2000. The effect of talk and writing on learning science: An exploratory study. *Science Education* 84 (5): 566–593.

Ruiz-Primo, A., M. Li, C. Ayala, and R. Shavelson. 1999. Student science journals and the evidence they provide: Classroom learning and opportunity to learn. Paper presented at the annual meeting of the National Association for Research in Science Teaching, Boston, MA (March 30). For information about the paper, contact Maria Araceli (Ayita) Ruiz-Primo at *aruiz@standford.edu.*

Ruiz-Primo, A., M. Li, and R. Shavelson. 2002. Looking into student science notebooks: What do teachers do with them. CRESST Technical Report 562. Los Angeles: CRESST.

Saul, W., J. Readon, C. Pearce, D. Dieckman, and D. Neutze. 2002. *Science workshop: Reading, writing and thinking like a scientist* (2nd ed.). Portsmouth, NH: Heinemann.

Schilling, M., L. Hargreaves, W. Harlen, and T. Russell. 1990. *Assessing science in the primary classroom: Written tasks.* London: Paul Chapman.

Shepardson, D. 1997. Of butterflies and beetles: First graders' ways of seeing and talking about insect life cycles. *Journal of Research in Science Teaching* 34 (9): 873–889.

Shepardson, D., and S. Britsch. 2001. The role of children's journals in elementary school science activities. *Journal of Research in Science Teaching* 38 (1): 43–69.

Songer, N. 2003. Persistence of inquiry: Evidence of complex reasoning among inner city middle school students. Paper presented at the annual meeting of the American Educational Research Association, San Diego, CA (April). For information on this paper, contact Nancy Songer at *songer@umich.edu.*

Songer, N., and P. Ho. 2005. Guiding the "explain": A modified learning cycle approach towards evidence on the development of scientific explanations. Paper presented at the annual meeting of the American Education Research Association, Montreal, Canada (April). For information on this paper, contact Nancy Songer at *songer@umich.edu.*

Vanosdall, R., M. Klentschy, L. Hedges, and K. Weisbaum. 2007. The effects of inquiry based science teaching in 5th grade. Paper presented at the annual meeting of the

American Educational Research Association, Chicago, IL. For information on this paper, contact Rick Vanosdall at *rvanosdall@coe.tsuniv.edu*.

Vitale, M., N. Romance, and M. Klentschy. 2005. Enhancing the time allocated to elementary science by linking reading comprehension to science: Implications of a knowledge-based model. Paper presented at the annual meeting of the National A ssociation for Research in Science Teaching, Dallas, TX. For information about this paper, contact Michael Vitale at *vitalem@mail.ecu.edu*.

Vygotsky, L. S. 1978. *Language and thought*. Cambridge, MA: MIT Press.

White, B., and J. Fredrickson. 1998. Inquiry, modeling, and metacognition: Making science accessible to all students. *Cognition and Instruction* 16 (1): 42–56.

## CHAPTER 8

Assessment of Laboratory Investigations

Arthur Eisenkraft
University of Massachusetts, Boston

Matthew Anthes-Washburn
Boston International High School

Equipped with his five senses, man explores the universe around him and calls the adventure Science. —Edwin Powell Hubble, astronomer, *The Nature of Science*, 1954

Laboratory investigations have been a staple of school science programs since the turn of the 20th century (Hofstein and Lunetta 2004). The most desired laboratory investigations are integrated into the science curriculum in a meaningful way to help students develop a deeper understanding of the science content, as well as an understanding of the nature of science, the attitudes of science, and the skills of scientific reasoning (Singer, Hilton, and Schweingruber 2006). Unfortunately, many of the laboratory investigations that take place in schools do not meet these criteria and, therefore, may not be meeting the goals we have for our laboratory programs.

How do we assess a laboratory investigation? The assessment has two distinct components. The first is an evaluation of the quality of the laboratory investigation itself: Does it meet the criteria for a desired laboratory experience? The second is an evaluation of the student performance and achievement from the laboratory investigation: Is the laboratory experience providing the intended results for the student?

Improving Laboratory Investigations

A useful preliminary task in this exploration is to ascertain what we presently know or believe about laboratory activities. In workshops that we have presented on this topic, we try to elicit participants' understandings by having them discuss a series of three questions. By using a "carousel"—each group answers a single question and then the groups move on to the next question, read the prior responses, and add additional responses (NCREL 1995)—we have found strong consensus on both the goals and characteristics of quality laboratory activities. The questions and selected participant comments (in italics) include:

- What student learning goals do we have for laboratory activities?
  - *Students develop an enthusiasm for research.*
  - *Students use evidence and claims to answer questions.*
  - *Students develop critical-thinking skills in analysis.*
  - *Students communicate their results and analysis.*
- What evidence will you accept that students have accomplished these goals?
  - *Students do the investigation effectively and safely.*
  - *Students design a follow-up experiment.*
  - *Students explain and present their results.*
  - *Students answer questions posed by peers, teachers, and experts in the field.*
  - *Students connect the laboratory activity to real life.*
  - *Students work effectively in teams.*
- What aspects of laboratory activities promote student learning?
  - *Student engagement*
  - *Inquiry-based*
  - *Collaboration with peers*
  - *Organizing and presenting data*
  - *Learning from success and failure*
  - *Connection to life and real-world problems*
  - *Raises new questions*
  - *Leads to design of new investigations*

Although there seems to be a consensus on the importance of inquiry, student engagement, safety, effectiveness of laboratory procedure, and the analysis and communication of results, the terms that are used can be interpreted differently by different people. We find, following the carousel, that it is worthwhile to engage in a participatory laboratory activity in order to make sense of the meanings of survey responses.

Everybody has used carbonless duplicate paper as forms for bank deposits, for purchase orders, or for telephone messages. How does this carbonless duplicate paper work? If given a piece of carbonless duplicate paper, how would you investigate the materials in order to better understand "how it works"? Participants at workshops were asked to explore the carbonless duplicate paper and to answer the following questions at the completion of this laboratory activity:

- What observations did you make?
- What data did you collect?
- How do you interpret your results?
- What is your model for how the carbonless duplicate paper works?

This is an example of an inquiry-based lab activity. Without instruction, participants developed various strategies for exploration and observation: drawing lines on the page of varying thickness, altering the pressure, reversing the order of the papers, using other papers, wetting the paper, smelling (and even tasting) the paper, heating the paper, and using other objects to create an image. Participants had to decide what data to collect and whether to make the observations qualitative or quantitative. When asked to summarize and interpret the results, participants had to decide how to organize their results and develop their own models to explain them.

Inquiry activities have the following essential features:

- Learners are engaged by scientifically oriented questions.
- Learners give priority to *evidence*, which allows them to develop and evaluate explanations that address scientifically oriented questions.
- Learners formulate *explanations* from evidence to address scientifically oriented questions.
- Learners evaluate their explanations in light of alternative explanations, particularly those reflecting scientific understanding.
- Learners communicate and justify their proposed explanations. (NRC 2000)

The carbonless carbon paper activity matches these features of inquiry.

Of course, the carbonless duplicate paper is not part of a physics or chemistry curriculum. It will not be found in textbooks, lab manuals, or on state frameworks. Although it may help students and teachers better explore the meaning of inquiry and the process of model building, most teachers have content needs and their limited laboratory time must be devoted to activities that help students understand the content of the course. How can we take a "traditional" laboratory activity and tweak it toward inquiry?

The "density lab" is a classic exploration found in the elementary school, middle school, and high school science curricula. A traditional textbook might introduce the topic of density through a brief definition: "Density is an important topic. It can help explain why iron sinks while wood floats. Density is defined as the ratio of the mass to the volume ($D = M/V$.)" The student is then invited to learn how to calculate density. A sample problem is provided where the mass of a substance is given as 5 g and the volume as 3 cm^3. The density is then calculated to be $D = M/V = (5 \text{ g})/(3 \text{ cm}^3) = 1.7$ g/cm^3. Needless to say, this treatment of the topic of density robs the student of an opportunity for inquiry and for appreciation of the way science understanding evolves.

An alternative approach to density that uses the lab as a vehicle for inquiry can be found in *Active Chemistry*, a high school chemistry curriculum developed with support from the National Science Foundation (Eisenkraft 2007). In the chapter where students are challenged to create a special effect for a movie, students engage in an activity titled "Mass and Volume." The activity begins by eliciting students' prior understanding by asking them to interpret what they see in a cartoon where a very muscular man is exerting himself while holding a lead object and a woman easily holds a similar aluminum object (Figure 8.1). Students are then asked to record the masses and volumes for different amounts of water. They are also asked to calculate the ratio of mass to volume for the water. They find that the ratio is always close to 1 g/cm^3. Students then conduct a similar experiment with isopropyl alcohol and find that the ratio is always close 0.8 g/cm^3. They also repeat the experiment with pieces of clay and find that the ratio is always close to 1.6 g/cm^3.

Figure 8.1 Mass and Volume

Source: Art by Tomas Bunk from Eisenkraft, A. 2003. *Active chemistry.* Armonk, NY: It's About Time. © It's About Time. Reprinted with permission.

Students then answer the question, "If you were told the mass and volume of an unknown substance, would you be able to identify whether the substance was water, alcohol, or clay?" Students successfully recognize from the outcome of their investigation that they can identify the substance by calculating the ratio of mass to volume and comparing it to water (1.00 g/cm³), alcohol (0.8 g/cm³), and clay (1.6 g/cm³). The text then explains how the importance of the mass-volume ratio that they have discovered warrants giving this ratio a specific name—density. The text goes on to explain how density is a characteristic property of matter and how different substances can be identified using a list of densities in much the same way that they did in this activity. Sample mathematical problems and homework are assigned. Of course, all students must learn about density because of its importance in science and nature and because the topic is found on all state frameworks. In addition, students in *Active Chemistry* also learn about density so that they can suspend solid objects of density 0.9 g/cm^3 in a water and alcohol mixture and film them as if they are suspended in air as part of a special effect for their movie.

In this example, one can see how the density lab is tweaked toward inquiry and the lesson follows a 7E instructional model (Eisenkraft 2003; Bybee 1997). The students are Engaged (equipment, activity, cartoon); student prior understandings are Elicited (cartoon analysis); the students Explore (the activity of measuring masses and volumes of water, alcohol, and clay); the students Explain (given the mass and volume, identify the substance); the text Explains (definition of density); the text Elaborates (density as a characteristic property of matter; density of various substances; density and its relation to floating and sinking); the students Extend (applying the concept of density to their movie special effect); and the teacher Evaluates (in all phases of the lesson by questioning students, by observing the students in labs, and by collecting and grading their lab reports).

The National Research Council (NRC) of the National Academy of Sciences published the study *America's Lab Report, Investigations in High School Science* (Singer, Hilton, and Schweingruber 2006), which surveyed the literature and research on laboratory investigations and summarized and made recommendations based on this research. The first task of the NRC committee yielded a definition of *laboratory*: "Laboratory experiences provide opportunities for students to interact directly with the material world (or with data drawn from the material world), using the tools, data collection techniques, models, and theories of science." The committee noted that this could include use of data found in climactic or astronomical databases but did not include previously collected data presented to students of a cart's motion down an incline since the students themselves can easily collect such data. Although the process of such data analysis may have value, such an exercise is not included as a lab investigation under this definition unless the students collect the data.

The goals of laboratory experiences (Singer, Hilton, and Schweingruber 2006) are as follows:

- Mastery of subject matter
- Developing scientific reasoning
- Understanding the complexity and ambiguity of empirical work
- Developing practical skills
- Understanding the nature of science
- Interest in science and science learning
- Developing teamwork abilities

We would add that an additional goal of laboratory experiences is to provide a common experience for each and every member of the classroom so that all students will know what is being discussed. With the diversity we now find in our schools and communities, we cannot expect that all students have similar backgrounds or experiences. For example, many students have never personally observed the waves at a beach. Rather than referring to waves at a beach in a book or in class, students should see waves in a tank as part of their classroom experience before reference is made to the waves at the beach. This is as much a pedagogical requirement as it is an equity issue.

The NRC study concluded that "typical lab" experiences were historically isolated from the flow of science teaching and, unfortunately, that this approach remains typical today. In contrast, the study also found evidence that some programs integrate labs with discussion and draw on the principles of learning from cognitive research studies. These "integrated instructional units" lead to better understanding of the nature of science, more interest in science, and development of more sophisticated aspects of scientific reasoning than the "typical labs." The study also concluded that there was inadequate evidence in the research literature of students' understanding of the complexity and ambiguity of empirical work, developing practical skills, or developing teamwork skills.

Improvement of lab investigations should begin with the use of design principles (Singer, Hilton, and Schweingruber 2006). Clear learning outcomes should guide all the instruction. Teachers can use the strategies found in *Understanding by Design* (Wiggins and McTighe 1998) as a template whereby a curriculum developer looks first at the big picture of what we want students to know, considers how achievement will be assessed, and then develops the best learning activities to move toward these goals. Teachers can also adopt a 7E instructional model (Eisenkraft 2003). The labs should then be thoughtfully sequenced into the flow of classroom science instruction. The lab activity should not be scheduled for Tuesday simply because that is when the equipment is available nor should the lab activity occur days, weeks, or months before or after the concept is introduced. The lab activity must be logically placed in the curriculum so that both content and learning about the processes of science are well integrated. Finally, student reflection and discussion time should be allocated in the design of the curriculum (NRC 2001; Black and Wiliam 1998).

It is disappointing to find that, too often, the quality of laboratory activity is poor. It is criminal that students in lower-level science classes and in schools with high concentrations of minorities spend little time in any kind of laboratory investigation whatsoever (Banilower, Green, and Smith 2004). If a goal of science education is to decrease the achievement gap of students, then all students must be provided with equal access to laboratory investigations.

Assessing Student Performance in the Laboratory

As we have seen, following design principles will improve the quality of laboratory investigations and providing lab time for all students will guarantee that all students have the opportunity to learn from the lab. Once we accomplish these changes, we need better methods of assessing how students perform in the laboratory.

Development of rubrics as a template for denoting criteria of performance has been shown to increase student achievement (Saphier and Gower 1997; Luft 1997; Siegel et al. 2006). A typical rubric clearly denotes each category to be evaluated and provides specific, required criteria for defining excellence, proficiency, and below-proficiency levels of performance. A sample rubric for evaluating a lab report is provided in Figure 8.2. In this rubric, the categories are Introduction; Materials, Methods, Roles; Results; and Discussion. In conducting workshops on the use of rubrics and evaluating their use in classrooms, we have generated three recommendations that will increase the value of such rubrics.

1. Assessment should directly address the goals of the laboratory investigation.
2. Students should participate in the assessment of their own performance.

3. Assessment should begin from a foundation of a proficient level of performance, providing ladders for students to achieve higher orders of thinking.

Figure 8.2 Rubric: Evaluating a Lab Report

	A level: Integrates, applies, innovates	B level: Connects, clarifies	C level: Identifies	Incomplete
Introduction • Interest in science and science learning	Completes requirements for B and C levels and defines several alternate hypotheses you will try to eliminate through your investigation.	Completes requirements for C level, and identifies the variables you will control and the variables you will test.	Introduction identifies what you initially believe about the topic of the investigation and lists the questions you initially have about how the project works.	Introduction is unclear with respect to your initial beliefs and the questions you have about the topic of the investigation.
Materials, Methods, and Roles • Developing practical skills • Developing teamwork abilities • Understanding the complexity and ambiguity of empirical work	Team designs its own investigation, tailored toward the questions they generated and designed to test the hypotheses they developed.	Completes the requirements for C level. Notes any troubles the team encountered carrying out the investigation and describes the solution developed by the team.	Materials, Methods, and Roles section identifies all the equipment used and outlines the methods by which the investigation was carried out. The author also outlines the role of each member of the team.	Materials, Methods, and Roles section is unclear with respect to the materials used, the methods used, or the role of each team member in the investigation.
Results • Developing scientific reasoning • Understanding the complexity and ambiguity of empirical work	Completes the requirements for B and C level and discusses the format used to communicate the results and the effectiveness of the choice.	Completes the requirements for C level, and describes significant features of the results.	Results section accurately communicates the results of the team's investigation.	Results of the team's investigation are incomplete, inaccurate, or unclear.

(continued)

Figure 8.2 *(continued)*

Discussion • Mastery of subject matter • Developing scientific reasoning • Understanding the complexity and ambiguity of empirical work • Understanding of the nature of science	Completes the requirements for B and C levels and answers the following questions: Why should we care? How can you apply your findings to your unit project?	Completes the requirements for C level and answers the following questions: Why do we believe? What evidence do you have that your findings hold true in the outside world?	Discussion explains the findings of the investigation, especially the following questions: What does it mean? What principles or generalizations can you infer from your investigation? How do we know? What empirical evidence supports your conclusion?	Discussion is unclear about the findings of the investigation.

First, as with any assessment, the assessment of student performance in laboratory investigation must match the goals of the task. In Figure 8.2, the goals of laboratory experience proposed by Singer, Hilton, and Schweingruber (2006) and described earlier are addressed in the categories of the rubric. In the Introduction, students catalog their initial beliefs and ideas, and develop questions they wish to investigate. This serves to address the goal that students develop interest in science and science learning, which can be assessed through use of the rubric. The Materials, Methods, and Roles category is an opportunity to assess students' progress in developing practical skills, developing teamwork abilities, and understanding the complexity and ambiguity of empirical work. Together, the Results and Discussion categories provide an instructor with evidence of students' progress in the mastery of subject matter, developing scientific reasoning, understanding the complexity and ambiguity of empirical work, and understanding the nature of science. Since each of these goals represents a process of continual intellectual development, one can also see in Figure 8.2 that the tiered approach of C level, B level, and A level represents increasing orders of thinking.

A second recommendation for the effective application of rubrics in assessment of student laboratory performance is to have students submit a completed rubric that serves as a self-evaluation of their work. Although this is no extra work for the instructor and still requires that the instructor grade the report using the rubric, comparison between the student and teacher rubrics can open up a valuable dialog.

In the best of all possible worlds, both the student and teacher agree that the grade on the report, based on the rubric, should be an A. In this case, no conversation is required. However, if two students each earn a C on their respective work, the necessary conversation between the teacher and each student will be markedly different if the self-evaluations of the students are different. If the first student's self-evaluation yields an A and the teacher finds the work deserves a C, then a discussion must ensue. This discussion can include the teacher asking specific questions and trying to ascertain if the student misunderstood elements of the rubric or whether the student misunderstood the quality required for an A. In another example, if the student self-evaluation yields a C and the teacher finds the work deserves a C, then a different discussion must ensue. In this case, the teacher may want to know why the student handed in C work. Did the student run out of time? Did the student not understand some of the concepts? Did the student not care about the work? Without having the evidence of the students' self-evaluation, the teacher cannot personalize the discussion to the individual student's needs.

The third recommendation requires a radical shift in the way some rubrics are generated and can raise student expectations, yield an improvement in student achievement, and more closely reflect the norms of the business and academic communities. In some rubrics, the original floor of expectations unfortunately evolves to become the ceiling of performance expectations. When rubrics are developed, a set of criteria are generated by the teacher (and, in the best cases, with input from the students). These criteria set the standard by which the students set expectations for their work. When the time to submit work arrives, the full rubric is presented. The criteria that were developed become the "excellence" standard. The proficient or below-proficient levels are generated by recognizing that the student may give fewer examples, may omit parts of the assignment, or may have not understood part of the assignment.

We propose that the criteria that are set forth initially should become the proficient level. Anything above proficient will require additional work. In this way, all students must strive to perform, at the very least, at the proficient level. This mirrors the expectation that we have for adults in society. All adults are expected to be proficient at their jobs. Less than proficient does not "cut it" in this competitive society. Students know this and operate on this principle in that they will not buy a shirt with a missing button or keep a CD that has one song that skips. In each of these areas, there is a minimum to what we are willing to accept, a level we refer to herein as "proficient." What has happened in some cases is that in setting our criteria for a project and then later building the rubric, we have inadvertently allowed the floor representing proficiency to become the ceiling representing excellence.

What will the improved rubric look like and how will it be created? The criteria for success on a project is agreed upon by teacher and students and this then is denoted as the proficiency or C level in the new rubrics. Anything less than this C level corresponds to incomplete work. If the student is able to meet the criteria as set forth and, additionally, connect and clarify, then the student reaches the B level. If, in addition to the criteria for proficient and good, the student also integrates, applies, and innovates, then the student reaches the A level. Each level above proficient encourages students to apply higher orders of thinking to the problem (Nunley 2006). The laboratory assessment rubric in Figure 8.2 demonstrates this approach. In order to clarify the difference in approaches that can inadvertently lead to the "floor" of expectations becoming the "ceiling" of some rubrics, a contrast of two rubrics is provided in Figures 8.3 and 8.4. These rubrics are for unit projects, in which the conceptual understandings students develop by synthesizing previous investigations are applied to original student work, in this case, the design of a sport that can be played on the moon. In Figure 8.3 you can see that "excellent" is where the bar is set for a good project. Because the excellent category was established first, and the other categories detail subsets of "excellent" that can earn a student a B or a C, the "minimal" category reads as a detail of what the student didn't do toward "excellent": "Attempts were made to describe physics. Need to clearly identify two or more principles and explain what they mean." In contrast, the newer rubric

(Figure 8.4, p. 158) sets "proficiency" at C level; anything less than that is incomplete, does not receive a grade, and requires further revision. Reading this column, one can see that proficiency is a positive goal to attain, more like the "excellent" in Figure 8.3: "Explains how and why pushing an object on the Moon is different from lifting it." Another notable difference between Figures 8.3 and 8.4 is that the latter is organized around Essential Questions, or "What should everyone learn in this activity?" (Wiggins and McTighe 1998), a strategy that ensures that the assessment matches the learning goals established at the outset of the activity.

Figure 8.3 Rubric: "Sports on the Moon"
(The description of what is acceptable represents "A level" in the rubric.)

Category	A level Excellent	B level Good	C level Minimal	Needs Improvement
Physics principles	Two or more principles described in the section. Each principle is clearly explained with examples that make the principle easy to understand.	Two or more principles described in the section. Needs to be explained in a clearer way. Perhaps some new examples will help.	Attempts were made to describe physics. Need to clearly identify two or more principles and explain what they mean.	Need to address physics principles.
Innovation	Was very interesting, kept on topic, very creative, and put much quality into the project.	Interesting but could be more creative. Possibly more quality could have been put into it.	Somewhat interesting, needs creativity and more attention to quality.	Not interesting— based too much on lists of facts and not on individual creativity or quality.
Depth	The reader is thoroughly convinced that the sport described should be played on the Moon. The proposal develops ideas thoroughly and discusses all the pros and cons of the project.	The proposal does a good job developing the sport. It needs a little more detail and a little more discussion of the pros and cons of the project.	The sport is somewhat developed, but the basis for making the decision lacks detail, such as discussing why the sport is well-suited for the Moon.	Development of sport superficial.

(continued)

Figure 8.3 (*continued*)

Appearance	Very well put together. Organized and stood out from other proposals. Unique.	Organized, well put together. Perhaps it could look more unique, or some of the parts might look a little more polished (e.g., typed, illustrated on high-quality paper).	Somewhat organized. Needs more work on the neatness (e.g., typing) and some details that will make the proposal unique.	Little attention paid to appearance. Proposal needs more attention to be presentable.
Timeliness	On time.	1 day late	2 days late	Proposals received after 2 days will be reviewed for comments, but will not earn physics dollars.

Figure 8.4 Rubric: "Sports on the Moon"
(The description of what is acceptable represents "C level" in the rubric.)

Category	A level Integrates, applies, innovates	B level Connects, clarifies	C level Identifies	Incomplete
	A level ideas: • Create a drawing, sketch, or cartoon with free-body diagram • Make an animation • Act it out • Use objects to make an example		• Demonstrate: a model of how it works on Earth/Moon • Make a movie • Write a song • Tell a story (analogy)	
1. How does an object fall on the Moon?	Gives quantitative analysis: **Calculates examples** (show work) of free-falling objects/projectiles and **how much time** it will take them to reach the ground. **Elaborates the nature of acceleration** due to gravity on the Moon by comparing it with the Earth's acceleration.	Explains the difference in **acceleration** on the Moon, using evidence from a laboratory activity, and accurately analyzes the effects it will have on the sport. **Makes appropriate changes to the sport** to accommodate different behavior of falling objects.	**Explains** that an object does not just float when released on the Moon, but it speeds up as it falls. Gives a value for the acceleration. **Explains** that objects accelerate at the same rate regardless of mass.	Claims that objects on the moon "float," or states that they fall slower without giving an explanation or value for the acceleration.

(*continued*)

Figure 8.4 (*continued*)

2. What is it like to push an object on the Moon?	Gives quantitative analysis: **Calculates examples** (show work) of several objects in the sport and how much force is required to lift them and to accelerate them. Compares Earth and Moon. **Original creation elaborates the nature of pushing vs. pulling** on the Moon. • See above for ideas	**Explains how and why** pushing an object on the Moon is different from lifting it, using evidence from a laboratory activity. **Makes appropriate changes to the sport** to accommodate the nature of pushing objects on the Moon versus lifting.	**Explains how and why pushing** an object on the Moon is different from **lifting** it.	Attempts to explain, but confuses the way an object is pushed on the Moon.
3. How does a thrown object behave on the Moon? (projectile motion)	Gives quantitative analysis: **Calculates examples** (show work) of several projectiles, their range and their time on the Earth compared with the Moon. **Original creation elaborates the nature of projectiles** on the Moon. • See above for ideas	**Explains how and why** there is a difference on the Moon for the **range** and **time** of a projectile, using evidence from a laboratory activity. **Makes appropriate changes** to the sport to accommodate the nature of projectiles on the Moon.	**Explains how and why** there is a difference on the Moon for the **range** and **time** of a projectile.	States that an object can be thrown farther on the Moon, but does not explain how or why.
4. How is friction different on the Moon?	Gives quantitative analysis: **Calculates examples** (show work) of several objects, their mass and weight, the coefficient of friction and amount of force they might experience on the Moon. **Original creation elaborates the nature of friction** on the Moon. • See above for ideas	**Explains how and why** the **friction** between a person or vehicle and the ground will be different on the Moon, using evidence from a laboratory activity. **Makes appropriate changes** to the sport to accommodate the nature of friction on the surface of the Moon.	**Explains how and why** the **friction** between a person or vehicle and the ground will be different on the Moon.	States that friction is different on the Moon, but does not explain how or why.

"We Are Not Science Students; We Are Student Scientists"

One of the purposes of laboratory investigations is to have students mirror the work of scientists. The phrase "creating student scientists not just science students" has been adopted by the Pathfinder Science group that guides teachers and students toward research projects (Pathfinder Science). A way in which to evaluate both the quality of the lab experience and the performance of students is to compare the cognitive processes that scientists use with those of our students. Chinn and Malhotra (2002) developed a theoretical framework by which such a comparison could be made. Here is their list of cognitive processes:

- Generating research questions
- Designing studies
- Making observations
- Explaining results
- Developing theories
- Studying research reports

Good laboratory investigations provide students with an opportunity to make observations. Although many laboratory investigations require students to explain results, they don't always mirror the way scientists view this process. Scientists explain results by making generalizations, transforming observations, finding flaws, and using indirect reasoning (Chinn and Malhotra 2002). Science students are often required only to report their results and state whether their observations match those of the textbook. If we are to help our science students to become student scientists, the breadth of opportunity to design studies, develop theories, study other research reports, and generate research questions must be increased and the assessment of laboratory performance must take these cognitive processes into account.

Scientists design studies. Discussions about inquiry make distinctions between guided inquiry and open inquiry (NRC 1996, 2000; Eick, Meadows, and Balkcom 2005.) The inherent danger in recognizing the value of students being involved in open inquiry lies in believing that all inquiry must be open inquiry. Returning to the carbonless duplicate lab activity, one wonders where participants would place this lab along the continuum of guided to open inquiry. Some people are quick to point out that the participants were given the carbonless duplicate paper—they didn't decide for themselves that this was something they wanted to study. If that is the defining criteria for inquiry, then it is quite understandable why teachers would be intimidated and shy away from inquiry. In reality, children must sometimes be guided toward questions.

In the inquiry addendum to the standards (NRC 2000), an example is provided in which students return from the playground with questions about the health of two trees. The teacher then runs with this idea and scientific investigations abound. Is this open inquiry because the students noticed the trees and the teacher apparently did not set the content domain? We would guess that some of the students also noticed that there was lots of noise when the fire engine passed by while others noticed that two of their classmates were having an argument about the rules of a game and still others noticed that it looked as if it might rain. The teacher could have picked up on any of these observations to have classroom discussions of fire safety, noise pollution, sociology of interactions, conflict resolution, or weather forecasting. When the teacher chose to focus on the question of the trees, it was similar to giving out carbonless duplicate paper. The teacher can guide the investigation and it will still be a suitable inquiry investigation.

Guided inquiry is also a necessary precursor of open inquiry. The scientist can engage in open inquiry because the scientist has the skills to make certain measurements and to use specific tools. The student who may want to know about floating and sinking will still require some guided inquiry to learn how to measure the mass and volume of liquids and irregular solids. As we encourage each other to transform our traditional labs to more inquiry-based lab experiences, we should agree that not all labs can be open-ended inquiry investigations and that guided inquiry has a valuable role in science education.

If we accept open inquiry to be one of the things professional scientists do, then some examples from the field of science may help clarify how a task can be high in inquiry without being devoid of directions. In field science, a student ecologist could easily be overwhelmed. Within the scope of her field of study are living organisms, nonliving elements of the environment, the characteristics and dynamics of populations, interspecific interactions, and the dynamics of energy and matter in the ecosystem (Ecological Society of America 2006). Where does an ecologist even start to observe and ask questions in such a complex system? She may be influenced by her previous observations and choose a "hot" area of study (such as global climate change) or she may even be influenced by the interests of an adviser or funder. Just as those studying the carbonless duplicate paper are influenced in their topic of study, all scientists experience outside influences on the area and scope of their investigations.

Even after choosing a question, a scientist is not without direction. She uses protocols and conventions established by her peers. For example, the field ecologist studying the biodiversity of a site will use established protocols for placing transects, data sampling, and analysis of that data, perhaps using the Shannon-Weaver (1949) index to quantitatively compare her biodiversity data to other data sets taken at other sites or at other times.

Without these protocols and conventions, our field ecologist would be lost. Having knowledge and an understanding of how to conduct these protocols does not hinder her inquiry, it enables it. In the same way, our students must learn the protocols and conventions of laboratory study necessary for them to be able to design their own investigations.

The Assessment Triangle

In both creating quality laboratory investigations and assessing performance on these investigations, we can use the assessment triangle of cog-

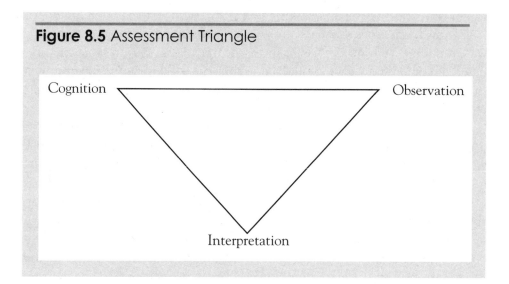

Figure 8.5 Assessment Triangle

nition, observation, and interpretation as a guide (Figure 8.5) (Pellegrino, Chudowsky, and Glaser 2001).

The purpose of assessments is to ascertain what a person knows—cognition. It would be wonderfully efficient if we could use an instrument to look into the brain and see if someone understands something. Since we are not able to physically evaluate the way the neurons are firing to assess understanding, we must make other observations. These observations often take the form of questions that the person answers or instructions that the person follows. The assessor must then interpret these observations to come to a conclusion about the person's cognition. If cognition, observation, and interpretation are not explicitly connected, the value of the inferences will be diminished.

A story will illustrate the components of the assessment triangle. A father took his eight-year-old daughter to the ophthalmologist to have her vision checked. The doctor asked the child to read the standard vision chart. She said that she couldn't. The doctor noted this and left the room for a few minutes, explaining that he would be back shortly. As soon as he was gone, the girl began to cry. When her father asked what was wrong, she remarked, "None of the words make sense." In this case, the doctor was trying to find out about the girl's vision (a physical attribute much easier to ascertain than cognition). He made an observation that she could not read the chart and interpreted this to mean that she had some visual impairment. The father, upon additional observation, found that she could see the letters but could not make sense of them. The father's interpretation is that the girl will not require glasses. The assessment triangle helps remind us that observations are imperfect as are interpretations.

When creating laboratory investigations, the assessment of the investigations should be of primary concern and not an afterthought. For example, when a student presents a data table in a lab report ("observation" in the assessment triangle) is that sufficient evidence for the teacher ("interpretation" in the assessment triangle) that the student understands how the data were collected, the limitation of the data, and how the data may change under different circumstances ("cognition" in the assessment triangle)? In the investigation, the importance of the big principles, content, and processes should be balanced with an equal weighting and attention given to the question, "What will we accept as evidence that the students understand the investigation?" (Wiggins and McTighe 1998). The assessment triangle can be a valuable guide in developing meaningful assessments of what students know and are able to do.

Conclusion and Next Steps

Assessing laboratory investigations requires an equal emphasis on assessing the quality of the investigation and assessing student performance. In reviewing the labs that are present in curricula or creating new labs or evaluating labs and programs during textbook adoption, stakeholders should pay particular attention to the points made in this chapter:

- Lab investigations should be an integral part of all science programs.
- All students should have access to laboratory investigations in their science classes.
- Lab investigations should be integrated with content and discussion and not be isolated from the stream of instruction.
- Lab investigations should be inquiry-based.
- Lab investigations should mirror the cognitive processes of scientists: Science students should become student scientists.
- Rubrics should be used to clarify the requirements of the lab investigation.
- Rubrics should be "built up" from agreed-upon criteria for success representing the "proficient" level of performance.

Improving the quality of laboratory investigations and the means by which we evaluate student performance in the lab requires professional development and is an evolutionary process. Teachers need to move toward increased use of inquiry in their classrooms and should develop strategies to change their teaching over time. Fortunately, there are footholds along the way that can help teachers improve their practices in incremental ways—by varying the degree to which inquiry is present (NRC 2000; Eick, Meadows, and Balkcom 2005). Progress can be measured by reviewing laboratory investigations to make sure they meet the above criteria and by evaluating student work using rubrics with clear expectations of student understanding.

References

Banilower, E. R., S. Green, and P. S. Smith. 2004. *Analysis of data for the 2000 National Survey of Science and Mathematics Education for the Committee on High School Science Laboratories* (September). Chapel Hill, NC: Horizon Research.

Black, P., and D. Wiliam. 1998. Inside the black box: Raising standards through classroom assessment. *Phi Delta Kappan* 80(2): 139–148.

Bybee, R.W. 1997. *Achieving scientific literacy.* Portsmouth, NH: Heinemann.

Chinn, C. A., and B. A. Malhotra. 2002. Epistemologically authentic inquiry in schools: A theoretical framework for evaluating inquiry. *Science Education* 86(2): 175–218.

Ecological Society of America. 2006. *www.esa.org/aboutesa/*

Eick, C., L. Meadows, and R. Balkcom. 2005. Breaking into inquiry. *The Science Teacher* 72(10): 49–53.

Eisenkraft, A. 2003. Expanding the 5E model. *The Science Teacher* 70(6): 56–59.

Eisenkraft, A. 2007. *Active chemistry.* Armonk, NY: It's About Time.

Hofstein, A., and V. N. Lunetta. 2004. The laboratory in science education: Foundations for the twenty–first century. *Science Education* 80: 28–54.

Luft, J. 1997. Design your own rubric. *Science Scope* (Feb.): 25–27.

National Research Council (NRC). 1996. *National science education standards.* Washington, DC: National Academy Press.

National Research Council (NRC). 2000. *Inquiry and the national science education standards: A guide for teaching and learning.* Washington, DC: National Academy Press.

National Resource Council (NRC). 2001. *Classroom assessment and the national science education standards.* Washington, DC. National Academy Press.

North Central Regional Educational Laboratory (NCREL). 1995. Pathways. Carousel Brainstorming. Learning Point Associates. *www.ncrel.org/sdrs/areas/issues/educatrs/profdevl/pd2reach.htm*

Nunley, K. 2006. Layered curriculum. *http://help4teachers.com*

Pathinder Science. *http://pathfinderscience.net/search/*

Pellegrino, J. W., N. Chudowsky, and R. Glaser, eds. 2001. *Knowing what students know: The science and design or educational assessment.* Washington, DC: National Academy Press.

Saphier, J., and R. Gower. 1997. *The skillful teacher: Building your teaching skills.* Acton, MA: Research for Better Teaching.

Shannon, C. E., and W. Weaver. 1949. *The mathematical theory of communication.* Urbana, IL: University of Illinois Press.

Siegel, M. A., P. Hydns, M. Siciliano, and B. Nagle. 2006. Using rubrics to foster meaningful learning. In *Assessment in Science: Practical Experiences and Educational Research.* Arlington, VA: NSTA Press.

Singer, S. R., M. L. Hilton, and H. A. Schweingruber, eds. 2006. *America's lab report: Investigations in high school science.* Washington, DC: National Academies Press.

Wiggins, G., and J. McTighe. 1998. *Understanding by design.* Alexandria, VA: Association for Supervision and Curriculum Development.

## CHAPTER 9

Assessing Science Knowledge: Seeing More Through the Formative Assessment Lens

Kathy Long, Larry Malone, and Linda De Lucchi,
Lawrence Hall of Science, University of California, Berkeley

Teaching elementary science has a lot in common with surfing. Both require your complete and undivided attention. You must continually assess exactly where you are and what is happening around you. At the same time, you have to glance ahead to anticipate what's in store, and you have to look over your shoulder from time to time to confirm your progress. Occasionally you miscalculate and wipe out. The wiser for it, you climb back on your board, catch the next wave, and negotiate the tricky crosscurrent with an intuitive change of direction. As your skill advances, the experience becomes increasingly exhilarating.

Assessment in Science Education

Developing skill in any practice is enhanced by meaningful information and timely feedback. The process of acquiring informative feedback is assessment. Assessment in science education is any process used to determine the quantity and quality of a student's knowledge and skill related to natural systems and the principles that govern their behaviors. Good science teaching includes good assessment practice.

Assessment of science learning can take many forms. One way to assess a fourth grader's knowledge of magnetism is to sit down with the student and discuss magnets and their interactions for 20 minutes, providing opportunities for the student to explain, demonstrate, and display (diagram, illustrate, list, graph, model) the scope and depth of his or her knowledge. Another way to assess the student's knowledge of the same content is to pose a problem, provide magnets and other materials, and observe his or her attempts to solve the problem. You could craft two or three constructed-response questions that require the student to generate explanations and support assertions with data. Or you might choose to have the student answer a set of multiple-choice questions on paper. These are a few of the many ways that a student's knowledge about magnets and their behaviors might be assessed.

The Role of Assessment

Assessment serves three prominent roles in science education. *Accountability assessments* get the greatest amount of public attention. Accountability assessments are driven by externally imposed criteria (standards) and examine the congruence between those criteria and student performance. The results of accountability assessments are used to evaluate students, teachers, and systems, and failure to meet achievement goals can have punitive consequences. Accountability assessments tend to limit the purview of science and encourage the teaching of right answers. Second, assessment can provide evidence of effectiveness. *Effectiveness assessments* are specific to a curriculum and answer questions about the quality of the curriculum materials and teaching methods used with students. If students achieve the outcomes that the curriculum purports to deliver, the program can claim to be effective.

For teachers, the most important form of assessment is classroom assessment. *Classroom assessments* traditionally fall into two categories, summative and formative. Summative assessment looks at end products of instruction and is used to assign value to student achievement, often taking the form of grades. Midterm and final exams are familiar examples of summative assessments.

In contrast, formative assessments are integrated into the teaching/learning process, are intimately associated with curriculum and instruction, and are primarily informational. Formative assessments provide diagnostic data about learning while learning is taking place. Formative assessments are used to monitor student knowledge and thinking, not to assign grades. Diagnostic information from formative assessments can be shared and processed by teachers and students alike. The knowledge and reasoning exposed by students in their responses to formative assessments are the raw materials teachers use to plan the next steps in instruction. Formative assessment is assessment *for* learning, not assessment *of* learning.

In this chapter we focus on classroom assessment practice, with emphasis placed on *embedded diagnostic assessment*, hereafter referred to simply as embedded assessment. The locus of activity for embedded assessment is the classroom, and the players are the classroom teacher and his or her 25–30 students. The goal of embedded assessment is to monitor and guide student progress toward understanding of complex scientific processes and ideas. Embedded assessment activities require teachers and students to rethink their responsibilities in the classroom learning community and to (ideally) redefine the teacher/student and student/student relationships.

Designing a Classroom Assessment System—ASK

Background

The traditional paradigm of classroom assessment is evaluation. The traditional instrument used to evaluate learning is the test administered after a unit of instruction. Right answers are tallied up, a metric is applied, and scores are recorded. Students look at their tests, review their successes and failures, feel a moment of satisfaction or regret, and then move on. The process of expressing knowledge and understanding is isolated from instruction. Traditional assessment has served little instructional function.

Thinking about the role of assessment began to change in the mid-1990s. The metaphor of the black box promoted by Black and Wiliam (1998) invited a renewed interest in the complex of interactions that occur between instructional practice and student learning.

Assessing Science Knowledge (ASK) was a four-year assessment research-and-development project funded by the National Science Foundation (2003–2007). The purpose of the project was to design and demonstrate a comprehensive assessment system that could be seamlessly integrated into a coherent, research-based upper-elementary science program. In the ASK Project, we endeavored to reduce the distance between instruction and assessment. We reasoned that knowing what students understand about a topic could inform instruction. This required thinking critically and precisely about what we wanted students to know as a result of instruction. We began by identifying and describing the big intellectual structures in a course—the constructs—and the major chunks of knowledge—the concepts—that we expected students to learn. The concepts provided a set of focusing "lenses" teachers could use to look for evidence of specific student thinking. With a coherent description of what students are expected to learn, it was then possible to develop high-quality assessment items to use throughout the course to monitor learning.

In ASK we developed assessment strategies, procedures, and instruments that teachers could incorporate into the flow of instruction to expose student thinking—specific observations of aspects of authentic activities, focused work in science notebooks, and explanations written on response sheets. An innovation in the use of these methods and materials occurred when students were invited to take part in the thoughtful processing of their own work on these assessments. Students cautiously accepted the opportunity to look critically at (i.e., to self-assess) their exposition of learning and understanding. They soon assumed responsibility for identifying flaws and inconsistencies in their responses. Identifying what they didn't know about a subject clarified where they needed to concentrate their thinking. As a result, the assessment practices became seamlessly integrated into instructional practices. The distinction between embedded assessment and instruction blurred until it virtually disappeared. The revised paradigm of assessment is one in which assessment is informative and an integral part of instructional practice.

Teachers told us they were dissatisfied with the traditional end-of-module summative assessment. They said they didn't want to wait until the end of the module to find out what their students were learning. In response, we introduced a mini-summative assessment, called an I-Check, after each investigation. The I-Checks provided summative data at major junctures throughout the module. The ASK staff had assumed at the beginning of the project that there was clear distinction between formative assessments, designed to yield diagnostic information, and summative assessment, designed to make judgments about the quantity and quality of learning that accrued as a result of instruction. Summative assessment was expected to provide effectiveness data only. As the project progressed, however, the function of the I-Checks changed. Teachers and students found that the summative data could also be instructive. Again, the I-Checks were used to identify inconsistencies and misconceptions in student understanding that needed to be addressed before moving on to new conceptual material. By using self-assessment methods, students have one more tool to help them improve their understanding of complex subjects. The dual function of assessment tools, serving both summative and formative functions, softens the traditional distinction between them.

ASK Project Organization

The two goals of the ASK Project were to (1) develop valid and reliable summative assessment tools that provide evidence that active-learning science programs teach science effectively, and (2) develop an embedded assessment system that allows teachers to monitor and analyze student progress with complex concepts and processes in science. For the purpose of this research we used the 16 Full Option Science System modules for grades 3–6 as the core science learning experience. FOSS was chosen because it is familiar and accessible to the ASK staff, it meets the standard of curriculum coherence and integrity, and it is widely implemented.

The project has nine field-test centers around the country. Each center has a director who coordinates 16 research classrooms during years 1 and 2, and 32 research classrooms during years 3 and 4. Data from all the teachers and students in 128 classrooms (years 1 and 2) and 256 classrooms (years 3 and 4) were analyzed to inform the evolution of the ASK assessment system. Year 4 of the ASK project started in the fall of 2006.

Theoretical Framework

The first phase of the ASK project involved enunciating a set of theoretical structures that would guide and frame the project activities. The starting point was the assessment triangle described in the National Research Council's *Knowing What Students Know: The Science and Design of Educational Assessment* (Pellegrino, Chudowsky, and Glaser 2001) The three vertices of the triangle identify the three major components of an assessment system: cognition, observation, and interpretation (Figure 9.1, p. 172).

Figure 9.1 Assessment Triangle

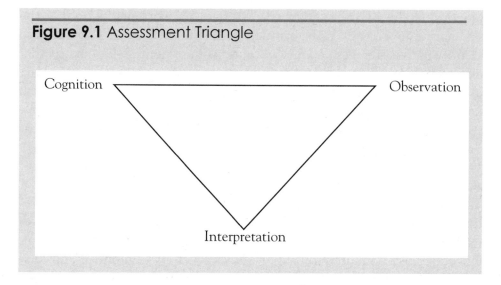

Cognition. The cognition vertex calls on assessment developers to think critically about the intended intellectual and behavioral outcomes of the curriculum. In the ASK system, the explanatory models that develop in the mind are called *constructs*. Each construct is then parsed into two or more *concepts*—the related, interacting pieces of knowledge that constitute the components of a construct. The specific pieces of knowledge, such as definitions, facts, and simple relationships, which combine to form concepts, are called *elements*.

The complexity of student understanding related to a construct (the concepts and elements to be learned) has to be consistent with the cognitive ability of the students at the grade level. Figure 9.2 shows examples of concepts and elements that contribute to the development of the construct related to understanding permanent magnetism. When the description of the thinking associated with a construct is complete and detailed, assessment of a student's progress toward full understanding of the construct can be precise and accurate.

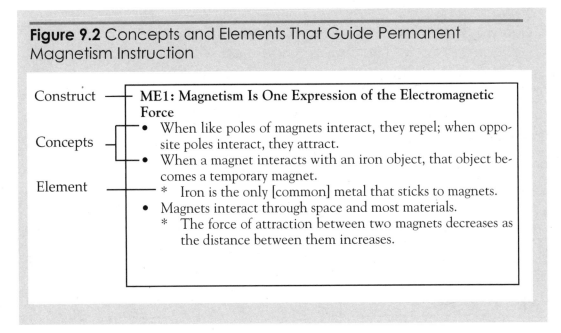

Figure 9.2 Concepts and Elements That Guide Permanent Magnetism Instruction

Construct —— **ME1: Magnetism Is One Expression of the Electromagnetic Force**

Concepts ——
- When like poles of magnets interact, they repel; when opposite poles interact, they attract.
- When a magnet interacts with an iron object, that object becomes a temporary magnet.

Element ——
  * Iron is the only [common] metal that sticks to magnets.
- Magnets interact through space and most materials.
  * The force of attraction between two magnets decreases as the distance between them increases.

Observation. Reasoning, knowledge, and understanding, the desired outcomes of science instruction, cannot be observed directly; they must be inferred from student-generated artifacts. ASK relies on two classes of student-generated artifacts: student work produced in the course of doing science and processing data, and student responses to specifically designed items on formal assessments. A large portion of the time spent developing the ASK system was directed to the exacting work of getting the assessment items right. An item that works well is (1) aligned with a concept or element, (2) equitable, providing universal access for all students, and (3) able to discriminate levels of understanding and expose misconceptions (see Figure 9.3, p. 174).

Figure 9.3 Samples of Items From an ASK Benchmark Exam Designed to Assess Constructs Associated With the Key Concept of Permanent Magnetism

I-CHECK
Investigation 1—The Force

ASK ID number
Date

me20. José discovered that he could move a magnet across the top of a wood table by moving another magnet under the table.

Explain why José is able to move the magnet on top of the table without touching it.

_____
_____
_____

? POLE   ? POLE   N   S

me68. Kari has a horseshoe magnet, but she isn't sure which pole is the north pole. She also has a bar magnet marked N and S.

How can Kari use her bar magnet to find out which pole of the horseshoe magnet is north?

_____
_____
_____

I-CHECK
Investigation 1—The Force

ASK ID number

me 5. Anne is investigating objects and magnets. She recorded this in her notebook.

I was surprised! A nail was stuck to the magnet. When I accidentally touched the nail to a paper clip, the paper clip stuck to the nail. I wonder what's going on.

a. Explain to Anne why the paper clip stuck to the nail.

_____
_____
_____

b. If Anne touched the nail to a copper penny, what do you think would happen and why?

_____
_____
_____

I-CHECK
Investigation 1—The Force

ASK ID number

me27. Keys can be made of iron or aluminum.

• If you want to find out if a key is made of iron, what can you do?

_____

• How will you know if the key is made of iron?

_____
_____

me28. Look at the picture on the right below. the bottom two magnets are stuck together.

a. Label the poles on each magnet.

These two magnets are attracting.

N

b. Explain why you labeled the poles the way you did.

_____
_____
_____

FOSS Magnetism and Electricity Module
© The Regents of the University of California
ASK Project use only. Do NOT duplicate.

Investigation 1 I-Check
4ME
Page 3

I-CHECK
Investigation 1—The Force

ASK ID number

me82. Look at the two pictures below.

Plastic container of iron filings   Magnet placed on top of the container

Explain what is happening when the magnet is placed on top of the container of iron filings.

_____
_____

me86. Think about the experiment you did in class using the balance shown in the picture.

a. What is the purpose of putting the spacers between the magnet in the cup and the magnet on the post?

_____
_____

b. What happens to the force between the magnets as more spacers are added?

_____
_____

FOSS Magnetism and Electricity Module
© The Regents of the University of California
ASK Project use only. Do NOT duplicate.

Investigation 1 I-Check
4ME
Page 4

Interpretation. Interpretation is making sense of student work and item responses. In ASK, teachers review students' science notebook entries frequently, making qualitative judgments of students' understanding. These quick assessments tap into the large currents that flow through an instructional sequence, suggesting where to slow down, when to revisit a concept in a modified manner, or when to move more quickly through the material.

Periodically it is useful to make more precise assessments of student progress. Carefully designed assessment tools (in ASK, items assembled into benchmark exams) elicit specific kinds of student responses. The responses are interpreted more rigorously using a process called coding. Once coding is complete for a set of items, the data can be aggregated to see how the student body is progressing with a given construct, or disaggregated to review an individual's performance. Multiple interpretations can be produced quickly and accurately when codes are entered into a computer program designed to manage assessment data (Figure 9.4).

Figure 9.4 Coding Guides for Two Items From an ASK Benchmark Exam

me68	Code	If the student...
	3	knows when opposite poles of magnets interact, they *attract*; when like poles of magnets interact, they *repel*. (ME1-ar)
		writes a procedure that uses the bar magnet in an appropriate way to apply the rule that opposite poles attract and like poles repel.
	2	writes a procedure that uses the bar magnet in an appropriate way but does not include how to tell which pole is north or south.
	1	provides any other answer.
	0	makes no attempt.

me28a	Code	If the student...
	2	knows when opposite poles of magnets interact, they *attract*; when like poles of magnets interact, they *repel*. (ME1-ar)
		labels all poles correctly.
	1	provides any other answer.
	0	makes no attempt.

me28b	Code	If the student...
	2	knows when opposite poles of magnets interact, they *attract*; when like poles of magnets interact, they *repel*. (ME1-ar)
		indicates that like poles repel and opposite poles attract.
	1	provides any other answer.
	0	makes no attempt.

The development of an assessment system is an iterative process. The constructs at the cognition vertex are refined in response to the student explanations (observation) and specific flaws in understanding (interpretation) that emerge. Then items (observation) change to reflect the restatement of the constructs (cognition). The coding guides (interpretation) change based on the revised items (observation), and so on. The goal is to reach homeostasis where the theory of learning, the items, and the evidence all complement one another and work in harmony.

ASK Project Activities

The ASK Project is about exposing student understanding of science content. Success depends on providing teachers with rock-solid tools for extracting the nuggets of scientific thinking from the prodigious quantity of collateral cognitive activity going on in students' heads. To access student scientific knowledge we needed good items (assessment questions), and lots of them. After identifying the learning outcomes we valued, we assembled the best items we could develop, based on experience and item-development theory, and sent them to our test centers for trial testing.

Each of the nine trial centers (years 1 and 2) tested two FOSS modules with revised assessments in the fall semester; four teachers using module 1 and four teachers using module 2. At the same time a second trial center tested the same two modules in a similar manner. Then in the spring semester, each trial center tested two different FOSS modules in eight different classrooms. This design provided data from eight classrooms for each module during both Year 1 and Year 2.

During Year 1, the four teachers testing any given module met in a study group six times. During these meetings they studied student assessment work, spending extensive time interpreting and categorizing student responses in a process called moderation. The moderation process required teachers to look critically at every student's response to an item and to make judgments about the quantity and quality of understanding it represented. They then generated a set of three to six statements that represented degrees of understanding—from *no knowledge*, through *incomplete or flawed understanding*, to *complete, accurate understanding*. The statements constituted the basis of a coding (scoring) guide for that one item. The process was then repeated for all of the other items.

All of the student work and teacher analyses were returned to the ASK Project staff for further analysis. Items were affirmed, discarded, or revised; new items were developed. The reworked item sets were subjected to additional scrutiny by our formative assessment partners at SRI International. Among other things, they checked for cognitive load, reading load, fairness, extraneous information, and bias. The assessments were revised and made ready for testing in the same classrooms in Year 2. Assessments were again revised, following the analysis of the teacher feedback and student work collected during Year 2.

In Years 3 and 4 the assessments were tested in twice as many classrooms at each of the (nine centers.) The item sets, now well along in the process of development, were analyzed for validity (assurance that the items actually measure what they are intended to measure) and reliability (statistical evidence that the items perform consistently in the hands of all users and in all circumstances). These determinations were made by our psychometric partners at the Berkeley Evaluation and Assessment Research Center (BEAR Center) at the Graduate School of Education at the University of California at Berkeley.

Assigning value to a student's test answers is usually referred to as scoring. In ASK, however, we do not do scoring per se. Instead we try to characterize the accuracy, completeness, and complexity of the student's knowledge. Throughout the ASK project we have struggled to describe the characteristics of different levels of understanding for all of the concepts taught. Each level is assigned a code, not a score. Teachers new to ASK want to add up the "scores" to see how much a student knows. Teachers need to realign their thinking to embrace a system of codes that describes a student's overall understanding of a concept, not an achievement score. To this end, the ASK staff is constantly tinkering with the coding guides to make them more precise and ever more informative with each revision of the system.

In Year 4, where we stand at the time of this writing, we are testing a computer-based analysis system developed by researchers at the BEAR Center, called ClassMap, which makes it possible for teachers to enter the assessment codes into the system and in moments have a comprehensive display of the progress made by the whole class or individuals for any of the constructs addressed in a module. This system will make it easy for a teacher to quickly identify content that needs general review by the whole class as well as individual students who are advanced or have deficits in their understanding of the content. The output of the ClassMap program is not just a display of student progress, it is also an analytical tool that presents an instantaneous nuanced snapshot of the cognitive currents running through the class. It is a digital lens that allows you to look more closely and with greater clarity at the learning taking place in your classroom.

Using ASK Embedded Assessments

A thoughtfully developed embedded assessment system is the sensory system of the curriculum. Embedded assessments keep information flowing from the learning environment to you at all times. The ASK Project has implemented a number of strategies and tools that facilitate the flow of information (Appendix, p. 190). Some of the strategies are suggested for use on an as-needed basis; others are considered critically important for adequately understanding student progress with difficult or complex concepts. Several tools and strategies have proven effective in the ASK work: quick writes, science notebook entries, and response sheets. We discuss each below; examples from the FOSS Magnetism and Electricity module for grades 3/4 will provide context for the discussion.

Quick Write. By third grade, students are typically able to express ideas in writing. The writing is not complex, but sound elements of a student's thoughts about a topic can be recorded on paper using words and labeled drawings.

Early in the exploration of a new topic, such as permanent magnetism, students can take two minutes to generate a quick write in response to a very general question (prompt): "What happens when a magnet comes close to a piece of metal or another magnet?" A quick-write prompt serves three functions. It activates students' prior knowledge, it documents students' entry-level understanding of the topic, and it produces an artifact that can be brought back after several more lessons for reflection. The quick-write exercise should be completed individually. The teacher collects the work and reviews it for ideas that correspond to the constructs that define the content of the curriculum. No marks or comments are applied to the quick writes; they are maintained in their original condition. These preinstructional ideas about magnetic interactions provide a number of markers that teachers note and use to put contours on their instruction: Students think magnets stick to all metals; students think magnets stick to each other; students don't mention poles, attract, repel, or interaction at a distance.

After students have had a number of opportunities to study the topic, the quick writes are returned and students self-assess their original work. Self-assessment adds power to the learning experience in several ways. Students can immediately determine that they have learned a lot. This is a valuable psychological boost for students. And students can review and revise their original responses to the prompt. One way to guide the process uses colored pens or pencils to code various classes of revision: green to underline statements that were correct in the original writing; red to modify original statements that were incorrect or deficient; and blue to add ideas or information that were absent in the original. Critical assessment and reorganization of ideas in this manner is a powerful learning strategy. Self-assessment is an example of assessment *for* learning.

Science Notebook Entry. A personal science notebook is an asset for learning and a valuable artifact for assessment. When students engage in active investigation, they interact with objects, organisms, and systems. During the investigation students record observations. These data take many forms, including metric measurements, lists, tables, and labeled illustrations. Subsequently, students organize, display, and analyze the data to discover relationships, make predictions, draw conclusions, and communicate explanations. The student's science notebook record of acquisition, management, and processing of authentic data constitutes a body of artifacts that can be assessed. Organizational and display skills (charts, diagrams, tables, graphs) can be monitored, and scientific knowledge can be assessed by reviewing students' predictions and explanations in the narrative sections of the science notebook.

The examples of one student's work in Figure 9.5, page 180, reveal that he is able to explain that south poles or north poles repel, and unlike poles attract. But he exposes confusion about the kinds of materials that stick to a magnet. His preliminary conclusion is that some of the metals tested are "real metals" and some are not real metals.

Figure 9.5 Samples of Student Science Notebook Entries

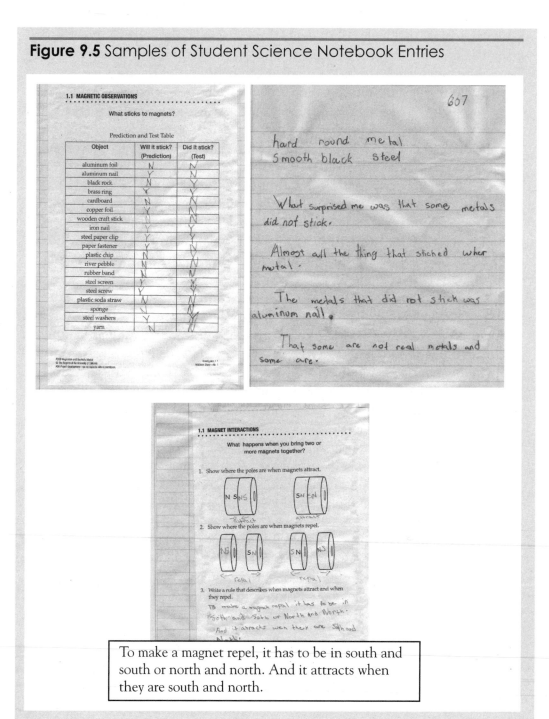

To make a magnet repel, it has to be in south and south or north and north. And it attracts when they are south and north.

This misconception is not uncommon among fourth-grade students. It will take a number of additional experiences with magnets and metals to understand that all metals share a set of properties, but that only a subset of those metals is magnetic. This concept will be assessed continually and progress toward understanding will be noted by the teacher.

Evidence suggests that one of the most effective ways to impact learning is to provide students with productive feedback on their work. Productive feedback does not include grades, check marks, stickers (stars and smiley faces), compliments ("good job"), or encouragement ("try a little harder"). Productive feedback requires getting into a student's reasoning and starting a dialog. Feedback should ask students to clarify, give examples, cite evidence, and provide more detail—for example, "I'm not sure what you mean by 'close to picking up the nail,' " "What other objects will the magnet stick to?" "Why do you say magnets can stick to rocks?" "I'm not sure how you got a nail to pick up a paper clip."

Feedback can be delivered verbally, written directly into the science notebook, or attached using a sticky note, depending on the rules of engagement established in the classroom. A student then responds to the feedback by restating his or her case or, if necessary, acquiring additional information to improve the statements. A quick review of the student's response to the feedback allows the teacher to better judge whether the deficiencies in the first effort were due to a failure to communicate or incomplete understanding.

Response Sheet. A response sheet is a designed embedded assessment tool. It assesses students' understanding of a major concept that has just been studied. It usually has one prompt that calls for a detailed constructed response. The prompt often takes the form of a disagreement between two people or a situation where two explanations or diagrams must be analyzed and discussed. Students are asked to sort out the scientific principles that are germane to the scenario (Figure 9.6, p. 182).

Figure 9.6 Sample of a Response Sheet

Max said, "Cardboard doesn't stop magnetism. Sponge does stop magnetism. I know because a paper clip doesn't fall when cardboard is put between a paper clip and a magnet. The paper clip does fall when sponge is put between a paper clip and a magnet."

Maxine said, "No, the paper clip falls because the sponge is thick. The paper clip is too far away from the magnet when the sponge is between them."

Who has a better idea? Explain why you think so.

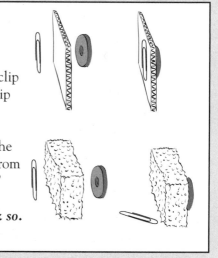

Because the scenario in Figure 9.6 is text heavy, it is appropriate for the teacher to hold up samples of the materials represented in the prompt and read the prompt aloud, slowly and precisely, to the class as students read along. Teachers should not interpret or "clarify" the prompt, simply read it to mitigate bias imposed by differences in reading ability. Students then work alone to write their best response to the scenario presented. The teacher collects and reviews the papers, making informal notes to himself or herself about specific misconceptions that recur frequently, as well as interesting individual responses. No marks are made on student papers, which are returned to students the next day for self-assessment. Before students critique their responses, there is a class discussion. The teacher calls on a student to state one element (fact or simple relationship) of a complete response to Max and Maxine. The statement goes on the board and is numbered. Then another student states a second element, and so on, until no student can offer any additional information.

1. *The paper clip has to be iron. Magnets don't stick to other metals.*
2. *Magnetic force gets weaker when it is farther away.*

3. *Max should try a thin sponge.*
4. *Max should try a stronger magnet (two magnets).*

In this case, students did not come up with the idea that magnetism goes through everything but iron, so it goes through sponge. This is valuable information—it suggests that there was a breakdown in instruction because not a single student brought this idea forward. The teacher has to decide whether to add this element to the numbered list or circle back later to reteach this important idea.

When the numbered list is complete, students review their responses and the list of elements that together constitute a complete response. They then self-assess their work by writing numbers by sentences in their original response that correspond to ideas on the board, and add or correct information as needed.

Using ASK Benchmark Assessments

The ASK Project uses benchmark (summative) exams before, after, and throughout a module. The survey (pretest) is administered before starting instruction and the posttest is administered following instruction. Together they provide a comprehensive assessment of students' overall progress related to knowledge and understanding of the constructs that are addressed in the science topic. In addition, I-Checks (I check my own work) are administered at the end of each coherent conceptual chunk throughout the curriculum. (In FOSS the major conceptual chunks are called investigations.) In the Magnetism and Electricity module, Investigation 1 concentrates on permanent magnetism. Three concepts and two elements (see Figure 9.2 earlier in this chapter) are addressed with multiple and varied exposures. At the conclusion of the instruction in Investigation 1, I-Check 1 is administered. It provides a measure of each student's overall understanding of the concepts taught in the preceding week or two, and a composite view of the performance of the class as a whole.

Coding guides have been developed for all benchmark items. A student's response to an item prompt is analyzed and, based on the germane elements in the response, may be coded between 0 and 4. The code corresponds to one or more elements of full understanding of a concept. But, remember, the code is not a score. This is a difficult concept for teachers. The temptation is to add up the code numbers to see what total an individual student got. The higher the number, the better the performance. The codes, however, only provide information about how completely and effectively the student is thinking about the concept being assessed *in the context of that one item*. The same concept will be assessed with two or three additional items, and it is the composite response that indicates the student's level of understanding (progress) with the construct being taught.

The teacher's review of the I-Checks will suggest one or two key items to process as a class. I-Checks are returned to students unmarked—no grades, checks, comments or other evaluative marks—so students can self-assess their responses as discussed earlier. This process provides students a "second chance" to clarify and enhance their answers before final evaluation of the responses. Where grades are obligatory, student performance on the I-Checks can be used as a component of a grade. But a more instructionally productive use of I-Check results is, again, as a source of information about student progress toward full understanding of complex ideas. I-Checks are designed to perform this formative function as well as the more obvious summative function.

Valuing Progress, Not Achievement

In the discussion of embedded assessment, we described single-item assessments: single-prompt quick writes, specific notebook entries, and narrowly focused response sheets. This kind of during-the-process monitoring of learning is very specific, targeting one concept while it is actively being taught. I-Checks are carefully designed, multiple-item assessments. They encompass two or more concepts. Each concept is addressed with two to four items. The several items that address a concept provide multiple points of reference regarding students' understanding of the concept. The end product of the assessment is a picture of the level to which students have progressed in their understanding of the construct.

Progress is the key. The ASK assessment system is designed to expose progress toward full understanding of complex ideas, not achievement of arbitrary objectives. Progress is the outcome that is valued. Thus, the learning expectation is the same for a student who enters with little or no prior knowledge and experience and a student who enters with substantial knowledge and experience. Both students are expected to advance their understanding, even though the actual knowledge at the conclusion of instruction may vary. An assessment system based on progress embraces and values the learning of every student, not just the most accomplished. An assessment system based on progress also values teachers who teach well, but may not bring all students to an externally imposed level of achievement at the same time.

New Assessment Paradigms Lead to New Classroom Practices—Implications for Professional Development

Incorporating diagnostic embedded assessments into the everyday practice of teaching science at the elementary level presents a number of challenges.

- It takes time to learn and time to enact. More time is required for professional development and more classroom time must be devoted to science teaching.
- Embedded assessment impacts the number of topics that can be covered.
- Teachers have to trust and act on the evidence of learning and understanding produced by students on assessments, not on intuitive beliefs of students' understanding.
- Teachers must be reflective and creative, not defensive or derisive, when assessment data show poor student understanding.
- Teachers must be ready to redefine the teacher/student and student/student relationships in the classroom culture.

Time. The preoccupation with reading and language arts instruction in U.S. schools today has eliminated science instruction from the curriculum in many elementary schools. Typically the 2.5–3 hours devoted to reading skill development and 1–1.5 hours on math skill development per day leave less than an hour for content subjects. Science has to compete for the few remaining minutes of instruction time with social studies, physical education, fine arts, and performing arts. It is ironic that the largest, most complex, most critically important body of knowledge for cogent engagement in contemporary society is excluded from the elementary school experience of the majority of American school children.

A major challenge facing educators who are ready to adopt a new assessment paradigm into their practice is finding the time to implement it. Time must be budgeted for teachers to learn the methods and skills associated with embedded assessment culture. Professional development includes time spent learning the theory and organization of the embedded assessment system and time working with experts and colleagues to learn how to interpret student work precisely and reliably. And finally, teachers need time to take risks in their classrooms as they try out new assessment strategies and practices with their students.

Coverage. Embedded assessment slows the rate of progress through a curriculum unit. Emphasis shifts from shallow coverage of many topics to deep and thorough understanding of fewer concepts. Embedded assessment provides continuous feedback to teachers and students concerning the quality and usefulness of the learning. The learning experience engages issues related to but distinct from the content, such as motivation to learn, responsibility for learning, critical analysis of learning, intellectual honesty, and a host of other cognitive processes associated with the inquiry process. Engaging topics deeply creates conflict with external directives that specify large numbers of topics to be covered.

A second challenge facing educators who want to bring embedded assessment practice into the classroom is the institutional will to break from the policies that mitigate against meaningful engagement with important ideas in science. Insightful administrators, those who value conceptual learning and understand how it empowers students to dispatch high-stakes tests routinely, enunciate the learning priorities that the teachers will strive to attain and create the environment where those priorities can be pursued. More administrators must find the courage to enact policies that invigorate teachers' creative energies and serve students' educational needs.

Trust. Teachers develop intimate relationships with their students. They know their strengths and weaknesses, personality traits, their most effective medium of expression, and so on. Teachers often develop a kind of sixth sense about how their students are advancing with their studies, sensing who is "getting it" and who is not. When confronted with a sample of student work that is confused or incomplete, a teacher will frequently say, "Oh, I know what he means," or "I know she knows that...she explained it to me yesterday." The teacher moves on to the next topic, confident that the students are competent with the subject matter, even though the student work did not support that conclusion.

This third challenge facing the effective use of embedded assessments is the innate nurturing nature of teachers. They often give students more credit for understanding than can be justified by objective analysis of the student work. It is difficult for teachers to fight back the impulse to say I know what he means, but a key to effective embedded assessments is to make judgments about learning based solely on the artifacts produced by students. When students are asked to clarify or add information, teachers are sometimes gratified to discover that their intuition was right, but more often they discover that the understanding they assumed to be there was not. Adopting this clinical element into the practice is critically important to effective assessment of learning.

Reflection. No teacher wants to face the fact that students don't understand the concepts he or she has just taught. Teachers often expect students to apprehend complex ideas fully with a single exposure. When students show that they don't have full understanding, teachers can get discouraged. The quickest way to avoid the "bad news" is to avoid assessing the students. When teachers take poor performance on the part of the students personally, they are not able to use classroom assessment effectively. Assessment results provide diagnostic information. This is the time when thoughtful reflection and creative solutions will move student learning forward.

Classroom culture. The ASK teachers were all highly qualified FOSS teachers before they self-selected to participate in the project. They had taught FOSS for several years and they were confident that their students achieved the goals and objectives of the FOSS curriculum. They were prepared to assume additional responsibilities as research partners in the project.

As the second year of the project progressed, grade-level teachers met monthly in small study groups and moderated (analyzed and coded) student responses to I-Check items. The student work showed that there were flaws and misconceptions in their understanding. The teachers returned the unmarked papers to their students and offered them the opportunity to review and rework their answers before the papers would be "graded." The proposition created disequilibrium in the students' minds. They were confused by and suspicious of the second chance to work on their answers. Students were told they could use all the resources available to them to improve their understanding and to better communicate that understanding. They could discuss items with other students, refer to their science notebooks, and review reading materials.

Students were motivated to enhance their understanding. At first the motivation was extrinsic: Take advantage of the opportunity to get a better grade. With more experience the motivation became intrinsic as students questioned, collaborated, and discussed science concepts as they strove for complete, classwide understanding. The perception of the role of assessment changed completely in the students' and teachers' minds. New teacher/student and student/student relationships evolved, resulting in a reformed classroom culture characterized by risk-free, open discussion of scientific ideas.

Conclusion

Transforming classroom culture is challenging. It involves fundamental rethinking of the roles and responsibilities of teachers and students. It involves navigating educational policy and community expectations. The new assessment paradigm is incompatible with an educational philosophy that promotes only the teaching of right answers and overt preparation for high-stakes tests. The new assessment culture requires teachers to relinquish authority and close the distance between teacher and student. The teacher enters into a partnership with students, and all members of the classroom learning community take responsibility for moving the science learning forward. The core element of the transformed classroom culture is the teacher's ability to use embedded assessments to expose student knowledge and thinking. How these data are used frames the challenge of reforming the classroom relationships that promote efficient science learning.

The new vision of science education assessment suggests a fundamental restructuring of learning values. Today we value achievement. Tomorrow we hope educational policy will evolve to value progress. A progress-based assessment system produces evidence that a student's understanding changed from a starting point to an end point as a result of an instructional episode. The initial level of understanding will vary from individual to individual, as will the final level. The important change is that students and teachers are valued for the advance, not the end point. Assessing progress is the most equitable way to produce evidence of effective educational practice and to evaluate all teachers for diligent attention to the educational needs of their students.

References

Black, P., and D. Wiliam. 1998. Assessment and classroom learning. *Assessment in Education: Principles, Policy, and Practice* 5: 7–73.

Pellegrino, J. W., N. Chudowsky, and R. Glaser, eds. 2001. *Knowing what students know: The science and design of educational assessment.* Washington, DC: National Academy Press.

Additional Reading

Atkin, J. M., P. Black, and J. E. Coffey, eds. 2001. *Classroom assessment and the national science education standards.* Washington, DC: National Academy Press.

Atkin, J. M., and J. E. Coffey, eds. 2003. *Everyday assessment in the science classroom.* Arlington, VA: NSTA Press.

Black, P., C. Harrison, C. Lee, B. Marshall, and D. Wiliam. 2003. *Assessment for learning: Putting it into practice.* New York: Open University Press.

Heritage, M. 2007. Formative assessment: What do teachers need to know and do? *Phi Delta Kappan* 89(2): 140–145.

Jorgensen, M. 2001. It's all about choices: Science assessment in support of reform. In J. Rhoton and P. Bowers (Eds.), *Professional development: Planning and design.* Arlington, VA: NSTA Press.

Stiggins, R. 2002. Assessment crisis: The absence of assessment for learning. *Phi Delta Kappan* (June): 758–765.

Wiliam, D., and M. Thompson. 2006. Integrating assessment with learning: What will it take to make it work? In *The future of assessment: Shaping, teaching, and learning,* Mahwah, NJ: Lawrence Erlbaum.

Wilson, M., and K. Sloane. 2000. From principles to practice: An embedded assessment system. *Applied Measurement in Education* 13(2): 181–208.

Appendix

Quick Reference Guide to Assessment Tools and Uses

Embedded Assessment Tool or Strategy	When	Purpose	Teacher Action	Kind of Information
Quick Write	Before starting instruction on a new concept.	Activate prior knowledge; determine entry-level knowledge; generate an artifact for future reflection.	Collect; use entry-level knowledge to guide instruction; return unmarked to students later for self-assessment.	Entry-level knowledge; academic vocabulary appropriate to the content at students' disposal.
Teacher Observation	During active investigations; during group discussions; during work in science notebooks.	Informal note of students' manipulative skills, development of ideas; participation in discussion; ability to plan investigations and conduct systematic investigations.	Cruise the classroom making informal notes about skill abilities and deficits; engage in 30-second interviews with individuals and groups.	Students' emerging ideas; students' abilities to work together; students' ability to record and organize data; students' participation in group discussions.
Notebook Entry	At various times during the active investigations; occasionally after the active investigation.	Assess student understanding of science facts and concepts; complete and accurate explanations of science principles.	Collect and review student notebooks and review specific student work for diagnostic purposes. Plan next steps in instruction.	Students expose elements of concepts; misconceptions; communication skills. Student work will reveal elements that need additional or alternative treatment in instruction.
Response Sheet	At a critical juncture in an investigation.	Assess the complexity of student understanding of concepts; assess how student can apply conceptual knowledge.	Collect and review Response Sheets. Make notes. Return them to students unmarked for self-assessment.	Completeness of expression of concept; ability to communicate understanding; ability to use knowledge in novel situation.
I-Check Exam	At the conclusion of each investigation.	Assess summary knowledge of concepts covered in an investigation. These summative assessments can be used formatively, too.	Collect and code the items. Use results to make judgments about quality and quantity of learning. Use results as a component of grade, when needed.	Reliable assessment of student knowledge based on multiple items addressing each concept. Results of assessment can be used to reteach content when deficits are identified.
Self-Assessment	After a constructed-response item has been coded and returned to students.	Let students revisit their work to improve their understanding. Students assess elements they got right, elements they can communicate better, and elements that need to be added.	Return student work to students, process students' responses in class, re-collect and review students' modified work for better understanding.	Refinement and enhancement of student understanding following a self-assessment cycle.

CHAPTER 10

Exploring the Role of Technology-Based Simulations in Science Assessment: The Calipers Project

Edys S. Quellmalz
WestEd

Angela H. DeBarger, Geneva Haertel, and Patricia Schank
SRI International

Barbara C. Buckley, Janice Gobert, and Paul Horwitz
Concord Consortium

Carlos C. Ayala
Sonoma State University

This material is based upon work supported by the National Science Foundation under Grant No. 0454772 awarded to SRI International, Edys Quellmalz, Principal Investigator, and transferred to WestEd under Grant No 0741029. Any opinions, findings, and conclusions or recommendations expressed in this material are those of the author(s) and do not necessarily reflect the views of the National Science Foundation. Contact: *equellm@wested.org*. Current versions of the assessments can be found at *http://wested. org/calipers*.

Requirements of the No Child Left Behind (NCLB) law for science testing at the elementary, middle, and secondary levels by the 2007–2008 school year are renewing scrutiny of available assessments as evidence of what students should know and be able to do in science. It is widely recognized that there is a major disconnect between the content and structure of large-scale accountability tests and classroom formative assessment practices. Traditional tests tend not to be aligned with the challenging goals set forth in the National Science Education Standards (NRC 1996) and state science standards and are too limited to capture the deep conceptual understandings at the heart of science reform.

A paramount lesson learned from earlier reform efforts is the need to align key components of the educational system—i.e., standards, curricula, and assessments (Quellmalz, Shields, and Knapp 1995; Smith and O'Day 1991). Effective reform programs promote student achievement by placing a greater emphasis on conceptual understanding and application to everyday situations, increasing use of technologies, and developing new forms of assessment (Shields, Marsh, and Adelman 1998). Several decades of science learning now inform our understandings of how to design assessments to probe what students know and can do (Bransford, Brown, and Cocking 2000; Pellegrino, Chudowsky, and Glaser 2001). Central to research-based test design is a shift from questions on discrete, factual content to questions that focus on relationships among concepts and tasks that require integration of reasoning and inquiry within significant, recurring, extended academic and applied problems. However, many assessments continue to rely on discrete items, primarily using the multiple-choice format. These tests still favor shallow content coverage (Quellmalz and Haydel 2002). Clearly, better methods for capturing compelling evidence of student science learning, both content knowledge and inquiry skills, must be made available.

The powerful capabilities of technology hold the key to transforming current assessment practices at both the state and classroom levels (Quellmalz and Haertel 2004). Currently, external, technology-based accountability assessments do not incorporate complex performance tasks, nor do technology-rich curricula yet employ principled assessment designs that provide student performance data that meet the standards of technical quality required for external assessments. What is needed, therefore, is development of assessment designs and examples that can take advantage of technology to bring high-quality assessments of complex performances into science tests with accountability goals and with formative goals. In this chapter, we describe a project funded by the National Science Foundation, "Calipers: Using Simulations to Assess Complex Science Learning."

Value and Uses of Simulations in Education

Increasingly, simulations are playing an important role in science and mathematics education. Simulations support conceptual development by allowing students to explore relationships among variables in models of a system. Simulations can facilitate knowledge integration and a deeper understanding of complex topics, such as genetics, environmental science, and physics (Buckley et al. 2004; Hickey et al. 2003; Krajcik et al. 2000; Doerr 1996). Moreover, simulations have the potential to represent content and relationships in ways that can reduce reading demands and allow students to "see" a variety of concepts and relationships (e.g., pictures, graphs, tables). Simulations are well-suited to investigations of interactions among multiple variables in models of complex systems (e.g., ecosystems, weather systems, wave interactions) and to experiments with dynamic interactions exploring spatial and causal relationships. Technology allows students to manipulate an array of variables, observe the impact, and try again. The technology can provide immediate feedback. Simulations also can make available realistic problem scenarios that are difficult or impossible to create in typical classrooms.

Simulations can allow students to engage in the kinds of investigations that are familiar components of hands-on curricula, but also to explore problems and discover solutions they might not be able to investigate in classrooms. They also allow experimentation with phenomena that are too large or small, too fast or slow, or too expensive or dangerous. In addition, simulations do not require the logistical planning involved in setting up equipment for hands-on science experiments.

Numerous studies have discussed the benefits of using simulations to support student learning. Model-It has been used in a large number of classrooms, and positive learning outcomes based on pretest-posttest data have been reported (Krajcik et al. 2000). Ninth-grade students who used Model-It to build a model of an ecosystem learned to create "good quality models" and effectively test their models (Jackson et al. 1995). After participating in the Connected Chemistry project, which uses NetLogo to teach the concept of chemical equilibrium, students tended to rely more on conceptual approaches than on algorithmic approaches or rote facts during problem solving (Stieff and Wilensky 2003). Seventh-, eighth-, and ninth-grade students who completed the ThinkerTools curriculum performed better than high school students on basic physics problems, on average, and were able to apply their conceptual models for force and motion to solve realistic problems (White and Frederiksen 1998). An implementation study of the use of BioLogica by students in eight high schools (Buckley et al. 2004) showed an increase in genetics content knowledge in specific areas, as well as an increase in genetics problem-solving skills. Studies conducted with BioLogica suggest that the activities maintain student engagement while also linking their explorations to underlying content in genetics (Horwitz and Christie 1999).

Calipers Project Goals

The Calipers project is a two-year demonstration project that aims to use technology-supported "benchmark assessments" to bridge the gap between external summative assessments of principled design and high technical quality and curriculum-embedded formative assessments.

The Calipers project has developed a new generation of technology-based science assessments that measure student science knowledge of the relationship of multiple components in a system and inquiry skills integrated throughout extended problem-based tasks. The Calipers simulation-based assessments are intended to augment available assessment formats; make high-quality assessments of complex thinking and inquiry accessible for classroom, district, program, and state testing; and reduce economic and logistical barriers that impede the use of rich science assessment. The Calipers project provides evidence of the feasibility, usability, and technical quality of the new simulation-based assessments. In addition, the project has prepared a plan for development of a larger pool of simulation-based complex assessments linked to key strands in the AAAS *Atlas of Science Literacy* (AAAS 2001) and core National Science Education Standards.

Development of the Calipers Assessments

The development of the Calipers assessments includes a principled approach to the assessment design, alignment of the assessments with key science standards and representative science curricula, pilot testing and revisions, and a plan for development of additional environments and assessments.

The Calipers assessments were designed to test science knowledge and inquiry strategies in two fundamental life and physical science areas. Life science standards related to populations and ecosystems were chosen for one of the simulation prototypes. Physical science standards related to forces and motion were selected for the second set of prototypes. For each area, Concord Consortium designed the model of the environment to be simulated, and SRI International and WestEd designed the assessment items and tasks related to the environments.

Design of the assessments followed a principled assessment design approach (Mislevy et al. 2003). The science knowledge and inquiry skills to be tested were specified. The evidence that would provide observations of achievement of the knowledge and inquiry was specified in terms of the types of student responses to be elicited and scoring criteria. Features of tasks and items that would elicit evidence of achievement were specified. For both of the content areas, the science knowledge and inquiry abilities were aligned with the AAAS Benchmarks and key ideas and the National Science Education Standards.

Guidelines for the Calipers assessment tasks included (1) specification of a driving, authentic problem, (2) design of items and tasks to take advantage of the simulation technology, (3) alignment with standards, and (4) alignment with the types of problems and activities presented in curricula.

Simulation-Based Assessments for Forces and Motion

The setting selected to simulate principles of force and motion included skiers and snowmobiles on a mountain. The driving problem was the need for a student dispatcher to coordinate the rescue of injured skiers by snowmobile units. The simulation engine developed by Concord Consortium built on its existing Dynamica engine. To demonstrate the flexibility of the environment for assessments at a range of levels of complexity, three assessments were developed to test concepts and inquiry strategies appropriate from the early middle school grades to grade 9 physical science. Students were asked to predict and explain what would happen to the snowmobile on varying terrain (e.g., sloped, frictionless). Student manipulations of the simulation included drawing force arrows and running the simulation.

Figure 10.1 presents a screen shot of a scene within one of the Mountain Rescue assessments. Students are asked to draw an arrow (see lower left) depicting the magnitude and direction of the friction force acting on the snowmobile and predict what will happen to the snowmobile. In a subsequent screen, after running the simulation to see if their prediction was correct, students are asked to explain to the rescue team why the snowmobile behaved as it did. Student manipulations of the simulation and responses to the question provide evidence of their knowledge of balanced and unbalanced forces on surfaces with and without friction. Other tasks and scenarios test inquiry skills for prediction, explanation, and interpretation of graphs. Questions related to simpler and more complex knowledge are asked in the three separate assessments, and additional inquiry skills, such as designing the experiment and communicating recommendations, are tested.

Figure 10.1 Force and Motion Assessment 1: Friction Force Drawing and Prediction Items

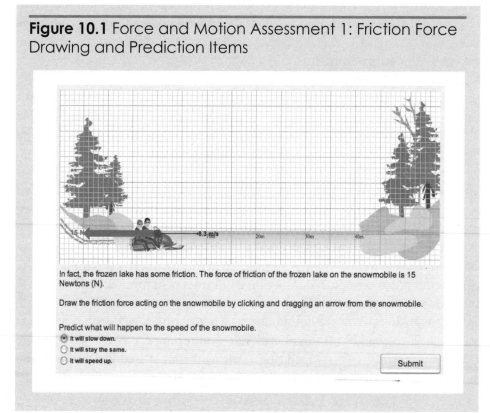

In fact, the frozen lake has some friction. The force of friction of the frozen lake on the snowmobile is 15 Newtons (N).

Draw the friction force acting on the snowmobile by clicking and dragging an arrow from the snowmobile.

Predict what will happen to the speed of the snowmobile.
- ⦿ It will slow down.
- ◯ It will stay the same.
- ◯ It will speed up.

Submit

As students participate in the force and motion assessments, the computer captures their answers to questions whether in the form of multiple choice, short answer, or essay. The computer records the magnitude and direction of arrows they draw and captures their manipulations of the simulations. When students experiment with the snowmobile speed to determine the best speed for getting to skiers on an icy hill, the computer captures the speed selected for each experimental trial. This information can be used to examine how each student in an entire class performs an experiment, a task that cannot be done in a classroom laboratory. We can determine if students have chosen experimental values that cover the range necessary and if they were systematic in exploring the range of values. Finally, we can determine if they were successful in accomplishing the task.

For many types of responses (e.g., multiple choice, drawing force arrows), the computer can automatically produce a score based on a rubric created by the Calipers assessment developers. To score multiple-choice questions, the computer identifies whether students selected the correct answer. Students' responses also can be automatically coded to facilitate the diagnosis of problem-solving strategies and types of errors in understanding. For example, in the first force and motion assessment students are asked to calculate how long it will take to travel a certain distance at a given speed. Students first select the correct formula for performing this calculation, then enter the values for distance and speed. The computer calculates the answer and students are asked to evaluate their answer. The computer automatically scores student responses using a rubric that awards 2 points for selecting the correct formula the first time, 1 point for selecting it on the second or third try, and 0 points for failing to select the correct formula within three tries. A similar scheme awards points for entering the correct values into the equation. If students accurately evaluate their answers, another point is awarded. In contrast to assessments that score only the final answer, this enables us to pinpoint where students have difficulty.

When students are conducting experiments to determine the best speed for the snowmobile to use to reach the skiers on the icy hill, the score is determined by examining if each experimental value entered is closer to or further away from the "correct" speed. Students receive one point for moving closer to the target speed. For the entire task, we average all the runs that a student makes. In addition, we take into account whether they bracket the target speed and whether they repeat any trials. Optimum performance on the experiment would include one run with a speed less than target but greater than the start speed of the team that failed at the task, one run with a speed greater than the target, and one at target.

For the constructed-response text-based questions, the computer captures the text exactly as the student types it. Another program displays the answers of the entire class, along with the question and the scoring rubric. The teacher or researcher reads the response, compares it to the rubric, and enters a score that the computer captures and integrates into the students' records.

When all of the responses have been scored by computer and humans, the results are placed in a database that can be explored in a variety of ways. A teacher or researcher can see how well students are performing on specific content or inquiry targets or how well students are performing on the assessment as a whole. Researchers can also compare how well students who are working with different curricula perform.

Simulation-Based Assessments for Ecosystems

The setting selected to simulate principles for populations and ecosystems is a newly discovered lake in the jungle. The driving problem is to explore the lake and describe its ecosystem. The simulation engine for modeling the ecosystem has been developed by Concord Consortium, building on its existing Biologica engine. To demonstrate the flexibility of the environment for assessments at a range of levels of complexity, three assessments were developed to test concepts and inquiry strategies appropriate from the early middle school grades to high school biology. Students are asked to identify the relationships of the fish and plant species and predict and explain the effects of introducing new fish species. Manipulations of the simulation include drawing food webs and varying the number of predator and prey.

Figure 10.2 presents a screen shot of a scene within one of the Fish World assessments, in which students observe species and draw a food web. Figure 10.3, p. 200, presents a screen shot of the population level of the ecosystem.

Figure 10.2 Analyzing the Relationships Among Organisms in the Ecosystem

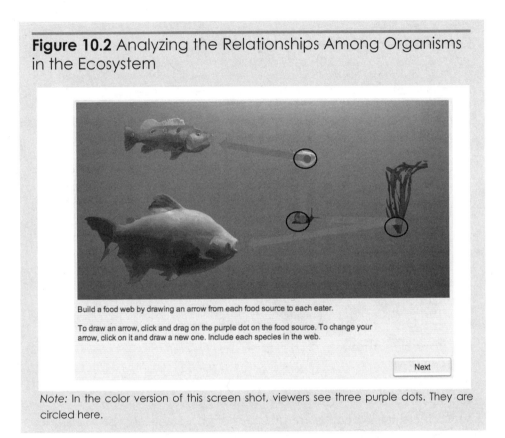

Build a food web by drawing an arrow from each food source to each eater.

To draw an arrow, click and drag on the purple dot on the food source. To change your arrow, click on it and draw a new one. Include each species in the web.

Next

Note: In the color version of this screen shot, viewers see three purple dots. They are circled here.

Several tasks have been designed using this layout—students modify variables and values that determine the size of populations of different organisms in the ecosystem over time. Changing population sizes are shown in a simulation and in a dynamically generated line graph.

As in the force and motion assessments, students' answers to the explicit questions and their actions manipulating the simulation are recorded by the computer and scored either automatically or by human scorers. The scores can be displayed by concept and inquiry skill, providing teachers and districts with standards-based feedback on the benchmark assessment. If the assessments were to be used for accountability, structured rater training and scoring sessions would produce interrater reliability data for the constructed-response items.

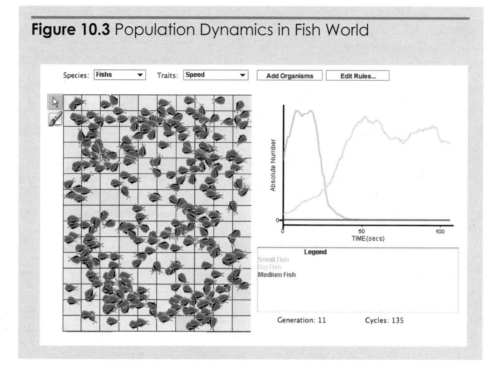

Figure 10.3 Population Dynamics in Fish World

Technical Quality of the Calipers Assessments

The Calipers assessments were first tested with small numbers of students for the feasibility of the navigation and questions. The assessments were then pilot tested in classrooms. Classes were selected that had completed units addressing ecosystems or force and motion. The classes varied in their prior use of technology. For the ecosystem assessments, students completed both a Calipers simulation-based assessment and a set of items developed for item clusters on the same content and inquiry by the AAAS project.

At the time of preparation of this chapter, the Calipers project had collected data from a variety of sources to document the assessments' technical quality. Reviews by external experts of alignment of the Calipers assessments with national science standards as well as the quality of the science content and items contributed evidence of the validity of the items. Cognitive analyses of students thinking aloud as they responded to the items contributed evidence of construct validity. Analyses of data from the pilot testing of the force and motion assessments indicated that the items discriminated between high and low science achievers and seemed to be working well as indicated by the spread of responses and the fit of the items to the IRT model, meaning that they were all contributing to the measurement of the force and motion content being tested. Similar data analyses are currently underway for the ecosystem assessments.

Promise of Simulation-Based Science Assessments

The Calipers demonstration project aimed to provide evidence of the feasibility, technical quality, and utility of simulation-based science assessments. The scientifically based principles underlying the simulation environments can be re-used for both assessment and instruction. For example, the ecosystem environment can be adapted for other aquatic (e.g., saltwater) or terrestrial (e.g., Arctic) biomes. The simulations can be used to design items testing factual content as well as interrelated knowledge of systems. Inquiry tasks asking students to design, conduct, analyze and interpret data, and communicate findings can be developed. Simulation environments developed for fundamental science systems can be re-used for elementary, middle, and secondary levels. Tasks and items developed in relation to the environments can be developed for curriculum-embedded and formative assessment activities or for external accountability. Reports linking students' scores to content and inquiry standards can provide valuable information about student progress. Most important, simulations can permit assessment of knowledge and standards not well measured by paper-based formats. The development of systematically designed science simulations promises to revolutionize both instruction and assessment.

References

American Association for the Advancement of Science (AAAS). 2001. *Atlas of science literacy*. Washington, DC: American Association for the Advancement of Science.

Bransford, J. D., A. L. Brown, and R. R. Cocking. 2000. *How people learn: Brain, mind, experience, and school*. Washington, DC: National Academy Press.

Buckley, B. C., J. D. Gobert, A. C. H. Kindfield, P. Horwitz, R. F. Tinker, B. Gerlits, U. Wilensky, C. Dede, and J. Willett. 2004. Model-based teaching and learning with Bio-Logica™: What do they learn? How do they learn? How do we know? *Journal of Science Education and Technology* 13: 23–41.

Doerr, H. 1996. Integrating the study of trigonometry, vectors, and force through modeling. *School Science and Mathematics* 96: 407–418.

Hickey, D. T., A. C. H. Kindfield, P. Horwitz, and M. A. T. Christie. 2003. Integrating curriculum, instruction, assessment, and evaluation in a technology-supported genetics learning environment. *American Educational Research Journal* 40: 495–538.

Horwitz, P., and M. Christie. 1999. Hypermodels: Embedding curriculum and assessment in computer-based manipulatives. *Journal of Education* 181: 1–23.

Jackson, S., S. Stratford, J. Krajcik, and E. Soloway. 1995. Model-It: A case study of learner-centered software for supporting model building. Paper presented at the Working Conference on Technology Applications in the Science Classroom, Columbus, OH.

Krajcik, J., R. Marx, P. Blumenfeld, E. Soloway, and B. Fishman. 2000. Inquiry-based science supported by technology: Achievement and motivation among urban middle school students. Paper presented at the annual meeting of the American Educational Research Association, New Orleans, LA (April).

Mislevy, R. J., N. Chudowsky, K. Draney, R. Fried, T. Gaffney, G. Haertel, A. Hafter, L. Hamel, C. Kennedy, K. Long, A. L. Morrison, R. Murphy, P. Pena, E. Quellmalz, A. Rosenquist, N. Songer, P. Schank, A. Wenk, and M. Wilson. 2003. *Design patterns for assessing science inquiry* (PADI Technical Report 1). Menlo Park, CA: SRI International, Center for Technology in Learning.

National Research Council (NRC). 1996. *National science education standards.* Washington, DC: National Academy Press.

Pellegrino, J., N. Chudowsky, and R. Glaser. 2001. *Knowing what students know: The science and design of educational assessment.* Washington, DC: National Academy Press.

Quellmalz, E. S., and G. Haertel. 2004. Technology supports for state science assessment systems. Paper commissioned by the National Research Council Committee on Test Design for K-12 Science Achievement (May).

Quellmalz, E. S., and A. M. Haydel. 2002. Using cognitive analysis to study the validities of science inquiry assessments. Paper presented at the annual meeting of the American Educational Research Association, New Orleans, LA.

Quellmalz, E. S., P. Shields, and M. Knapp. 1995. *School-based reform: Lessons from a national study.* Washington, DC: U.S. Government Printing Office.

Shields, P. M., J. A. Marsh, and N. E. Adelman. 1998. *Evaluation of NSF's Statewide Systemic Initiatives (SSI) program: First year report.* Menlo Park, CA: SRI International.

Smith, M. S., and J. O'Day. 1991. Systemic school reform. In S. Fuhrman and B. Malen (Eds.), *The politics of curriculum and testing: The 1990 yearbook of the Politics of Education Association* 33: 93–108.

Stieff, M., and U. Wilensky. 2003. Connected Chemistry—Incorporating interactive simulations into the chemistry classroom. *Journal of Science Education and Technology* 12: 285–302.

White, B. Y., and J. R. Frederiksen. 1998. Inquiry, modeling, and metacognition: Making science accessible to all students. *Cognition and Instruction* 16: 3–118.

## CHAPTER 11

Using Standards and Cognitive Research to Inform the Design and Use of Formative Assessment Probes

Page D. Keeley and Francis Q. Eberle
Maine Mathematics and Science Alliance

Assessment is a ubiquitous part of classroom practice (NRC 2001). Science educators agree that accurate and dependable assessments are integral to measuring and documenting student achievement as well as informing teaching and learning. However, in the current climate of high-stakes testing and accountability, the balance of time, resources, and looking at student work has been tilted considerably toward the formal, summative assessment side. This emphasis on summative assessment reduces the time classroom teachers spend on understanding their students' thinking before and throughout instruction and on using this information to adjust their teaching to promote deeper conceptual learning.

Research in the learning of science reveals that students come to the classroom with a variety of preconceptions about the natural world (Bransford, Brown, and Cocking 2000). Many of these preconceptions are held by students everywhere, regardless of their grade level, geographic location, or academic performance level. If these commonly held ideas are not identified and used by the teacher to build a bridge between students' preconceptions and the scientific ideas, students may retain their naive ideas about how the natural world works, despite the fact that they can play back the correct answers on formal tests (Black 2003). If that is the case, how much can teachers really glean from summative assessments to inform teaching for conceptual change? In response to the need for classroom assessments that uncover students' ideas in science and inform instruction, we have developed a process that links commonly held ideas identified through cognitive research to key concepts in state and national learning goals. We have used the results of this linking process to develop a bank of formative assessments we call "probes" to help teachers examine student thinking. In addition to providing a source of assessment material, we have designed professional development materials that build teachers' capacity to develop their own probing assessments. It is this informal, naturally embedded form of assessment *for* learning that balances the opportunity to learn and promote student thinking (Black et al. 2003) with assessment *of* learning that we highlight and discuss in this chapter.

Context for Our Work—Curriculum Topic Study

There are many resources and professional development opportunities that help teachers design and use standards-aligned summative assessments for accountability purposes and informing programmatic changes. However, there are fewer resources and professional development opportunities that help teachers design and use assessments that both align with standards and link to the research on student learning for the purpose of uncovering students' ideas. In our work with teachers, we use a process called Curriculum Topic Study (CTS) developed through our National Science Foundation–funded project, "Curriculum Topic Study—A Systematic Approach to Utilizing National Standards and Cognitive Research," to design standards-based and research-informed assessment probes (Keeley 2005). CTS begins with a thorough study of learning goals, instructional implications, and research on students' ideas in a topic area of science. Figure 11.1 shows an example of a CTS guide used to study the topic of conservation of matter. Each of the six sections in the left-hand column of the CTS guide has a specific purpose that is linked to the vetted resources on the right-hand side. Using the CTS guide, teachers read, study, and identify the key concepts and ideas related to a curricular topic (Section III and V) and the research on students' ideas (Section IV) using the national standards and cognitive research resources listed on the guide. The full process for using curriculum topic study guides is described in more detail in *Science Curriculum Topic Study: Bridging the Gap Between Standards and Practice* (Keeley 2005). Here we will focus on how we use parts of the CTS process to develop CTS assessment probes.

Figure 11.1 Standards- and Research-Based Study of a Curricular Topic—"Conservation of Matter"

CONSERVATION OF MATTER

Section and Outcome	Selected Sources and Readings for Study and Reflection Read and examine *related parts* of:
I. Identify Adult Content Knowledge	**IA:** *Science for All Americans* ▸ Chapter 5, *Flow of Matter and Energy*, pages 66–67 ▸ Chapter 10, *Understanding Fire*, pages 153–155
II. Consider Instructional Implications	**IIA:** *Benchmarks for Science Literacy* ▸ 4D, *Structure of Matter* general essay, page 75; grade span essays, pages 76–79 ▸ 10F, *Understanding Fire* general essay, page 249; grade span essays, pages 250–251 **IIB:** *National Science Education Standards* ▸ Grades K–4, Standard B essay, pages 123, 126 ▸ Grades 5–8, Standard B essay, page 149 ▸ Grades 9–12, Standard B essay, page 177; Standard F, essay, pages 193, 197
III. Identify Concepts and Specific Ideas	**IIIA:** *Benchmarks for Science Literacy* ▸ 4D, *Structure of Matter*, pages 76–80 ▸ 10F, *Understanding Fire*, pages 250–251 **IIIB:** *National Science Education Standards* ▸ Grades K–4, Standard B, *Properties of Objects and Materials*, page 127 ▸ Grades 5–8, Standard B, *Properties and Changes of Properties in Matter*, page 154 ▸ Grades 9–12, Standard B, *Structure and Properties of Matter*, pages 178–179
IV. Examine Research on Student Learning	**IVA:** *Benchmarks for Science Literacy* ▸ 4D, *Conservation of Matter*, pages 336–337 **IVB:** *Making Sense of Secondary Science: Research Into Children's Ideas* ▸ Chapter 8, *Conservation of Matter*, page 77; *Mass*, pages 77–78 ▸ Chapter 9, *The Solid State*, page 79; *The Liquid State*, pages 79–80; *The Gaseous State*, page 80; *Melting*, page 80, *Evaporation*, page 81; *Dissolving*, pages 83–84 ▸ Chapter 10, *Combustion*, pages 87–88, *Conservation of Matter Through Change*, pages 88–89
V. Examine Coherency and Articulation	**V:** *Atlas of Science Literacy* ▸ *Conservation of Matter*, pages 56–57
VI. Clarify State Standards and District Curriculum	**VIA:** *State Standards:* Link Sections I–V to learning goals and information from your state standards or frameworks that are informed by the results of the topic study. **VIB:** *District Curriculum Guide:* Link Sections I–V to learning goals and information from your district curriculum guide that are informed by the results of the topic study.

Visit www.curriculumtopicstudy.org for updates or supplementary readings, Web sites, and videos.

Source: Keeley, P. 2005. *Science curriculum topic study: Bridging the gap between standards and practice.* Thousand Oaks, CA: Corwin Press, p.163. Reprinted with permission.

The Upfront Part of Backward Design

The critical and often overlooked first step in designing assessments involves having a clear understanding of the specific ideas students need to learn and the pedagogical implications for how students learn them. Children at a young age already begin to develop conceptual models of how things in the natural world work. These ideas become their explanations and are highly resistant to change. Scientific models that explain phenomena can be complex and demanding and they sometimes vary based on changing conditions or context (Stewart, Cartier, and Passmore 2005). Science educators have the important yet often daunting task of building a bridge between students' alternative conceptions and the scientific view of the natural world. A systematic and deliberate study of a science topic that examines standards and research on learning clarifies the "end in mind" and provides a framework that takes into account students' ideas and their developmental readiness to give up those ideas in favor of new ones. We consider this type of study to be the "upfront part of backward design" of an assessment probe.

What Is an Assessment Probe?

An assessment probe is a type of diagnostic assessment that provides information to the teacher about student thinking related to a concept in science. A diagnostic probe becomes formative when the information goes beyond merely knowing what students think, to using the information about student thinking to guide instruction. This use of assessment is supported by research on how students learn. "Students come to the classroom with preconceptions about how the world works. If their initial understanding is not engaged, they may fail to grasp the new concepts and information that are taught or they may learn them for purposes of a test, but revert to their preconceptions outside the classroom" (Bransford, Brown, and Cocking 2000, p. 14).

The selection of the word *probe* rather than *task* is intentional. Consider two dictionary definitions:

> **probe**—a usually small object that is inserted into something so as to test conditions at a given point; a device used to penetrate or send back information; a device used to obtain specific information for diagnostic purposes
>
> **task**—a usually assigned piece of work often to be finished within a certain time; something hard or unpleasant that has to be done

Probe connotes an exploration or investigation into student learning, as opposed to a *task*, which has a defined end point. Probes are concerned less with the correct answer or quality of the student response and focus more on what students are thinking about a concept or phenomenon and where their ideas may have originated. Tasks focus more on the extent to which a student has met an assessment target by examining the accuracy and quality of his or her response. While tasks can be used for formative purposes, our use of the word *probe* rather than *task* signifies the need to go deeper into students' thinking and use assessment as a way to examine science teaching and learning.

Contrasting a CTS-Developed Probe With a Traditional Assessment Item

The water cycle assessment item in Figure 11.2 is a typical traditional assessment that often appears on a state or local standardized test. The assessment item is intended to assess "water cycle–related ideas." However, all it really assesses is students' recognition of the term *evaporation*. Furthermore, students have seen multiple water cycle diagrams with arrows (labeled "evaporation") that point from a water source up to a cloud. The item essentially is a recall of terminology used in a diagrammatic representation. It can be answered with very little conceptual understanding of the processes that make up the water cycle. Using an item such as this does little to inform instruction that will help students understand what happens to water during evaporation.

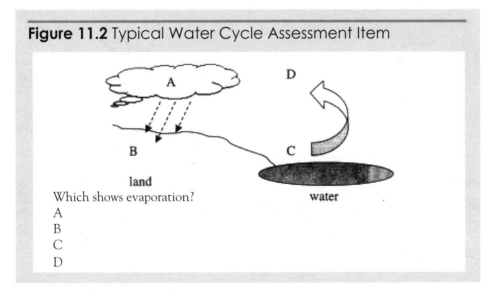

Figure 11.2 Typical Water Cycle Assessment Item

land

Which shows evaporation?
A
B
C
D

water

CHAPTER

11

SECTION 2: PROBING STUDENTS' UNDERSTANDING

Contrast the water cycle item in Figure 11. 2 with the "Wet Jeans" probe in Figure 11.3. This probe was developed using the CTS process, which carefully examines ideas in the standards and research on student learning before developing an assessment item. This assessment item uses a familiar phenomenon to ask students what happened to the water in a pair of wet jeans an hour after they dried. Technical terms like *evaporation* and *water vapor* are intentionally not used; the assessment seeks to examine students' thinking about the phenomenon, not their recall of terminology or the features of a diagram. Students can score correctly on the item in Figure 11.2 yet choose a distracter (an incorrect choice) in Figure 11.3, supported by an explanation that reveals their "private" misconceptions.

Figure 11.3 CTS Probe for Water Cycle Ideas

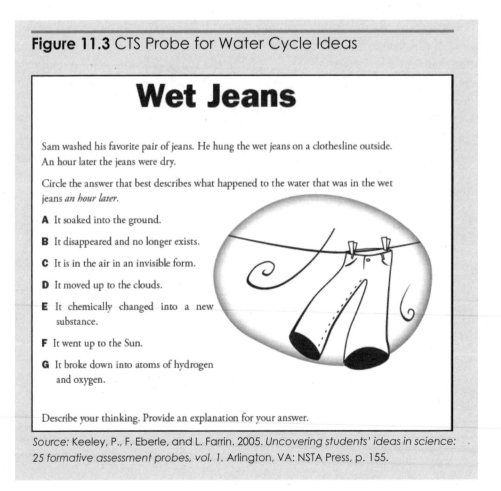

Wet Jeans

Sam washed his favorite pair of jeans. He hung the wet jeans on a clothesline outside. An hour later the jeans were dry.

Circle the answer that best describes what happened to the water that was in the wet jeans *an hour later*.

A It soaked into the ground.

B It disappeared and no longer exists.

C It is in the air in an invisible form.

D It moved up to the clouds.

E It chemically changed into a new substance.

F It went up to the Sun.

G It broke down into atoms of hydrogen and oxygen.

Describe your thinking. Provide an explanation for your answer.

Source: Keeley, P., F. Eberle, and L. Farrin. 2005. Uncovering students' ideas in science: 25 formative assessment probes, vol. 1. Arlington, VA: NSTA Press, p. 155.

In field-testing these probes, we administered them to multiple classes across different grade levels. In a class of 22 sixth graders who had previously "learned" about the water cycle in fifth grade, the number of students who selected the correct response (C) and distracters broke down into: A- 2, B- 3, C- 6, D- 7, E- 0, F-1, G-3. Only about 27% of the students chose the correct response. Six out of seven of the students who chose distracter D used a form of the word *evaporation* in their explanation, such as "The water evaporated and went up into the sky to form clouds." Four drew pictures with the arrow pointing up to a cloud. Even choosing a correct response may reveal that students have only a partial understanding. For example, one student who selected the correct response, C, explained it as "The water turned into air when it evaporated."

To help us better understand why recognizing the difference in these two items is critical to using assessment to design and improve instruction, one has to consider the specific idea(s) that make up a learning goal in the standards and the research that describes students' thinking related to those ideas. Using the Curriculum Topic Study Guide, "The Water Cycle," a careful examination of the Benchmarks for Science Literacy shows that by the end of grade 5, students should know that "when liquid water disappears, it turns into a gas (vapor) in the air..." (AAAS 1993, p. 68). This idea is a precursor to the middle school National Science Education Standards (NSES) and Benchmarks learning goals that suggest that by the end of eighth grade students should know that when "water evaporates from the surface of the earth, [it] rises and cools..." (AAAS 1993, p. 69; NRC 1996, p. 160). Examining the grades 3–5 *Benchmark* idea alerts teachers to the importance of making sure students first understand that during the process of evaporation, water changes form and goes into the air, even though we cannot see it. This explains many related phenomena such as humid weather or where the water comes from that ends up on the outside of a cold beverage glass on a hot day. Furthermore, examining the research on students' commonly held ideas indicates that some students believe that the water simply ceases to exist after evaporation (Driver et al. 1994).

The research summary in Chapter 15 of *Benchmarks* also points out that by the end of fifth grade, students should be able to accept air as the location for evaporating water if instruction was specifically targeted toward this idea. If assessments that probe for this idea are not used, teachers may never be aware that their own students have all kinds of alternative ideas about where the water goes, even though they can describe and name the processes associated with the water cycle. The variety of student responses to the probe in Figure 11.3 indicates the need for targeted instruction that is explicitly tied to the goal of developing an understanding that there is water we cannot see in the air around us.

Because standardized test items, teacher-made assessments, and commercial classroom tests often do not include distracters that are based on students' commonly held ideas, such as the example in Figure 11.3, they often are not useful in eliciting students' preconceptions or uncovering misconceptions even after instruction. Some large-scale-test developers intentionally avoid items that include distracters based on common misconceptions. When test developers analyze distracters, they examine who is choosing each option for an item by dividing students who were tested into low-scoring and high-scoring groups and looking at which distracters each group chose. High-scoring students are more apt to pick the correct response and low-scoring students often pick a distracter. If high-scoring students are picking distracters that low-scoring students are picking, that is often a cue to the test developer to revise or reject the item. Something about the distracter is attracting high-performing students to pick it. This is because high-performing students often hold the same misconception as low-performing students. Thus, an item that shows that a significant number of high-performing students have a misconception is often not used on a standardized test. When test items are developed using this process, they do not effectively serve the purpose of gauging the extent and type of students' misconceptions that the CTS process points out are likely to be commonly held ideas.

This raises the issue of using released test items or other forms of summative assessment during professional development. If teachers primarily use released test items or develop similar items to prepare students for high-stakes, external assessments, how much of this effort contributes to truly understanding how students think and learn? How much can teachers really glean from summative assessments to inform teaching for conceptual change? It is this dichotomy between the purpose and use of summative and formative assessments that we have addressed in our work. While some narrowly focused, high-stakes tests have a negative effect on teaching practice and yield little information on students' misconceptions, we acknowledge that there are some well-developed summative test items that can be used to inform instructional practices. Additionally, if students do not perform well in a particular area, the results from a summative assessment can be used to select formative assessment probes that may reveal root causes for poor performance on a summative item.

In response, we have developed banks of formative assessment items for NSTA's *Uncovering Student Ideas in Science* series, such as the example in Figure 11.3, that probe for conceptual understanding and provide feedback to the teacher and student on student thinking before new ideas are introduced and that may used as checkpoints and as rich discussion starters throughout instruction. In addition, we have developed professional development materials to help teachers design similar types of assessments.

The CTS Assessment Probe Development Process

After teachers have been introduced to the CTS process and used a variety of CTS-developed probes, they learn how to apply CTS in developing their own formative assessment probes. Figure 11.4, page 212, shows a scaffold we developed for teachers to use with the CTS formative assessment design process (Mundry, Keeley, and Landel, In Press). A group of elementary and middle school teachers we worked with in an urban district in Massachusetts used the scaffold to develop probes related to the topic of the water cycle. They began by using "The Water Cycle" CTS study guide. They used section III of the CTS process to group ideas from the national standards under the concept categories of Evaporation, Condensation, Precipitation, Cycling, and Clouds and Fog. They then examined the research for commonly held ideas that fell under these concept categories. From their two lists they chose the *Benchmarks* idea "Clouds and fog are made of tiny droplets of water" to focus on for the development of their assessment probe (AAAS 1993, p. 68).

Section IV of the CTS guide, which identifies selected readings from the research on students' learning, directed the teachers to the water cycle section in Driver et al.'s book (1994). From reading and studying that section they learned that some students think that clouds are made of smoke or cotton wool, that they are created by steamlike vapor that comes from kettles or seawater heated by the Sun, or that they come from water evaporated from puddles (Driver et al. 1994). These ideas from research were linked to the cloud composition idea from the standards to create a probe. The teachers decided to use a concept cartoon format, a visual depiction of con-

Figure 11.4 Scaffold for Designing CTS Formative Assessment Probes

STEP 1: Identify the related CTS guide(s) for the topic you want to assess. Decide whether to study the entire topic, a subset of ideas within the topic, K–12, or a grade span.

STEP 2: Examine CTS sections III and VI. List each *specific idea* found within a national, state, or local learning goal. If you are listing more than one grade level, record the grade level. (Optional: Group clusters of ideas by concept.)

STEP 3: Examine Section IV and/or the narrative research notes in Section V. List the major *research findings* about students' ideas. Also include any context issues, representations or phenomena, developmental considerations, or strategies that could inform the design of your probe. (Optional: Group clusters of research ideas by concept.)

STEP 4: Look for matches between *ideas* or a *concept* in Step 2 (CTS sections III and VI) and *research findings* from Step 3 (CTS section IV).

STEP 5: Select one of the matches from Step 4. Choose an appropriate format, such as justified multiple choice or a justified list, that would best elicit students' conceptions of the *idea* or *concept* you selected.

STEP 6: Develop the prompt to target the *idea* or *concept* from Step 5. Choose selected responses (including distracters) to specifically target the *research findings*. Use distracters that come from the research findings. In addition, you may include student ideas you have observed through your own practice.

STEP 7: Share a draft version of your probe(s) with your colleagues in order to gather constructive feedback, pilot the probe with a sample of students, and modify as needed.

STEP 8: Administer the assessment probe and organize and analyze the student data. What do the responses tell you about students' thinking? What will you do with the information?

STEP 1
Identify the CTS guide that addresses the ideas you want to examine.
⬇

STEP 2
List the *specific ideas* contained in the standards.
⬇

STEP 3
List *findings* from the research on student learning.
⬇

STEP 4
Match the *ideas* or a *concept* from the standards with *findings* from the *research* on learning.
⬇

STEP 5
Select a format that would best elicit students' conceptions.
⬇

STEP 6
Develop the prompt and selected responses.
⬇

STEP 7
Get feedback from colleagues, pilot with students, and modify.
⬇

STEP 8
Give probe, analyze data, and use results to inform actions taken.

text and children's ideas, to elicit their students' ideas (Naylor and Keogh 2000). Figure 11.5 shows the probe the teachers created after using the CTS process to inform the prompt and selected responses. Feedback from colleagues led to a few suggested revisions such as changing the distracter "they look like big cotton balls to me" to "the cloud material is like a cotton ball" in order to focus the probe more on the material of clouds. The probe was piloted with students, and assessment data were analyzed. Results from the probe informed the teachers of the need to design a lesson to explicitly address the composition of clouds.

Figure 11.5 Teacher-Created Cloud Probe

In addition to building capacity to use formative assessment, teachers are able to contribute to the development of a bank of formative assessment probes to share with other teachers. Teachers describe the assessment probe development process as a powerful form of professional development. They see the value in taking the time to identify and clarify learning goals and examine what the research has to say about students' ideas in science. Discussing ideas for assessment probes that have the potential to reveal similar, commonly held ideas among their own students has led teachers to examine their own thinking, sometimes bringing their own misconceptions to the surface.

Uncovering Students' Ideas in Science—Formative Assessment Probes

Although the process of developing probes is arduous and demanding and requires a lot of practice, it is still a worthwhile professional development experience. In addition to having teachers design their own probes, we have used CTS to develop and publish a collection of science formative assessment probes. These probes, published through NSTA, provide a ready source of research-based, field-tested assessment probes that target ideas in the standards across multiple grade levels (Keeley, Eberle, and Farrin 2005; Keeley, Eberle, and Tugel 2007; Keeley, Eberle, and Dorsey 2008).

The probes have a two-tiered format. Tier I includes the prompt that asks questions about common, everyday phenomena. The prompt format can involve students in making predictions that can be tested, have students analyze an argument between people who hold different ideas, or have students use a justified list to see if they are bound by the context or examples in which ideas were learned. Figure 11.6 shows an example of a justified-list format. The prompt is followed by a selected-response section based on a list of items, several of which are identified in the research as problematic for students. For example, many students from all grade levels fail to identify air as matter. Tier 2 asks students to explain their thinking and provide a "rule" or reasons for their selections.

The CTS guide shown in Figure 11.1 was used to develop a cluster of conservation of matter–related probes. A "crosswalk" between the standards and research, as shown in Figure 11.7, page 216, shows the results of using CTS to match related concepts and ideas from the related national standards to concepts and ideas from the related research findings. The shaded areas show the link between the ideas in the standards and the research findings that were then used to develop the probe "Ice Cubes in a Bag" shown in Fig-

Figure 11.6 "Is It Matter?" Probe

Physical Science Assessment **Probes**   10

Is It Matter?

Listed below is a list of things that are considered matter and things that are not considered matter. Put an X next to each of the things that you consider to be matter.

X rocks   X salt

X baby powder   X Mars

X milk   X Jupiter

___ air   ___ steam

___ light   X rotten apples

___ dust   ___ heat

___ love   ___ sound waves

X cells   X water

X atoms   ___ bacteria

___ fire   ___ oxygen   ___ gravity   ___ dissolved sugar

___ smoke   ___ stars   ___ magnetic force X electricity

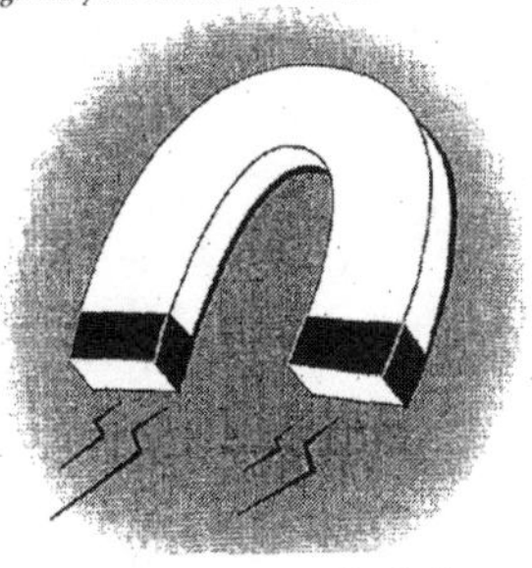

Explain your thinking. Describe the "rule" or reason you used to decide whether something is or is not matter.

I decided the things that I thought were matter, because I knew matter had to do with mass, because mass, how much matter is in something.

Uncovering Student Ideas in Science   79

ure 11.8, page 217. The probe is used by teachers to find out if their students think the mass of ice changes after it melts. It also shows whether students can use the idea of systems in their thinking. Since the bag is sealed, nothing can get in and nothing can get out, thus the mass stays the same. Teachers often recognize that this latter idea is seldom used by students. Additional linkages were made between the standards and research to develop a cluster of additional probes that targeted other conservation of matter–related ideas as seen in Figure 11.9, page 217.

Figure 11.7 Crosswalk Between Grades 3–8 Conservation of Matter–Related Concepts and Ideas and Research on Students' Ideas in Science

Science Concepts and Ideas	Research Findings
Properties ▪ Objects have many observable properties, including size, weight, and shape. Those properties can be measured using tools such as rulers and balances. (*NSES* K–4, p. 127) ▪ Materials can exist in different states—solid, liquid, and gas. (*NSES* K–4, p. 127) ▪ Air is a substance that surrounds us, takes up space, and whose movements we feel as wind. (*BSL* 3–5, p. 68) **Physical and Chemical Change** ▪ Water can be a liquid or solid and can go back and forth from one form to another. If water is turned into ice and then ice is allowed to melt, the amount of water is the same as it was before freezing. (*BSL* K–2, p. 67) ▪ No matter how parts of an object are assembled, the weight of the whole object made is always the same as the sum of the parts; and when a thing is broken into parts, the parts have the same total weight as the original thing. (*BSL* 3-5, p. 77) ▪ Substances react chemically in characteristic ways with other substances to form new substances with different characteristic properties. In chemical reactions, the total mass is conserved. (*NSES* 5–8, p. 154) **Interactions in a Closed System** No matter how substances within a closed system interact with one another, or how they combine or break apart, the total mass of the system remains the same. (*BSL* 6–8, p. 79) **Particulate Matter** The idea of atoms explains the conservation of matter: If the number of atoms stays the same no matter how they are rearranged, then their total mass stays the same. (*BSL* 6–8, p. 79)	**Matter and Its Properties** ▪ Students need to have a concept of matter in order to understand conservation of matter. (*BSL*, p. 336) ▪ Students need to accept weight as an intrinsic property of matter to use weight conservation reasoning. (*BS,L* p. 336) ▪ Confusion between weight and density contributes to difficulty understanding conservation of matter. (*BSL*, p. 336) ▪ The concept of mass develops slowly. *Mass* is often associated with the phonetically similar word *massive* and thus may be equated with an increase in size or volume. (Driver et al., p. 78) ▪ The idea that gases possess material character is difficult. Students may not regard gases as having weight or mass. Until they accept gas as a substance, they are unlikely to conserve mass in changes that involve gases. (Driver et al., p. 80) **Physical and Chemical Change** ▪ There is often a discrepancy between weight and matter conservation with dissolving. Some students accept the idea that the substance is still there but the weight is negligible, is "up in the water," or it no longer weighs anything. (Driver et al., p. 84) ▪ Some students believe one state of matter of the same substance has more or less weight than a different state. (Driver et al., p. 80) ▪ In changes that involve a gas, students are more apt to understand that matter is conserved if the gas is visible. (*BSL*, p. 337) ▪ Weight conservation during chemical reactions is more difficult for students to understand, particularly if a gas is involved. (*BSL* p. 337) ▪ Many students do not view chemical changes as interactions. They have difficulty understanding the idea that substances can form from a recombination of the original atoms. (*BSL*, p. 337) ▪ Students have more difficulty with the quantitative aspect of chemical change and conservation. (Driver et al., p. 88) ▪ The way a student perceives a chemical or physical change may determine whether he or she understands that matter is conserved. For example, if it looks as if something has disappeared or spread out more, then the student may think the mass changes. (Driver et al., p. 77) **Particle Ideas** ▪ Newly constructed ideas of atoms may undermine conservation reasoning. For example, if a material is seen as being dispersed in very small particles, then it may be regarded as having negligible weight or being more spread out and less heavy. (Driver et al., p. 77)

Note: NSES= *National Science Education Standards* (NRC 1996); BSL= *Benchmarks for Science Literacy* (AAAS 1993); and Driver et al. = *Making Sense of Secondary Science: Research Into Children's Ideas* (Driver et al. 1994).

Source: Modified from Keeley, P., F. Eberle, and L. Farrin. 2005. *Uncovering students' ideas in science: 25 formative assessment probes, vol. 1.* Arlington, VA: NSTA Press, p. 6.

Figure 11.8 "Ice Cubes in a Bag" Probe

Ice Cubes in a Bag

You are having an argument with your friend about what happens to the mass when matter changes from one form to another. To prove your idea, you put three ice cubes in a sealed bag and record the mass of the ice in the bag. You let the ice cubes melt completely. Ten minutes later you record the mass of the water in the bag. Which of the following best describes the result? Circle your prediction.

A The mass of the water in the bag will be less than the mass of the ice in the bag.

B The mass of the water in the bag will be more than the mass of the ice in the bag.

C The mass of the water in the bag will be the same as the mass of the ice cubes in the bag.

Describe your thinking. Provide an explanation for your answer.

Source: Modified from Keeley, P., F. Eberle, and L. Farrin. 2005. *Uncovering students' ideas in science: 25 formative assessment probes, vol. 1.* Arlington, VA: NSTA Press, p. 49.

Figure 11.9 Conservation of Matter Cluster

A. Lower Level of Sophistication—Objects	B. A Different Type of Physical Change
Imagine you have a whole cookie. You break the cookie into tiny pieces and crumbs. You weigh all of the pieces and crumbs. How do you think the weight of the whole cookie compares to the total weight of all the cookie pieces and crumbs? A. The whole cookie weighs more. B. The whole cookie weighs less. C. The whole cookie weighs the same as all of the cookie crumbs and pieces. Provide an explanation for your answer.	A glass of unsweetened lemonade weighs 255 grams. A spoonful of sugar is weighed before stirring it into the lemonade. The sugar weighs 25 grams. What will the weight of the sweetened lemonade be after the sugar is added? A. It will weigh slightly less than 255 grams. B. It will weigh slightly more than 255 grams. C. It will weigh 230 grams. D. It will weigh 280 grams. E. It will weigh the same: 255 grams. Provide an explanation for your answer.

(continued)

Figure 11.9 (*continued*)

C. A Chemical Change	D. Changing Context—Life Science

Serena put a spoonful of baking soda inside an empty balloon. She slipped the end of the balloon over a flask containing vinegar. She made sure the baking soda did not fall into the flask. She recorded the mass of the balloon, flask, vinegar, and baking soda. She then held up the end of the balloon so all the baking soda fell into the flask. The liquid fizzed, causing the balloon to expand. How would you compare the total mass of the materials before and after the balloon expanded?
 	
    A. The mass was less before it expanded.
    B. The mass was more after it expanded.
    C. The mass stayed the same.
 	
Provide an explanation for your answer.
	A sealed jar contains 5 bean seeds, air, and a moist paper towel. Nothing can get in or out of the jar. The total mass of the jar and its contents is 500 grams. After 12 days, the seeds sprouted into seedlings that were 6 cm long. The total mass of the sealed jar and its contents was recorded again. Which best compares the total mass of the jar and its contents before and after the seeds sprouted?
	
	    A. The mass of the jar with seedlings was less.
	    B. The mass of the jar with seedlings was more.
	    C. The mass of both jars was the same.
	
	Provide an explanation for your answer.

Source: Modified from Keeley, P., F. Eberle, and L. Farrin. 2005. *Uncovering student ideas in science: 25 formative assessment probes, vol. 1.* Arlington, VA: NSTA Press.

Analyzing a Probe

The probes can be used and disaggregated to give quick, targeted feedback to teachers. Results from our evaluations of teachers' use of the probes reveal how much they value the amount and type of data they get back. The data often surprise them and are used to inform their teaching. For example, one of the teachers we work with who used the "Ice Cubes in a Bag" probe, Figure 11.8, found that more students in her seventh-grade class chose the incorrect response than the number of students who selected the correct response, with justifications such as "The water spreads out more so it must be heavier" or "Ice floats [or is less dense] so it must be lighter than the water." A correct response—C: "I think the water would weigh the same because it's the same as taking some paper, you rip it up but it still remains the same"—reveals that the student has a basic notion of matter conservation during a physical change but fails to use more sophisticated ideas—such as the number of atoms are the same when the form changes—or fails to recognize that the substance is in a closed system—that is, nothing can get in, nothing can get out. The two ideas from the Benchmarks—the number of atoms remaining the same when the form changes and the concept of a closed system—are ideas that students should be able to use by the end of eighth grade to explain conservation-related phenomena.

The responses from the probe were used by the teachers to make sure these ideas were addressed during instruction. The teachers recognized the need to help students (a) distinguish between changes in density that happen during water's change in state and conservation of matter during a change in state, (b) recognize that the presence of air in a closed system does not decrease the mass of a substance, (c) know that changing form and shape does not affect the total amount of matter, and (d) understand that the same numbers of atoms and the idea of a closed system are ideas that can be used to explain conservation of matter. The probe also revealed to teachers that students may need further instruction in distinguishing between mass and volume.

As teachers analyze their students' responses through the lens of the research on common misconceptions, teachers learn how to separate idiosyncratic ideas from commonly held ideas cited in the research. An idiosyncratic idea tends to be an outlying idea unique to a particular student and not cited in the research. For example, a student who selected B explained that "Gravity affects the water more. The water has more mass because gravity presses down on it making the water more spread out and heavier." This student's response is not common and may point to the need to differentiate instruction for this particular student without addressing the idea with the entire class.

Deconstructing CTS-Developed Assessment Probes

Deconstructing a probe using CTS is a professional development activity that helps teachers learn how to clarify the science content, identify the targeted idea or concept, examine the level of sophistication of the idea, identify the related research findings, and consider instructional implications. A comparison across several related ideas in a topic and the research findings points out the need to use more than one assessment probe for a concept or topic. For example, Probe A in Figure 11.9 may be used to assess a lower-level idea: conservation of matter using objects rather than substances. Probe B can be used to find out if students generalize the conservation principle to other types of physical changes besides a change in state. Probe C addresses both the middle school idea in the NSES and Benchmarks on conservation of matter during a chemical change. Students may recognize that matter is conserved during a physical change but exhibit different ideas when a chemical change is involved. Probe D changes the context from a physical science example to a life science example involving a transformation of matter. It may be used to see if high school students can apply previously "learned" conservation ideas in a biological context. *How People Learn* describes the importance of transfer of learning: "Students' abilities to transfer what they have learned to new situations provides an important index of adaptive, flexible learning; seeing how well they do this can help educators evaluate and improve their instruction" (Bransford, Brown, and Cocking 2000, p. 235).

Further Inquiry Into Student Ideas

Teachers often raise new questions when examining student work from the probes. For example, students' explanations from the conservation of matter probes have led some teachers to wonder if their students have a correct conception of matter. Do they know what matter is? The "Is It Matter?" probe in Figure 11.6 shows an example of a student response that fails to recognize forms of matter that one cannot see or detect easily, such as air, dust, steam, bacteria, oxygen, and dissolved sugar, as being matter, even though the explanation says that mass has something to do with being matter. These types of responses to this probe have alerted teachers to the need to explicitly address the idea that gases, dissolved substances, and microscopic matter are still regarded as matter and why.

Are "Right Answers" Enough?

We also encourage teachers to look carefully at correct responses and discuss whether they feel their students have a full understanding of the scientific idea. For example, three ninth graders' responses to the "Apple in the Dark" probe shown in Figures 11.10.A, 11.10.B, and 11.10.C, developed with the CTS guide "Visible Light, Color, and Vision," address the NSES grades 5–8 idea that "Something can be 'seen' when light waves emitted or reflected by it enter the eye" (NRC 1996). However, an examination of student responses reveals different understandings about how light inter-

Figure 11.10 A, B, C "Apple in the Dark" Probe

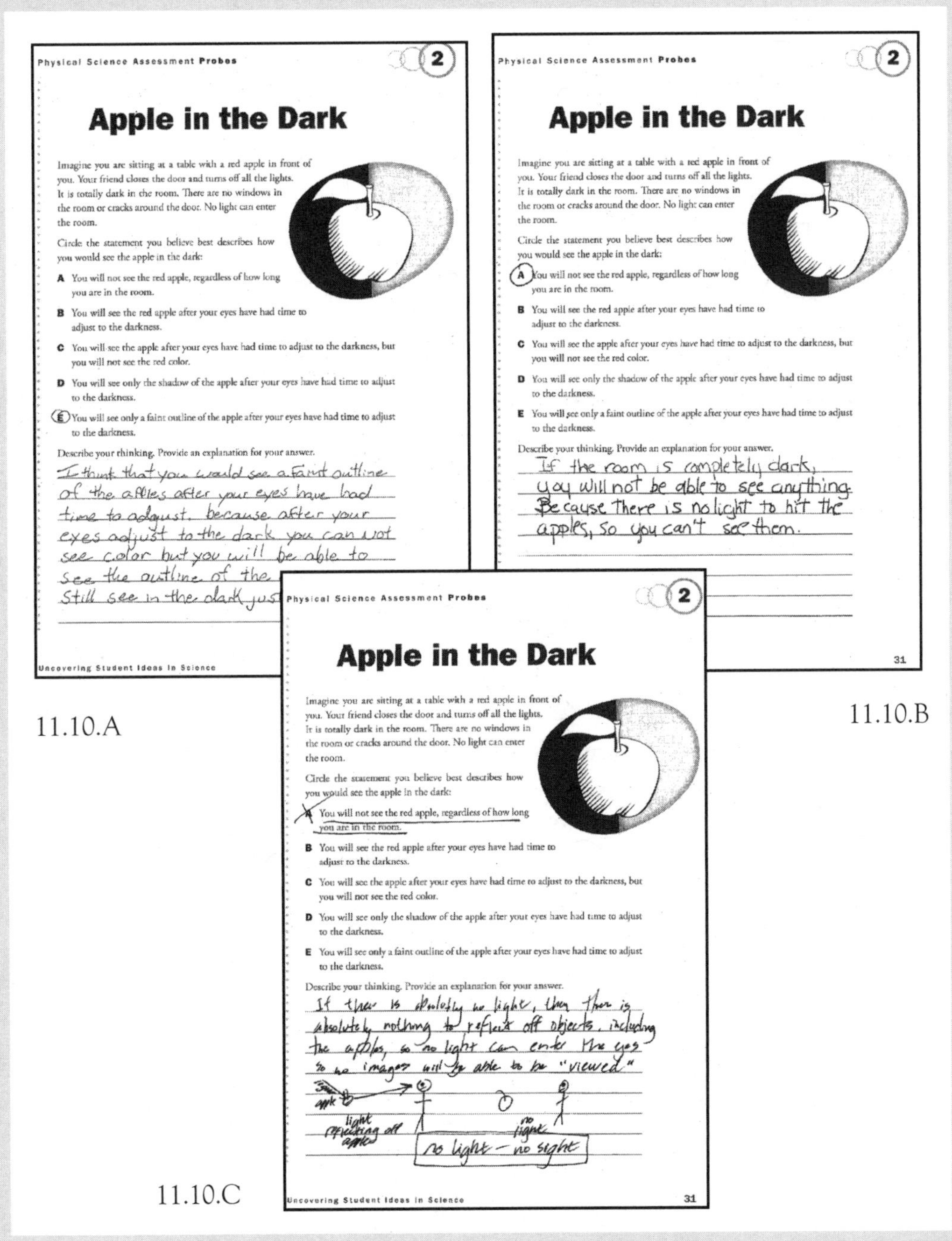

Physical Science Assessment **Probes**　　②

Apple in the Dark

Imagine you are sitting at a table with a red apple in front of you. Your friend closes the door and turns off all the lights. It is totally dark in the room. There are no windows in the room or cracks around the door. No light can enter the room.

Circle the statement you believe best describes how you would see the apple in the dark:

A You will not see the red apple, regardless of how long you are in the room.

B You will see the red apple after your eyes have had time to adjust to the darkness.

C You will see the apple after your eyes have had time to adjust to the darkness, but you will not see the red color.

D You will see only the shadow of the apple after your eyes have had time to adjust to the darkness.

E You will see only a faint outline of the apple after your eyes have had time to adjust to the darkness.

Describe your thinking. Provide an explanation for your answer.

I think that you would see a faint outline of the apples after your eyes have had time to adjust. because after your eyes adjust to the dark you can not see color but you will be able to see the outline of the ... Still see in the dark jus...

Uncovering Student Ideas in Science

11.10.A

Physical Science Assessment **Probes**　　②

Apple in the Dark

Imagine you are sitting at a table with a red apple in front of you. Your friend closes the door and turns off all the lights. It is totally dark in the room. There are no windows in the room or cracks around the door. No light can enter the room.

Circle the statement you believe best describes how you would see the apple in the dark:

A You will not see the red apple, regardless of how long you are in the room.

B You will see the red apple after your eyes have had time to adjust to the darkness.

C You will see the apple after your eyes have had time to adjust to the darkness, but you will not see the red color.

D You will see only the shadow of the apple after your eyes have had time to adjust to the darkness.

E You will see only a faint outline of the apple after your eyes have had time to adjust to the darkness.

Describe your thinking. Provide an explanation for your answer.

If the room is completely dark, you will not be able to see anything. Because there is no light to hit the apples, so you can't see them.

31

11.10.B

Physical Science Assessment **Probes**　　②

Apple in the Dark

Imagine you are sitting at a table with a red apple in front of you. Your friend closes the door and turns off all the lights. It is totally dark in the room. There are no windows in the room or cracks around the door. No light can enter the room.

Circle the statement you believe best describes how you would see the apple in the dark:

A You will not see the red apple, regardless of how long you are in the room.

B You will see the red apple after your eyes have had time to adjust to the darkness.

C You will see the apple after your eyes have had time to adjust to the darkness, but you will not see the red color.

D You will see only the shadow of the apple after your eyes have had time to adjust to the darkness.

E You will see only a faint outline of the apple after your eyes have had time to adjust to the darkness.

Describe your thinking. Provide an explanation for your answer.

If there is absolutely no light, then there is absolutely nothing to reflect off objects, including the apples, so no light can enter the eyes so no images will be able to be "viewed"

light reflecting off apple　no light　no sight

11.10.C

Uncovering Student Ideas in Science

31

acts with objects. CTS was used to identify the linked research finding that students of all ages have difficulty understanding that light must enter the eye in order to see an object. When students do not understand that light is something that travels away from its source to another place, they have difficulty explaining reflection of light off objects and fail to recognize that light must enter the eye in order to see something (AAAS 1993).

The student response in Figure 11.10.A reflects the research finding that some students believe if you stay in the dark long enough your eyes will adjust and you will be able to see an object, although you will not be able to see its color (Guesne 1985). Figure 11.10.B shows the correct choice from the selected responses, but the explanation does not reveal whether the student understands that light must reflect off the apple and enter the eye in order to see the apple. The student's explanation "no light to hit the apples so you can't see them" does not reveal whether the student uses a model of light reflecting off an object or the model that involves light illuminating an object without being reflected. Additional probing by the teacher would be needed to find out what model of light and vision the student is using. Conversely, Figure 11.10.C shows a conceptual understanding of how objects are seen by their reflected light entering the eye.

With each of these responses the probes revealed different "understandings" students have of how light interacts with objects, resulting in our ability to see them. It is how this information is used to inform teaching and learning that makes this assessment inextricably tied to instruction. In many cases teachers recognize that students have had no opportunity to experience total darkness and need to address in their instruction what "darkness" means. They also recognize the consequences of using activities such as examining how one's pupils dilate in the dark and contract in the light if the learning goal of the activity is not explicitly tied to the idea of the pupil dilating to let more light in when light conditions are dim versus being able to "see in the dark." If there is no light, the eye will not see an object regardless of how much the pupil dilates.

Reflection From a Middle School Teacher

We recently held an online workshop for teachers using the assessment probes along with the CTS process. Each teacher selected a probe and made a prediction regarding how he or she thought his or her students would respond. A middle school teacher, who we will call Andre (not his real name), selected the "Wet Jeans" probe shown in Figure 11.3. Andre's prediction about students' ideas and his online reflection provide insights into the power of using assessment probes:

Andre's Selection of a Probe, Modification, and Prediction

I'm choosing the "Wet Jeans" probe because we started the water cycle unit before I started this online course. We're still in it, and I haven't mentioned the words water cycle *or* evaporation *yet—so it will help me as we proceed through the week. The students started by having to prove that the water on the side of a cold glass came from the air surrounding the glass or came through spaces in the glass. They used cobalt chloride paper, balances, and liquids other than water to make their point. We're finishing those experiences this week—will do the probe and then try to figure out how the water vapor gets into the air. The probe will help me plan the rest of the unit (I really like this—the suspense is killing me).*

I think it must be OK to use a probe in midstream. I'll be making an administrative modification by adding two more distracters to the seven already listed: (H) The water changed into a new material in the jeans and is still there in another form and (I) The water particles changed to air particles. I'd like to add letter H because of the reference from Driver on page 159 that students thought the water changed into the container object and letter I because I've heard students in the past say that water and air are interchangeable.

My prediction of responses from my students is: A 2% B 16% C 60% D 10% E 2% F 0% G 4% H 2% I 4%

Andre's Analysis of Results

Oh my!! My prediction was way off. The results for the probe with 95 students responding to the probe are: A 2%, B 0%, C 24% (correct response), D 27%, E 5%, F 1%, G 13%, H 2%, I 26%. (Note: H and I were added distracters.)

The A responses said the water dripped to the ground and then evaporated. Hmmm. Most of the C responses said the water vapor jumped into the spaces between air particles. Some stated that if water appears (condenses) on a cold glass then it must be in the air somewhere. About 10 D responses stated that the water evaporated and went up to the clouds and about 10 responses said they've studied the water cycle and this is what happens—otherwise it would never rain. An interesting response was that clouds attract water. Four of the 5 E responses saw water vapor as a different substance than water. At least 8 of the G responses

saw the Sun tearing apart or somehow causing the water to break into hydrogen and oxygen.

The H answers were the most unique. They saw the water soaking or evaporating into the jeans. One said that once the water was absorbed into the jeans they were dry. The I answers included liquid changes to a gas—therefore water to air. At least 10 responses said the water evaporated—[these students'] definition of evaporation seemed to exclude the idea that water could exist as a vapor. The I answers were the most convoluted—some matter of fact and others stating that water gets heated up, evaporates, and becomes air/water particles or air particles. This presents an interesting challenge. I'm having fun with it so far.

Andre's Reflection

I was surprised by the number of students choosing the clouds. I might try the drawing-a-picture strategy the next time I do it. About 10–15 students choosing the cloud response said in their explanations that the water evaporated and moved up to the cloud only to come back as rain. I'm not discounting the possibility of the cloud diagram influence because about five responses indicated exactly that. I wonder if fewer cloud responses would happen if the question read "It moved directly to the clouds."

I haven't given my students the answer yet. Here is what I'm going to do in class tomorrow based on reading the responses and a gut feeling I have about the cloud responses:

The students in each class will be sorted into groups of three or four, with a variety of opinions in each group. A short time for discussion will start and then I'll round robin one student from each group in one direction to a different group and another student again from each group in the opposite direction. This will add new ideas to each group. I will remind students that no one has been told the scientific best answer. Students will use red pens to change their responses, if they want to. They will also write an explanation (in red) for staying with their original or why they changed to a different distracter. By moving from group to group, I think I'll get a feel for (a) how solid the responses are, (b) whether changes are being made to go along with the "smart" kids (at least two of the "smart" kids chose G in every class), (c) whether cloud responses were influenced by water cycle diagrams, and (d) what's with the 23 responses that see water particles changing to air particles. I need to hear more about these responses! Had the distracter said the water particles mixed with the air particles, I could understand because that would be very similar to correct response.

I don't plan to take a long time with this—maybe 10 minutes tops. Once the red pen responses are in, I can make a decision on the next move. The probes and this course are very energizing and providing ideas to use for years to come. I'm interested if you or others feel I am trying to squeeze too much out of the "Wet Jeans" probe.

Andre's Final Comments

Have the probes changed my teaching practices? Definitely. The "Wet Jeans" probe that I didn't get a chance to finish before vacation will be revisited tomorrow. Guess what? I can't wait to get back to it. Someone in this course stated that letting ideas brew or stew for a while is fine. The probes have presented me with a palate of refreshing strategies to choose from to help me attempt to understand and encourage students' ideas.... I've only used one so far, but students were excited because they had positive experiences with them in a previous teacher's class, approached the task seriously when I gave them my first one, and even argued for their favorite answer out the door and all the way to the lunchroom.... At times parents wonder what we do. It feels good to have the strategies, research, and tools to explain our purpose and defend our practices. They will become an integral part of every unit I teach.

The probes are equalizers and I like to project the idea to students that no one has the right answer all the time. Using inquiry strategies through the years has been rewarding as the "listeners" in the class become more confident and start talking. Everyone benefits from using these probes.

Concluding Thoughts

The use of CTS-developed assessment probes helps teachers gain insight into their students' understandings that would be difficult to glean from large-scale, standardized test results. While knowing what students think is important information in its own right, it is not formative unless it is used to modify the teaching and learning activities students are engaged in (Black and Wiliam 1998). The formative assessment probes developed through the CTS process of linking standards and research also provide evidence for adapting and differentiating teaching repertoires to fit the needs of students. These actions can only be taken when the quality and saliency of the assessment information enables teachers to make appropriate decisions that have positive effects on student learning.

References

American Association for the Advancement of Science (AAAS). 1993. *Benchmarks for science literacy*. New York: Oxford University Press.

Black, P. 2003. The importance of everyday assessment. In J. M. Atkin and J. E. Coffey (Eds.), *Everyday assessment in the science classroom*. Arlington, VA: NSTA Press.

Black, P., C. Harrison, C. Lee, B. Marshall, and D. Wiliam. 2003. *Assessment for learning*. Berkshire, UK: Open University Press.

Black, P., and D. Wiliam. 1998. Inside the black box: Raising standards through classroom assessment. *Phi Delta Kappan* 80(2): 139–148.

Bransford, J., A. Brown, and R. Cocking. 2000. *How people learn*. Washington, DC: National Academy Press.

Driver, R., A. Squires, P. Rushworth, and V. Wood-Robinson. 1994. *Making sense of secondary science*. New York: Routledge.

Guesne, E. 1985. Light. In R. Driver, E. Guesne, and A. Tiberghien (Eds.), *Children's ideas in science*. Buckingham, UK: Open University Press.

Keeley, P. 2005. *Science curriculum topic study: Bridging the gap between standards and practice*. Thousand Oaks, CA: Corwin Press.

Keely, P., F. Eberle, and C. Dorsey. 2008. *Uncovering student ideas in science: Another 25 formative assessment probes, volume 3*. Arlington, VA: NSTA Press.

Keeley, P., F. Eberle, and L. Farrin. 2005. *Uncovering student ideas in science: 25 formative assessment probes, volume 1*. Arlington, VA: NSTA Press.

Keeley, P., F. Eberle, and J. Tugel. 2007. *Uncovering student ideas in science: 25 more formative assessment probes, volume 2*. Arlington, VA: NSTA Press.

Mundry, S., P. Keeley, and C. Landel. In Press. *A leader's guide to science curriculum topic study: Designs, tools, and resources for professional training*. Thousand Oaks, CA: Corwin Press.

National Research Council (NRC). 1996. *National science education standards*. Washington, DC: National Academy Press.

National Research Council (NRC). 2001. *Classroom assessment and the national science education standards*. Washington, DC: National Academy Press.

Naylor, S., and B. Keogh. 2000. *Concept cartoons in science education*. Chesire, UK: Millgate Publishing.

Stewart, J., J. Cartier, and C. Passmore. 2005. Developing understanding through model-based inquiry. In S. Donovan and J. Bransford (Eds.), *How children learn science in the science classroom*. Washington, DC: National Academy Press.

High-Stakes Assessment: Test Items and Formats

S ection 3 reviews the format and cognitive demands of high-stakes test items, recommendations for the design of state-level tests, and lessons learned from literacy tests.

Discussion Questions

- What are the critical features of well-designed high-stakes test items?
- How can analyzing test questions help teachers become more proficient in classroom-based assessment?
- What communication skills are needed for students to demonstrate their understanding on high-stakes tests?
- What insights and recommendations are assessment experts contributing to the national dialogue on science assessment? What are the areas of consensus and of divergence?

Chapter Summaries

The opening chapter in this section presents an analysis by assessment experts on strategies they use to ensure that test items both align with standards-based learning goals and are devoid of bias that might influence some test takers. Authors George DeBoer and Cari Hermann Abell of Project 2061 at AAAS, Arhonda Gogos at Sequoia Phamaceuticals, An Michiels in Leuven, Belgium, Thomas Regan at the American Institutes for Research, and Paula Wilson in Kaysville, Utah, describe the item development approach advocated by Project 2061, fundamental to its online collection of test items AAAS is developing for middle and secondary classrooms. The authors emphasize the roles of both student feedback and psychometric analysis in their evaluation items for this collection. The also emphasize the importance of students' ability to apply their knowledge to explain phenomena and make predictions in unfamiliar contexts.

Researchers Audrey Champagne and Vicky Kouba from the State University of New York at Albany and their colleague Linda Gentiluomo, a professional development specialist in Schenectady, New York, analyze the purpose and design of extended constructed-response items in large-scale assessments such as NAEP and TIMSS. These items require students to answer an initial question and elaborate in written responses to subsequent related questions. The authors' goals are to help teachers become better informed about the rigorous research underlying these items and better able to use this information to improve their own instruction and test-writing skills. While the extended-format makes a greater cognitive demand on students, it reveals more about their understanding. In the chapter, which includes several examples of extended response items and their grading rubrics, the authors examine the design of these items, the content they test, and the cognitive demands they make on students. The chapter also includes a discussion of the complex scoring guides used to evaluate student responses.

Marian Pasquale and Marian Grogan also examine the cognitive demands of high-stakes test items. The authors, both from the Center for Science Education at EDC in Boston, recommend helping students succeed in high-stakes tests by exposing them to similar items in classroom-based assessments. This chapter, drawing on professional development programs the authors developed for middle school teachers, presents several examples of released test items and analyzes their format and the information they provide about student understanding. The authors suggest using these as models for creating classroom assessments that are aligned with their high-stakes counterparts and offer similar cognitive demands.

Meryl Bertenthal and colleagues Mark Wilson, Alexandra Beatty, and Thomas Keller summarize the recent National Research Council (NRC) study of and recommendations for high-stakes state science assessment systems. The NRC report identifies several concerns about individual states' standards and testing—in particular, content standards that include large numbers of disconnected topics and are used to guide the development of state-level tests. As the authors report, the NRC study suggests reorganization of state standards to emphasize the big ideas and supporting them with well articulated learning progressions. They further emphasize that the report advocates a multi-component state assessment system rather than a state test. Readers will want to compare some of the goals for classroom-based assessments listed in previous chapters to the following NRC goals for state systems: (1) monitoring student performance and informing teaching and learning; (2) monitoring schools and administrators for accountability purposes; and (3) certifying that students have met the state standards.

Peter Afflerbach, an expert in the teaching and assessment of reading from the University of Maryland, shares his impressions about high-stakes assessment in the field of literacy. His chapter addresses two questions: (1) How might assessment practice from other disciplines inform high quality science assessment? and (2) What are the challenges and accomplishments of student assessment in different content domains? In his discussion, Afflerbach notes that current assessment practices have several imbalances due to policies that favor particular assessment formats. He suggests that these imbalances prevent teachers from using assessments that would best serve their students.

## CHAPTER 12

Assessment Linked to Science Learning Goals: Probing Student Thinking Through Assessment

George E. DeBoer and Cari Herrmann Abell
Project 2061, American Association for the Advancement of Science

Arhonda Gogos
Sequoia Pharmaceuticals

An Michiels
Leuven, Belgium

Thomas Regan
American Institutes for Research

Paula Wilson
Kaysville, Utah

S tandards-based reform of K–12 science education is built on the idea that fundamental improvement begins with the development of a well-articulated and coherent set of learning goals to guide instruction and assessment. This vision for reform has been supported for over a decade by Project 2061, the education initiative of the American Association for the Advancement of Science (AAAS), through its *Science for All Americans* (1989) and *Benchmarks for Science Literacy* (1993), and by the National Research Council (NRC) through its *National Science Education Standards* (1996).

The focus of this chapter is on how to design science assessment items that are linked to the content standards in *Benchmarks for Science Literacy* and the *National Science Education Standards*. It outlines AAAS Project 2061's approach to the design of assessment items that can be used with students in middle and early high school. The chapter describes (1) our criteria for determining that the assessment items test the targeted ideas and not some other ideas, (2) precautions we take to ensure that each assessment item is a fair and accurate measure of student knowledge, and (3) the use we make of student feedback during item development rather than relying solely on psychometric features to judge the suitability of the items. Drawing on a set of example items and students' responses to them, we show how data obtained from students through pilot-testing provide important insights about what students do and do not know and how those insights can be used to improve the effectiveness of each assessment item as a measure of student learning.

Background

With the adoption of the federal No Child Left Behind Act (NCLB) in 2002, the need for assessments that are well aligned to content standards became greater than ever. NCLB placed a new emphasis on standards-based accountability for students, teachers, schools, school districts, and states. Beginning in the 2007 academic year, testing in science became part of the NCLB mandate, and each state is now required to assess students' achievement of its science standards at least once at the elementary, middle, and high school levels. This emphasis on accountability through assessments linked to content standards has the potential to bring all parts of the instructional system in science together with the common goal of improving student learning in science. The importance of science content standards is also noted in a 2006 report from the National Research Council:

> To serve its function well, assessment must be tightly linked to curriculum and instruction so that all three elements are directed toward the same goals. Assessment should measure what students are being taught, and what is taught should reflect the goals for student learning articulated in the standards. (p. 4)

But, for standards-based accountability to produce the desired results, assessment and instruction must in fact be aligned to the most important ideas in science, and assessment must in fact probe student understanding of those ideas. If that can be accomplished, assessment can provide direction for instruction and provide valuable feedback to teachers about their teaching and their students' learning. However, there is growing concern on the part of policy makers, educators, and the public about the quality of state assessments, which are key to the success of the NCLB strategy. A 2006 study from the American Federation of Teachers, for example, concludes that for reading, mathematics, and science, the three subject areas that are the focus of NCLB testing, only 11 states have strong content standards and tests that are aligned to them.

Although much of the debate about assessment has focused on the high-stakes testing required at the state level by NCLB, the tests that teachers have available to them also warrant attention. Classroom testing should enable teachers to probe students' understanding of the science ideas specified in the content standards and get feedback about their own teaching of those ideas. According to the National Research Council,

> ...classroom teachers are in the position to best use assessment in powerful ways for both formative and summative purposes, including improving classroom practice, planning curricula, developing self-directed learners, reporting student progress, and investigating their own practices.... Teachers need not only...interpret the assessment-generated information, they also must use the information to adapt their teaching repertoires to the needs of their students. (2001, p. 15)

Whether formative or summative in nature, classroom-based or statewide, high-quality assessments can contribute to student learning and promote science literacy for all as it is envisioned in the national and state science standards. At present, however, most educators agree that there are too many poorly written assessment items that do not align properly with the content standards. These problems are prevalent in other content areas as well, not just in science. A recent report by the National Mathematics Advisory panel (2008) notes that there are many flawed items on existing tests and suggests that "careful attention must be paid to exactly what mathematical knowledge is being assessed by a particular item and the extent to which the item is, in fact, focused on that mathematics" (p. 60).

To help address these problems in the context of science assessment, AAAS's Project 2061 is developing test questions for middle school and early high school science that are aligned to core ideas in national and state content standards and designed to be highly effective probes of students' understanding of those ideas. Because these assessment items are aligned to the ideas in the content standards and not to any particular curriculum material, to get an item correct students must demonstrate real understanding of those ideas rather than merely reciting words they have memorized from their textbooks or heard in their classrooms. The items are filling other needs as well:

- They enable *teachers* to keep track of their students' understanding of specific ideas over time and to conduct classroom research on the effects of various instructional strategies on student learning of those ideas. Many of the assessment items that are being developed expect students to use their knowledge to explain and predict phenomena that they may not have encountered before in school. Items that are embedded in real-world contexts accessible to students but different from those commonly used in textbooks or classroom lessons enable teachers to gauge more precisely their students' knowledge of the science ideas. The items also provide diagnostic information to help teachers determine what misconceptions or other problems may be impeding their students' learning.

- They provide *test developers and test administrators*, particularly those at the state and district levels, with models for items that are well aligned to the science ideas targeted in state and national standards and that also conform to rigorous psychometric, linguistic, and cognitive requirements. Tests that are developed from such items can reliably inform education policy and decision making and ensure that the consequences of decisions made based on those tests are fair for students, teachers, administrators, and schools.

- They provide *curriculum developers and researchers* with high-quality assessment items that are aligned to content standards to compare the effectiveness of various instructional materials objectively. Existing assessment items are not focused enough on the specific ideas in the content standards to provide precise and replicable measures of student understanding of the ideas and skills included in those content standards. High-quality assessment items linked to content standards can also be integrated into instructional materials themselves. Although instructional materials often include embedded questions and summary assessment activities, they are rarely linked to specific content standards and are rarely presented as probes to help teachers uncover their students' thinking so that instruction can be adjusted based on how students respond.

- They give *parents* and other members of the public specific information about what it is that children, teachers, and schools are being held accountable for with respect to the content standards of their state and local communities and what alignment of assessment to those content standards means. Clear statements of the standards themselves, as well as assessment items that measure understanding of the ideas in the standards, are essential if parents are to contribute meaningfully to their children's education.

Developing Assessment Items Aligned to Standards

The AAAS Project 2061 procedure for developing assessment items involves three stages: (1) clarifying the targeted content standard, (2) designing assessment tasks that are precisely aligned to the specific ideas in the targeted content standards, and (3) using data derived from one-on-one interviewing and pilot-testing items with students to improve the items' effectiveness. By focusing on the ideas that an item targets and on the item's likely effectiveness as an accurate probe of students' knowledge of those ideas, the process helps to articulate what is being tested by a particular item, thus improving the validity of interpretations that can be made from test results.

1. Clarifying the Content Standards

Both *Benchmarks for Science Literacy* and *National Science Education Standards* are organized around ideas and skills that all students should have learned by the end of each grade band if they are to achieve the goal of science literacy by the time they graduate from high school. The learning goals in these documents, and the organization of these learning goals, provide guidance for developing instruction and curriculum as well as assessment.

Key ideas. Although state and national content standards provide important guidance to assessment developers regarding what students should know in science, to increase the precision of the content alignment, we further subdivide the content standards into finer-grained statements of knowledge, or key ideas. We then clarify each key idea by indicating what it is that we expect students to know about that idea and what the boundaries of that knowledge are for purposes of assessment. Consider the following key idea for a benchmark from the topic of plate tectonics:

The outer layer of the Earth—including both the continents and the ocean basins—consists of separate plates.

Clarification statements. Clearly, there are concepts in this statement about Earth's plates that need to be elaborated. What knowledge, for example, should students have of what a plate is? Our clarification specifies the following:

Students are expected to know that the solid outer layer of the earth is made of separate sections called plates that fit closely together along the entire surface where they are in contact such that each plate touches all the plates next to it. They should know that any place where two plates meet is called a plate boundary. They should know that plates are continuous solid rock, miles thick, which are either visible or covered by water, soil, or sediment such as sand. They should know that the exposed solid rock of mountains is an example of plate material that is visible. Students are not expected to know the term bedrock. Students should know that there are about 12–15 very large plates, each of which encompasses large areas of the earth's outer layer (e.g., an entire continent plus adjoining ocean floor or a large part of an entire ocean basin), which together are large enough to make up almost the entire outer layer of the earth.

.....They should also know that there are additional smaller plates that make up the rest of the outer layer, but they are not expected to know the size of the smaller plates or how many there are. Students are expected to know that the boundaries of continents and oceans are not the same as the boundaries of plates. They should know that some boundaries between plates are found in continents, some in the ocean floors, and some in places where oceans and continents meet. Students are not expected to know the names of specific plates or the exact surface areas of plates. Students are not expected to know the terms lithosphere, crust, or mantle; the difference between lithosphere and crust; or that a plate includes the crust and the upper portion of the mantle. (DeBoer 2007)

This clarification statement was written in response to three questions that are central to the design of assessments that target a key idea: (1) Is the description of plates that is specified here what is needed for students of this age to form a mental model that allows them to predict and explain phenomena involving plates? (2) Is the description of plates that is specified here needed for students to understand *later ideas* and the accompanying phenomena that they will encounter? (3) Will the specified terminology contribute enough to students' ability to communicate about the targeted ideas to make that terminology worth learning?

In the case of Earth's plates, we judged that students should know what the plates are made of, approximately how thick they are, approximately how many there are, that the plates are not all the same size and shape, and that each plate directly touches all the plates next to it. These elaborations of the term *plate* are needed so that students can be helped to develop a mental model of a plate that enables them to understand ideas about plate motion and the consequences of plate motion. We made a judgment that although the term *lithosphere* is often used when communicating about plates, neither the term nor the concept (i.e., that it is made of both crust and the upper portion of the mantle) contributes enough to explaining phenomena related to plate motion to hold students accountable for knowing the term. At other times, we may judge that the idea behind the term is important even if the term itself is not. We recognize that individual teachers may choose to show students the relationship between lithosphere, upper mantle, and plates, but our assessment items will not include the term for reasons explained above.

Misconceptions. Research on student learning also plays an important part in helping us clarify the learning goals. It provides information about the age-appropriateness of the ideas we are targeting and the level of complexity of the mental models we can expect students to develop. The research on student learning also identifies many of the misconceptions that students may hold, which we include as distracters in the items so that we can test for these ideas alongside the ideas we are targeting.

Connections among ideas. Once the ideas to be assessed have been identified and clarified, our next step is to see how those ideas relate to other ideas within a topic and across grade levels. The objective here is to be as clear as possible about what ideas we are explicitly testing and what prior knowledge we can assume students will have. For example, if we expect students to know that digestion of food involves a process in which the atoms of molecules from food are rearranged to form simpler molecules, should we assume that students already know that molecules are made of atoms? If not, are test questions on chemical digestion really just testing whether students know the relationship between atoms and molecules?

We use the conceptual strand maps published in the *Atlas of Science Literacy* (AAAS 2001; 2007) to make judgments about when students can be expected to know certain ideas along a K–12 learning progression for a particular topic. The *Atlas* map for the topic of natural selection in Figure 12.1, for example, has four strands: changing environments, variation and advantage, inherited characteristics, and artificial selection. The interconnections among the ideas in these strands are visually represented—or mapped—to show the progression of ideas within each conceptual strand through four grade bands as well as the links between ideas across strands. In the *variation and advantage* strand, a benchmark at the 6–8 grade level says: "In all environments…organisms with similar needs may compete with one another for resources, including food, space, water, air, and shelter" (AAAS 2001, p. 83). This is preceded on the *Atlas* map by a benchmark at the 3–5 grade level that says: "For any particular environment, some kinds of plants and animals survive well, some survive less well, and some cannot survive at all" (AAAS 2001, p. 83). In testing whether students know that organisms in ecosystems compete with each other for resources, we would assume that students already know that not all organisms in an ecosystem survive. As a rule, unless we have reason to believe otherwise, we assume that students already know the ideas listed at an earlier grade band and use the ideas and language from those earlier ideas in item development for the grade band that follows.

2. Aligning Assessment Items to Content Standards

We use two criteria to determine whether the content targeted by an assessment item is aligned to the content in a particular key idea. The *necessity* criterion addresses whether the knowledge specified in the learning goal is *needed* to successfully complete the task, and the *sufficiency* criterion addresses whether the knowledge specified in the learning goal is *enough by itself* to successfully complete the task (Stern and Ahlgren 2002). If the targeted knowledge is not needed to answer the question, then the item is obviously not a good indicator of whether or not students know that

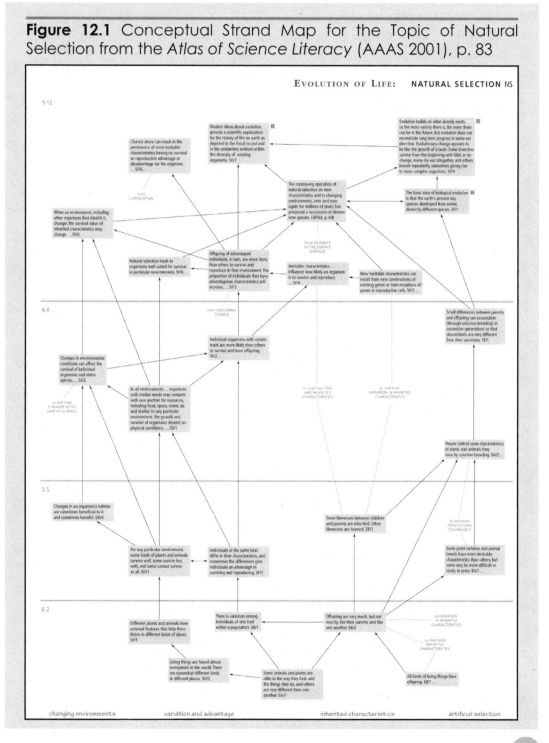

Figure 12.1 Conceptual Strand Map for the Topic of Natural Selection from the *Atlas of Science Literacy* (AAAS 2001), p. 83

targeted idea. Also, if additional knowledge is needed, it is difficult to know if an incorrect response is due to students' not knowing the targeted idea or that additional idea. The purpose of such careful alignment is to help reduce errors in interpreting students' correct and incorrect responses. When items are well aligned to the targeted content, students' responses are more likely to provide accurate insights into their understanding of that content. But improving content alignment is not enough. There are other factors that can also affect the validity of an item, and Project 2061's assessment analysis procedure takes those factors into account as well (AAAS 2008).

Improving validity. Test items should be written in such a way that teachers and researchers can draw valid conclusions from them about what students do and do not know about the ideas being tested. Unfortunately, many test items have features that make it difficult to determine if a student's answer choice reflects what the student does and does not know about an idea. When an item is well designed, students should choose the correct answer only when they know an idea, and they should choose an incorrect answer only when they do not know the idea. They should not be able to answer correctly by using test-taking strategies that do not depend on knowing the idea (a false positive response) or be so confused by what is being asked that they choose an incorrect answer even when they know the idea being tested (a false negative response). To improve an item's validity, we identify and eliminate as many problems with comprehensibility and test-wiseness as we can.

3. Using Student Data to Improve Items

Rigorously applying a set of criteria to determine the alignment of test items to learning goals and identifying features that obscure what students really know are important steps in the development of items that accurately measure the knowledge we want students to have. But findings from our research indicate that this approach works much more effectively when used in combination with one-on-one interviews with students or pilot tests of items in which students' answer choices are compared to the explanations they give for their answers (DeBoer and Ache 2005).

By comparing the answer choices that students select with their oral or written explanations, we determine if an assessment item is measuring what we want it to measure or if something about the item makes it more likely for students to give false negative or false positive responses to the item (DeBoer et al. 2007, 2008). In the pilot-testing, students are asked the questions that appear in Figure 12.2. For questions 3 through 6, students are asked to explain why an answer choice is correct or not correct or why they are "not sure."

Figure 12.2 Questions Posed to Students in the Pilot Testing of Test Items

1. Is there anything about this test question that was confusing? Explain.

2. Circle any words on the test question you don't understand or aren't familiar with.

3. Is answer choice A correct?   Yes   No   Not Sure

4. Is answer choice B correct?   Yes   No   Not Sure

5. Is answer choice C correct?   Yes   No   Not Sure

6. Is answer choice D correct?   Yes   No   Not Sure

7. Did you guess when you answered
   the test question?   Yes   No

8. Please suggest additional answer choices that could be used.

9. Was the picture or graph helpful? If there was no picture or
   graph, would you like to see one?   Yes   No

10. Have you studied this topic in school?   Yes   No   Not Sure

11. Have you learned about it somewhere else?   Yes   No   Not Sure
    (TV, museum visit, etc.)? Where?

The following six examples illustrate the kinds of information we are able to derive from pilot testing the items. The pilot tests were carried out in both urban and suburban middle and high schools serving a wide range of students. The examples show how we use what we have learned to improve the items' alignment to the key ideas and their validity as measures of student learning.

Example 1: Matter and Energy Transformations in Living Systems

*Key Idea: Organisms use molecules from food to make complex molecules
that become part of their body structures.*

This item was intended to find out if students know that food contains
molecules used by organisms to make other molecules that become incorporated into their body structures. The distracters test commonly held misconceptions about what happens to the food that organisms consume.

When a baby chick develops inside an egg, the yolk in the egg
is its only source of food. As the chick grows, the yolk becomes
smaller. Why does the yolk become smaller?

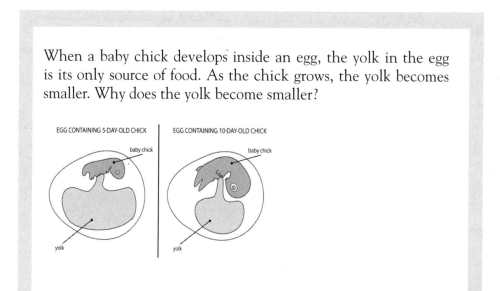

A. The yolk enters the chick, but none of the yolk becomes part
   of the chick.
B. The yolk is broken down into simpler substances, some of
   which become part of the chick.
C. The yolk is completely turned into energy for the chick.
D. The yolk gets smaller to make room for the growing chick.

Pilot testing with middle school students showed that 21.6% of those
who were tested got this question correct. Only 4 of the 16 students who
chose the correct answer (B) gave any indication that they knew that the
yolk becomes incorporated into the chick's body. Those students said that

the yolk "becomes the chick" or "becomes part of the chick," but none of those students used the word *molecule* in their explanation or in any other way suggested that they know that there is a chemical transformation of the yolk into new substances that become part of the chick's body. Other students said the chick "needs" the yolk to grow or said that the yolk "helps" the chick to grow.

Students who chose each answer:

	A	B	C	D	Not Sure/Blank	Total
#	8	16	23	20	7	74
%	10.8	21.6	31.1	27.0	9.5	100

Based on these results, the following revisions were considered:
1. Change answer choice B to read: "The yolk gets smaller because some of the atoms of the molecules from the yolk are assembled into new molecules that become part of the chick's body."
2. To more explicitly test if students have the less sophisticated idea that food is *needed* but not that it becomes incorporated into a growing animal's body, include the following distracter: "The yolk gets smaller because some of the molecules from the yolk are used by the chick to live and grow, even though none of the atoms of the molecules from the yolk become part of the chick's body."

These changes should make it more likely that students will not get the test item correct without understanding the idea being tested. In addition, the suggested revisions remove the idea that there is an intermediate stage in which "simpler substances" are formed.

Example 2: Atoms, Molecules, and States of Matter

Key Idea: For any single state of matter, increasing the temperature typically increases the distance between atoms and molecules. Therefore, most substances expand when heated.

This item was written to test whether students know that molecules get farther apart when they are heated and that this molecular behavior explains why most substances expand when heated. The item also includes common misconceptions related to thermal expansion and the behavior of molecules, especially the existence of "heat molecules."

The level of colored alcohol in a thermometer rises when the thermometer is placed in hot water. Why does the level of alcohol rise?

A. The heat molecules push the alcohol molecules upward.
B. The alcohol molecules break down into atoms which take up more space.
C. The alcohol molecules get farther apart so the alcohol takes up more space.
D. The water molecules are pushed into the thermometer and are added to the alcohol molecules.

Pilot testing showed that 25.9% of the students answered this question correctly. The most common response was that heat molecules push the alcohol molecules upward. Pilot testing also revealed that eight of the students were not familiar with the terms *alcohol* or *colored alcohol*, at least not in the context of a thermometer. A number of students wrote comments that confirmed that they held the misconception that there are *heat molecules*.

Students who chose each answer:

	A	B	C	D	Not Sure/Blank	Total
#	48	7	28	5	20	108
%	44.4	6.5	25.9	4.6	18.5	100

Based on the results of pilot testing, the following revisions were considered:

1. Because answer choice A is the only one that has the word *heat* in it, students may choose answer choice A because they connect the liquid rising in the thermometer with heat rising. Therefore, add *heat* to one or more of the answer choices.
2. Change the word *alcohol* to *liquid*.

These changes remove a word that students may find confusing in the context of thermometers. The changes also make it less likely that students will be drawn to answer choice A, in which they associate the liquid *rising* with their knowledge that "heat *rises.*"

When we interviewed students about this item, we found that they had difficulty reconciling what they expected to be a very small expansion of the liquid in the bulb into what appears to be a very large expansion of the liquid in the narrow tube of the thermometer. One student who knew that substances expand when heated did not believe the liquid could expand that much and, therefore, chose A. Even though her commitment to "heat molecules" did not appear to be strong during the interview, it seemed to her that something besides thermal expansion had to explain such a large increase. Because of developmental issues regarding younger children's ability to engage in proportional reasoning, the thermometer context may be a difficult context for general testing of beginning middle school students' understanding of thermal expansion. But these results also point to an opportunity to provide focused instruction to help students see that a small change in the volume of a liquid is amplified in a narrow tube.

Example 3: Atoms, Molecules, and States of Matter

Key Idea: All atoms are extremely small.
This item was developed to test students' knowledge of the size of atoms. The clarification of this idea says that students should know that atoms are millions of times smaller than other small things such as cells or the width of a hair.

> Approximately how many carbon atoms placed next to each other would it take to make a line that would cross this dot? •
>
> A. 6
> B. 600
> C. 6000
> D. 6,000,000

The results of our pilot testing with middle and early high school students showed that 22.8% of the students answered this item correctly. Approximately 25% of the students said they did not know what a carbon atom was and a number of students had difficulty imagining a line of carbon atoms across the dot.

Students who chose each answer:

	A	B	C	**D**	Not Sure/Blank	Total
#	19	23	30	51	101	224
%	8.5	10.3	13.4	22.8	45.1	100

Based on the results of pilot testing, the following revisions were considered:
1. Change "carbon atom" to "atom" so that students who do not yet know about specific atoms will not be disadvantaged.
2. Because imagining a line of carbon atoms across a dot may create an unnecessary cognitive load on students, try to create a simpler context for this item.

Example 4: Force and Motion

Key Idea: If an unbalanced force acts on an object in the direction opposite to its motion, the object will slow down.

This item was developed to test students' understanding of the relationship between forces on an object and the object's motion, focusing specifically on what happens when a force acts in the direction opposite to the object's motion.

> A ball is kicked straight up. The ball's speed decreases as it moves upward. Why does the ball's speed decrease?
> A. Because the ball gets heavier as it gets farther away from the ground.
> B. Because the force of the kick diminishes as the ball moves upward.
> C. Because the force of the ball's motion decreases.
> D. Because the force of gravity is in the opposite direction to the ball's motion.

Results of pilot testing the item with middle school students showed that 76.9% of the students got this item correct. However, 42% indicated that they were not familiar with the word *diminished*.

Students who chose each answer:

	A	B	C	D	Not Sure/Blank	Total
#	1	1	2	20	2	26
%	3.8	3.8	7.7	76.9	7.7	100

Based on the results of pilot testing, the following revision was considered:
Change the wording of answer choice B from "…the force of the kick diminishes as the ball moves upward" to "…the force of the kick runs out as the ball moves upward."

Example 5: Plate Tectonics

Key Idea: The solid outer layer of the Earth—including both the continents and the ocean basins—consists of separate plates.

Two items were piloted to test students' knowledge of the term *bedrock* to help decide if it should be used in assessment items. The two items are identical except that one uses the term *bedrock* and the other uses the descriptive phrase *solid rock*.

Item without the term *bedrock*.
Which of the following are part of Earth's plates?
A. Solid rock of continents but not solid rock of ocean floors.
B. Solid rock of ocean floors but not solid rock of continents.
C. Solid rock of both the ocean floors and the continents.
D. Solid rock of neither the ocean floors nor the continents.

Students who chose each answer:

	A	B	C	D	Not Sure/Blank	Total
#	5	3	19	0	6	33
%	15.2	9.1	57.6	0.0	18.2	100

Item with the term *bedrock*.
Which of the following are part of Earth's plates?
A. Bedrock of continents but not bedrock ocean floors.
B. Bedrock of ocean floors but not bedrock of continents.
C. Bedrock of the ocean floors and the continents.
D. Bedrock of neither ocean floors nor continents.

Students who chose each answer:

	A	B	C	D	Not Sure/Blank	Total
#	1	2	17	3	11	34
%	2.9	5.9	50.0	8.8	32.4	100

The results show that 50% of the middle school students were able to correctly answer this question when the term *bedrock* was used compared to 57.6% of the students when *solid rock* was used. However, there were also a greater number of *not sure* responses when *bedrock* was used (32.4% compared to 18.2%). In addition, 32 of the 34 students wrote responses indicating that they did not know what bedrock is. Without understanding the term, students were apparently translating *bedrock* to mean *rock* even though they were not sure what *bedrock* is. Based on the results of pilot testing, it was decided not to include the term *bedrock* in assessing student knowledge about the composition of Earth's plates.

Example 6: Control of Variables

Key Idea: If more than one variable changes at the same time in an experiment, the outcome of the experiment may not be clearly attributable to any one of the variables.

The following item was developed to determine if students understand that the way to determine if one variable is related to another is to hold all other relevant variables constant. The item was also designed to test a number of common misconceptions that students have regarding the control of variables, including the idea that all of the variables should be allowed to vary in a controlled experiment.

A student wants to test this idea: The heavier a cart is, the greater its speed at the bottom of a ramp. He can use carts with different numbers of blocks and ramps with different heights.
Which three trials should he compare?

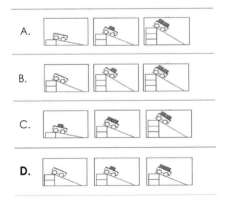

Students who chose each answer:

	A	B	C	D	Not Sure/Blank	Total
#	20	1	6	41	8	76
%	26.3	1.3	7.9	53.9	10.5	100

The results of pilot testing the item showed that 53.9% of the students answered correctly. In addition 26.3% chose answer choice A, which targets the misconception that both variables should vary at the same time. Answer choices B and C, however, were less successful distracters. Answer choice B was chosen by only one student. Of the six students who chose C, three students said they rejected answer choices A and B because there were no weights in one of the carts for those answer choices. Also, three students thought the word *trials* in the stem referred to the answer choices and circled three answer choices as correct. Six students (even some of those who chose the correct answer) thought that the word *blocks* in the stem referred to the parts of the ramp rather than the weights in the cart.

Based on the results of pilot testing, the following revisions were considered:
1. Replace the blocks with metal balls, and increase the number of balls in each cart by 1 so that there are no empty carts.
2. Replace answer choice B with three carts with the same weights on different height ramps.
3. Replace the question in the stem with: "Which 3 sets of experiments should the student do?" Label the 3 images for each answer choice "Experiment 1," "Experiment 2," and "Experiment 3."

Conclusion
As the examples above illustrate, many factors can affect how well a test item measures what students know about a particular idea. In our work, we have analyzed hundreds of items covering more than a dozen science topics. The items have come from a wide range of sources including international, national, and state tests; curriculum materials; and a variety of item banks. Nearly all of these items had problems that compromised how well they measure what students know. Most item developers depend on sophisticated quantitative analyses to judge the suitability of their test items. The

items are selected if they produce reliable results and discriminate among categories of test takers. This approach is inadequate, according to the National Mathematics Advisory Panel (2008), which called for items that are also designed to measure "specified constructs" as a way to reduce the number of flawed items (p. 61). This recommendation makes sense in science as well, and we offer intensive qualitative analysis described in this chapter, along with more quantitative approaches, as a way to produce items that are effective measures of what we want to measure.

References

American Association for the Advancement of Science (AAAS). 1989. *Science for all Americans*. New York: Oxford University Press.

American Association for the Advancement of Science (AAAS). 1993. *Benchmarks for science literacy*. New York: Oxford University Press.

American Association for the Advancement of Science (AAAS). 2001. *Atlas of science literacy*. Washington, DC: AAAS and NSTA.

American Association for the Advancement of Science (AAAS). 2007. *Atlas of science literacy, Vol. 2*. Washington, DC: AAAS and NSTA.

American Association for the Advancement of Science (AAAS). 2008. Project 2061's approach to assessment alignment: Assessment analysis utility. Retrieved April 12, 2006, from *www.project2061.org/research/assessment/assessment_form.htm*.

American Federation of Teachers. 2006. Smart testing: Let's get it right (Policy Brief No. 19). Washington, DC: American Federation of Teachers (July).

DeBoer, G. E. 2007. Developing assessment items aligned to middle school science learning goals. Presented at the Knowledge Sharing Institute of the Center for Curriculum Materials in Science. Washington, DC, July 22–25.

DeBoer, G. E., and P. Ache. 2005. Aligning assessment to content standards: Applying the Project 2061 analysis procedure to assessment items in school mathematics. Paper presented at the meeting of the American Educational Research Association, Montreal, Canada (April). Retrieved September 29, 2006, from *www.project2061.org/research/assessment/aera2005.htm*

DeBoer, G. E., N. Dubois, C. Herrmann Abell, and K. Lennon. 2008. Assessment linked to middle school science learning goals: Using pilot testing in item development. Paper presented at the annual conference of the National Association for Research in Science Teaching, Baltimore, MD (March).

DeBoer, G. E., C. Herrmann Abell, and A. Gogos. 2007. Assessment linked to science learning goals: Probing student thinking during item development. Paper presented at the annual meeting of the National Association for Research in Science Teaching, New Orleans (April).

National Mathematics Advisory Panel. 2008. *Foundations for success: The final report of the National Mathematics Advisory Panel.* Washington, DC: U.S. Department of Education.

National Research Council (NRC). 1996. *National science education standards.* Washington, DC: National Academy Press.

National Research Council (NRC). 2001. *Classroom assessment and the* national science education standards. Washington, DC: National Academy Press.

National Research Council (NRC). 2006. *Systems for state science assessments.* Washington, DC: National Academies Press.

Stern, L., and A. Ahlgren. 2002. Analysis of students' assessments in middle school curriculum materials: Aiming precisely at benchmarks and standards. *Journal of Research in Science Teaching* 39(9): 999–910.

CHAPTER 13

Assessing Science Literacy Using Extended Constructed-Response Items

Audrey B. Champagne and Vicky L. Kouba
University at Albany, State University of New York

Linda Gentiluomo
Schenectady, New York, School District

Our chapter addresses what teachers can learn from extended constructed-response items[1] that they can apply to strengthening their students' science literacy. Our goal is to provide teachers with information that will make them better formative and summative assessors; better judges of the quality of state, national, and international assessments that their students experience; and better interpreters of assessment data and its applications for educational decision making.

Three large-scale assessments—one national, the National Assessment of Educational Progress (NAEP) (National Assessment Governing Board 2005), and two international, Trends in International Mathematics and Science Study (TIMSS) (Mullis et al. 2005) and the Programme for International Student Assessment (PISA) (Organisation for Economic Co-operation and Development [OECD] 2003)—have fueled U.S. science education debates influencing federal education policies, the status of U.S. education worldwide, and the practice of science education in the United States.

---

[1]The science framework for the 2009 National Assessment of Educational Progress distinguishes two types of constructed-response items. Short constructed-response items "generally require students to supply the correct word, phrase, or quantitative relationship in response to the question given in the item" while extended constructed-response items require more extensive responses (National Assessment Governing Board 2005, pp. 95-96). The PISA 2003 framework refers to items requiring students to construct answers as open-constructed-response items (p. 52) while the TIMSS 2007 assessment framework refers to this item type as a constructed-response item (p. 102).

The student knowledge and abilities assessed by NAEP, TIMSS, and PISA are contained in frameworks for each assessment as are examples of items and student responses to those items. These frameworks are sources of information valuable to classroom teachers because they include examples of the kinds of items that appear on the assessments, alignment of the items with the science content the items are designed to measure, and scoring guides for the items.

Constructed-Response Items on International Assessments

Constructed-response items contribute from one-third to one-half of students' scores on these influential assessments. This emphasis on constructed-response items will trickle down to state assessments and influence classroom science assessments and science teachers' responsibility for developing their students' ability to compose science text.

In this section, we take a close look at examples of constructed-response items: one from PISA, the other from TIMSS. Our goals are to (1) compare the different formats of constructed-response items, (2) examine the consistency of the item with the scoring guide and the science content the item purports to measure, and (3) estimate the breadth of knowledge and mental processing involved in students' successful performance of the item.

While we understand that teachers may be skeptical about the relevance of information about items from international tests, believing their time is better spent carefully reviewing their students' responses to items they have designed, we believe there are good reasons for teachers to review items from PISA and TIMSS. Items that appear on these assessments have undergone a rigorous development process, including review by content experts and experienced teachers and field testing. Furthermore, they represent international perspectives on the science all students should know and the ways in which student understanding should be assessed. Analysis of items from external assessments helps teachers to focus on the characteristics and quality of items apart from how their students have responded to the items and to review the items more objectively than they might items of their own design.

Our analysis is based on descriptions of the science knowledge and processes contained in the 2009 NAEP Science Framework (National Assessment Governing Board 2005). The knowledge to be assessed is stated as principles. An example of a principle is "Properties of solids, liquids, and gases are explained by a model of matter that is composed of tiny particles in motion" (p. 30). Processes, called *practices* in the NAEP framework, describe science knowledge in action. "Using Science Principles" and "Using Scientific Inquiry" are examples of science practices. Each practice is elaborated in terms of component practices. "Explaining observations" is a component practice of "Using Scientific Inquiry." Performances describe the integration of science principles and practices.

The term *cognitive demands* refers to the knowledge and mental processes hypothesized to be necessary for a student to perform a practice that places a science principle in action. For example, consider an extended-response item that asks students to explain how clothes are dried when placed in an electric clothes dryer. This item is aligned with the performance "Explain observations of phenomena using science principles from the content statements" (National Assessment Governing Board 2005, p. 65) and the content statement "One way to change matter from one state to another and back again is by heating and cooling" (National Assessment Governing Board 2005, p. 31). A student responds: "Clothes get dry in an electric dryer because the clothes are heated, the water in the clothes evaporates, and the fan blows the water vapor out of the dryer." On the basis of this response, we make inferences about the student's knowledge or, more precisely, we make inferences about what is in the student's mind. The converse of this process is to generate hypotheses about cognitive demands based on the item and its scoring guide or our expectations for an adequate student response. In the first case, student responses are the data on which the cognitive demands are based. In the second the cognitive demands are based on the item, its scoring guide, or our expectations for student performance.

The science framework for the 2009 NAEP provides an organizing structure for the cognitive demands we hypothesize are required to meet a performance expectation for an item or the inferences we make about a student's knowledge from that student's response to the item. Figure 13.1, page 256, contains a summary of four knowledge types—declarative, procedural, schematic, and strategic—that define cognitive demand. While the NAEP framework defines cognitive demands primarily in terms of science practices and principles, Figure 13.1 broadens the definitions to include knowledge related to science literacy more broadly defined, including understanding science text and composing explanations and descriptions that

Figure 13.1 Four Knowledge Types That Define Cognitive Demands

	Cognitive Demands
Declarative Knowledge (Knowing *that*)	Basic knowledge about the world in general and science in particular, including facts, principles, definitions of terminology, science representations (such as formulas and equations), and attributes of science communication (such as scientific explanations and descriptions of investigations or experiments).
Procedural Knowledge (Knowing *how*)	Knowing how to and being proficient at standard procedures and processes such as inquiry, design, implementation of investigations and laboratory procedures, reading science text for understanding, and writing science descriptions and explanations.
Schematic Knowledge (Knowing *why*)	Knowledge contained in mental structures organized in systematic ways that have broad application for explaining the natural world. Knowledge about broadly applicable concepts such as system, conservation, structure and function, and theories such as atomic and kinetic molecular theory are contained in mental structures called schema.
Strategic Knowledge (Knowing *how to plan*)	Knowledge that enables the analysis of situations and tasks for the purpose of planning. It includes approaches to understanding the situation or task and developing a plan to approach the task in a meaningful way. Strategic knowledge plays an important role in successful test taking, in general, and understanding science items and how to construct responses to them, in particular.

are components relevant to composing responses to constructed-response items.

Science content and practices are two sides of the assessment coin. Typically, practices are assessed in a science context. Students are seldom asked to describe how to assess the validity or the design of science investigations in the abstract. Rather, items tend to describe an investigation and ask students to evaluate the validity and design. Items designed to assess students' understanding of science principles require coordination of the principle with a practice. Consequently, sorting out whether an item is assessing a principle or a practice generally is not possible. Although we presented the clothes dryer example as an assessment of the practice, a correct response is evidence for understanding of at least that portion of the science principle related to evaporation, that is, understanding the principle that heating changes water from the liquid to the vapor state.

The National Science Education Standards (NRC 1996), the Benchmarks for Science Literacy (AAAS 1993), and the TIMSS and PISA assessment frameworks do not describe in detail the knowledge implied in statements such as "One way to change matter from one state to another and back again is by heating and cooling." Consequently, a student's understanding of the clothes dryer requires more knowledge and reasoning abilities than are evident in either the practice or the content statement.[2]

Analysis of an assessment item to identify *all* that a student needs to know to write a satisfactory explanation of the phenomenon (clothes drying in the electric dryer) reveals the quantity and complexity of knowledge and reasoning required. The results of the analysis provide teachers with a valuable template against which to evaluate students' explanations and to identify what makes one explanation satisfactory and another not.

If we only think about student performance on an assessment item in terms of whether an answer is right or wrong, we may underestimate the challenge of producing a correct response to the item. When a student is asked by a classroom teacher to explain how clothes dry in an electric dryer, some students have a good idea of what the teacher expects regarding the science information to be used in the explanation and the structure of explanations that the teacher will find acceptable. On assessments designed by their teachers, students know the audience for their writing and write to that audience, the teacher. When responding to an item contained on an external test, however, students are not aware of the writers' expectations for their responses or how the responses will be scored. In the case of the clothes dryer example, is the expectation that the response will be at the molecular level or simply relate heating to increased evaporation rate? Are students expected to apply the systems concept, noting that the water vapor must be blown out of the dryer (otherwise the clothes would not dry)? What are the characteristics of satisfactory explanations? What is the acceptable content of sentences in an explanation? In what order should the sentences appear?

Beyond the science knowledge and proper forms for explanations, the context in which the item is framed is also important for composing a response to the item. A student in a third world country might be able to compose a properly structured science explanation and know that heating increases the rate

---

[2]The American Association for the Advancement of Science's *Atlas of Science Literacy* (AAAS 2001) illustrates the complexity of declarative knowledge contained within the National Science Education Standards and the Benchmarks.

of evaporation but may not be able to answer the question as posed because he has no familiarity with electric clothes dryers. This is a simple but real example of knowledge necessary to respond correctly to an item that might be missed by a student in a third world country. Although identification of *all* the knowledge necessary to respond to an item is an impossible task, serious attention to the knowledge required to respond to the task is valuable when trying to understand why students are not performing well on items.

PISA Constructed-Response Items

We turn now to the analysis of two sets of constructed-response items: one from PISA, the other from TIMSS. The PISA item set is about immunity and is situated in an historical context. The TIMSS item set is about the density of metals and is situated in the context of identifying the metal content of a crown. In addition to the item sets as they are presented to students, we have information about the science content the item sets are designed to assess and scoring guides for each item in each of the sets.

Figure 13.2 contains an item set from PISA. The item set represents an international perspective on the form of constructed-response items and how responses to the items will be scored. PISA is administered to 15-year-olds worldwide and has been designed to assess literacy "broadly defined"—meaning students' science knowledge and abilities in situations they would likely meet in their everyday lives. Figure 13.2 contains two items, scoring information, as well as the science process and concept being measured. Both items are open constructed-response items set in a life and health situation measuring students' understanding of scientific investigation related to the concept of human biology.

Reading and Writing Knowledge and Skills Required for Successful Performance of This Item Set

Successful performance of the item in Figure 13.2 requires students to read and understand a text passage. The language in the passage is straightforward but has some vocabulary that may not be familiar to students—for instance, *onslaught* and *pulverized*. The passage sets the item in a historical context and thus contains information irrelevant to the ideas that Boylston was trying to test and his approach to testing them.[3] This may be a distraction to students who are having difficulty understanding the historical content if they spend time trying to understand the context and its relationship to the science being assessed.

Figure 13.2 PISA "Stop That Germ!" Item Set

Stop That Germ![a]

Item

As early as the 11th century, Chinese doctors were manipulating the immune system. By blowing pulverized scabs from a smallpox victim into their patients' nostrils, they could often induce a mild case of the disease that prevented a more severe onslaught later on. In the 1700s, people rubbed their skins with dried scabs to protect themselves from the disease. These primitive practices were introduced into England and the American colonies. In 1771 and 1772, during a smallpox epidemic, a Boston doctor named Zabdiel Boylston tested an idea that he had. He scratched the skin on his six-year-old son and 285 other people and rubbed pus from smallpox scabs into the wounds. All but six of his patients survived.

1.1 What idea might Zabdiel Boylston have been testing?
1.2 Give two other pieces of information that you would need to decide how successful Boylston's approach was.

Scoring and Comments on Science Example 1.1[b]

Full Credit
Code 2: Answers with reference both to the idea that infecting someone with smallpox will provide some immunity AND to the idea that by breaking the skin, the smallpox was introduced into the blood stream.
Partial Credit
Code 1: Answers that refer to either of the above points.
No Credit
Code 0: Other responses.

Item type: Open constructed response
Process: Understanding scientific investigation (Process 2)
Concept: Human biology
Situation: Science in life and health

Scoring and Comments on Science Example 1.2

Full Credit
Code 2: Answers that provide the following TWO pieces of information: the rate of survival without Boylston's treatment AND whether his patients were exposed to smallpox apart from the treatment.
Partial Credit
Code 1: Answers that provide either of the above points.
No Credit
Code 0: Other responses.

Item type: Open constructed response
Process: Understanding scientific investigation (Process 2)
Concept: Human biology
Situation: Science in life and health

───────

[a]Science Unit 1: Stop That Germ! *PISA 2003 Assessment Framework: Mathematics, Reading, Science, and Problem Solving Knowledge and Skills.* © OECD. Reprinted with permission.

[b]Science Examples 1.1 and 1.2. *PISA 2003 Assessment Framework: Mathematics, Reading, Science, and Problem Solving Knowledge and Skills.* © OECD. Reprinted with permission.

Both items 1.1[4] and 1.2 in Figure 13.2 ask students to provide information without any reference to how that information should be structured. Each scoring guide contains two phrases containing the information that will be given full credit. Thus it seems that student scores will not depend on the form of the response but only on the information contained in it. Figure 13.3 contains performance expectations for both examples, illustrating possible forms that student responses might take. Three of the examples reference the question that is being responded to. Some teachers require that students include the question in a constructed response. This form for the response reminds the student of the purpose of the response and reminds him or her to answer the question. It also makes the response more complete. The second response for Example 1.2 does not contain a statement of the question and, strictly speaking, does not respond to the

---

Figure 13.3 PISA "Stop That Germ!" Performance Expectations (Possible Forms That Student Examples Might Take)

Performance Expectations Example 1.1

The idea Boylston was testing whether rubbing pus from smallpox on scratched skin keeps people from dying from smallpox.

When Boylston rubbed pus from smallpox on scratched skin, he was testing to see if getting smallpox pus in the human body kept people from dying from smallpox.

Performance Expectations Example 1.2

To decide for sure if rubbing pus on scratched skin kept people from dying from smallpox you would need to know that everyone of the people he scratched really got pus in their bodies and that they had a chance to get smallpox. Did any people that he didn't scratch who had a chance to get smallpox die?

Boylston did not do a good test of his idea. He did not have a control group. He did not make sure that all the people he scratched were coughed on by people with smallpox.

---

[3]The American Association for the Advancement of Science's *Atlas of Science Literacy* (AAAS 2001) illustrates the complexity of declarative knowledge contained within the National Science Education Standards and the Benchmarks. Some students, seeing the lengthy text, might choose not to read the text but go right to the first question and, armed with the question, search the text for the answer, ignoring the irrelevant historical information. This would be an example of strategic knowledge in action. These students have analyzed the situation and devised an efficient plan to complete the task that does not involve following a set procedure of reading the item from top to bottom as it appears in the test booklet.

[4]The American Association for the Advancement of Science's *Atlas of Science Literacy* (AAAS 2001) illustrates the complexity of declarative knowledge contained within the National Science Education Standards and the Benchmarks. Notice that Example 1.1 in Figure 13.3 asks for an idea, but the scoring guide requires two ideas for full credit.

question. The statement that Boylston did not have a control group suggests the writer understood the need for information about the survival rate for individuals who were exposed to smallpox but were not treated using Boylston's method.

Science Knowledge and Practices the Items Are Assessing

According to the assessment framework, these items are assessing the concept *human biology* and the process *science investigation*. What is not evident from the scoring and comments for the item set are the components of human biology being assessed. The scoring guides are vague with regard to how exact the match must be between a student's response and the phrases in the scoring guide for the student to receive full credit. For 1.1 in Figure 13.2, will a response that does not use the word *immunity* but says that "the treatment protected the person against smallpox" or repeats the text by saying "the treatment prevented a more severe onslaught of smallpox" get a score point? Does *protection* mean not getting smallpox, getting a less severe case, or simply surviving until the epidemic is over? Is it necessary that the response mention the bloodstream, or would a score point be given to the response "the smallpox got under the skin"? Notice that the scoring guide equates smallpox, the disease, with the virus that causes the disease. Is this distinction important? Or is it acceptable to conflate the disease with the virus?

A literal interpretation of the scoring guide suggests that students are not being assessed on their ability to use the terms *infection* and *immunity* appropriately or to be explicit about the necessity of the smallpox virus to get into the bloodstream to produce partial immunity. Nor does the scoring guide seem to expect that students should distinguish the disease from the microorganism that causes it. This suggests that 1.1 in Figure 13.2 is not designed to measure the students' understanding of immunization or diseases such as smallpox caused by microorganisms.

"Understanding scientific investigation" is the process the item set is designed to assess. In 1.1 in Figure 13.2, students are expected to infer the purpose of an investigation from its description. In 1.2 in Figure 13.2, students are required to identify variables to assess the validity of the investigation. Judging whether an investigation is well matched to the question it means to answer and identifying variables to be controlled so that valid conclusions can be drawn are components of scientific investigation.

The claim that this item set measures understanding of scientific investigation might be challenged and an alternative claim made that 1.1 measures only reading comprehension and 1.2 critical-reasoning skills.

Cognitive Demands

Figure 13.4, page 262, contains some of the cognitive demands posed by the germ item set based on the analysis of the items and scoring guides. Because it is not entirely clear from the scoring guides how much students are expected to know about acquired immunity or how the Boylston strategy might have produced it, the declarative and procedural knowledge is tentative and can only be verified by the item developers.

Figure 13.4 PISA "Stop That Germ!" Cognitive Demands

Knowledge Type	Knowledge

Declarative	Human Biology Vocabulary
	    pulverized
	    onslaught
	    epidemic
	    immune system
	    smallpox

Procedural	Scientific Investigation Reading for Understanding Critical Reading Writing Responses to Questions

Schematic	Human Immune System
	    natural immunity
	    acquired immunity
	Germ Theory of Disease
	Science Investigations

Strategic	Test Taking

The text describes Boylston's treatment without reference to (1) the possibility that he might have been conducting an investigation, (2) any variables he might have controlled, or (3) whether his outcome variable was whether the individuals he treated got smallpox, got severe cases of smallpox but did not die, got mild cases of smallpox, or just died of whatever. Neither do the questions make reference to scientific investigation. Without any direct reference to scientific investigation, students may not have called on their schematic or procedural knowledge of the characteristics of scientific investigations to answer the questions but rather applied their critical-reading skills to the text.

Our discussion above suggests that placing the investigation in context may be a distraction to the student who carefully reads the text and attempts to understand the historical context before reading the questions. Other students may take a different and more efficient approach to the item: skip the lengthy text, read the first question, and search the text for information relevant to the question while ignoring the irrelevant historical information. We might infer from this approach that students taking the second approach have strategic knowledge. These students analyzed the situation and devised a more efficient plan to complete the task. The approach does not involve following a set procedure for reading the item from top to bottom as it appears on the page of the test booklet.

Cognitive demands based on items and scoring guides are only hypotheses about the cognitive demands items actually placed on students. Cognitive demands can best be inferred by observing students doing the items using thinking aloud protocols or by the analysis of students' responses to the items.

TIMSS Constructed-Response Item

We turn now to an item set ("Metal Crown") from the grade 8 science TIMSS. The six constructed-response items as they are presented to the students and guides used to score student responses are contained in the Appendix on page 273. Brief descriptions of the items are contained in Figure 13.5. All of the items require constructed responses: one a short constructed response, the other five extended constructed responses.

Figure 13.5 TIMSS "Metal Crown" Item Set

Item Number	Item Description

SO32709	*Calculate the density of a block*
	    Short constructed response
	    Content Domain: Physical Science/Density
	    Cognitive Domain: Recall/Recognize, Define, Compare/
	    Contrast/Classify

(continued)

Figure 13.5 *(continued)*

SO32711	*Describe a procedure to measure volume of an irregular object (crown)*
	    Extended constructed response
	    Content Domain: Physical Science/Density
	    Cognitive Domain: Design or Plan an Investigation

SO32712A	*Explain why scientists do multiple measures*
	    Extended constructed response
	    Content Domain: Physical Science
	    Cognitive Domain: Design or Plan an Investigation

SO32712B	*Show how scientists obtained the density of the crown*
	    Extended constructed response
	    Content Domain: Physical Science/Density
	    Cognitive Domain: Analyze/Solve Problems

SO32713A	*Use a table of density and the density of the crown to identify the metal the crown is made of*
	    Extended constructed response
	    Content Domain: Physical Science/Density
	    Cognitive Domains: Interpret Information, Draw
	        Conclusions

SO32713B	*Use a given density for the crown and a table of metal densities to identify the metal or mixture of metals the crown is made of*
	    Extended constructed response
	    Content Domain: Physical Science
	    Cognitive Domains: Interpret Information, Draw
	        Conclusions

Source: TIMSS Grade 8 Science Concepts and Science Items. *http://necs.ed.gov/timss/pdf/ timss8_Science_ConceptsItems.pdf* ©TIMSS & PIRLS International Study Center. Reprinted with permission.

Reading and Writing Knowledge and Skills Required for Successful Performance of This Item Set

This item set is introduced with a story (Appendix, p. 273) describing a king's suspicions about the composition of the metal composing a crown a jeweler made him. The king asks his scientists to investigate his suspicions. The text introducing each item in the set describes the scientists' approach to determining if the king's suspicions are reasonable. The language structure of the introductory text is straightforward and the vocabulary related to the context is familiar. Because kings, jewelers, and crowns are familiar to most grade 8 students, they are unlikely to find the context of the item difficult to understand.

Pure metal, mixtures of metal, weight/mass, gram, density, and *volume* are the science-specific vocabulary in the introductory text. The words are familiar, but students' understanding of the concepts is limited. Understanding that density is a property that can be used to identify metals and mixtures of metals is essential to understanding the premise of the item set that density could be used to investigate the king's suspicions.

With the exception of item 09, items in this set require students to compose written responses of different types. Item 11 requires the students to write a description of a laboratory procedure. Item 12A requires writing an explanation. Item 12B requires students to demonstrate their understanding of the scientists' work. Items 13A and 13B asks the students to explain the reasoning leading to their answers about the composition of the crown.

The scoring guides describe the information required for scored points and, in some instances, examples of student responses that would be given score points and examples of student responses that would not earn score points. The guides are clear about the information that should be included in student responses but do not specify the form in which that information should appear. For instance, the scoring guide for Item 11 does not suggest the characteristics of a well-written description of a laboratory procedure. As such, the scoring guide sets a standard for performance much lower than most science teachers set for their students. Science teachers are likely to expect descriptions in paragraph form including the purpose of the procedure, and details critical to its success, as, in this case, the importance of starting with clean and dry equipment.

Performance expectations such as those in Figure 13.6, page 266, provide examples of responses that meet teachers' expectations. These illustrate both the information that should be included in the response and the structure of the response.

Item 12A requires students to write an explanation. The scoring guide lists information that will earn a score point and examples of student responses that contain the information. The student responses are one or two sentences long with no reference to what is being explained. Do the

Figure 13.6 Performance Expectations for TIMSS Item 11

Performance Expectation A

To measure the volume of the crown, the scientists should put the crown in the glass container and use the graduated cylinder to add water to the container until the top of the crown is just covered. They need to write down how much water it took to cover the crown and mark how high the water is in the glass container. Then dump the crown and water out of the container and dry the container. Now use the graduated cylinder to add water to the mark to show how high the water was in the container and write down how much water it took to fill the container to the mark. When the scientists subtract the volume of water it took to fill the container to the line from the volume of water it took to fill the container to the line when the crown was in the container, they will get the volume of the crown.

Performance Expectation B

The scientists should put the glass container in the dry plastic tray and fill the glass container to the top. Then put the crown in the water. They should put the water that spills out in the graduated cylinder. The level of the water in the cylinder is the volume of the crown.

explanations meet the science education community's expectations? Should the explanation (1) include a description of what is being explained, (e. g., why scientists do multiple measures); (2) explain why a different volume may be measured each time (e. g., spilling some of the water, not drying the graduated cylinder thoroughly before it is used, or misreading the meniscus); and (3) identify other sources of error (e. g., inaccurate recording of readings or incorrect arithmetic calculation)?

Item 12B requires students to describe a procedure for calculating the mean of five volumes. Example responses show either the calculations or a verbal description of the calculations. None of the responses makes any reference to the fact that the mean of the volumes is being calculated.

Responses to items 13A and B follow a similar pattern consisting of single declarative statements in the form of conclusions without reference to the information on which the conclusion is based or reference to the situation to which the conclusion refers.

Scoring guides for the Crown items that call for explanations are comprised of one or two sentences or phases that, taken out of context, would not be recognized as an explanation.

Are explanations contained in structured paragraphs what is called for in national standards? If so, what are the characteristics of explanations that make them good science? Are the "explanations" in the scoring guides examples of good science explanations?[5]

Science Knowledge and Practices the Items Are Assessing

The set *engages* the students in the process of analyzing a problem, the composition of the crown, and a strategy for its solution. This item set *measures* students' knowledge of density as a property of substances and objects and density's measurement. While the students are not posed the problem and asked to develop a plan for solving it, they are required to demonstrate their ability to perform component parts of the solution, including calculating the volume of a regular geometric figure, a cube, and measuring the volume of an irregular object, a crown. They use the volumes to calculate the densities of the metal block the king gave to the jeweler and the crown the jeweler delivered to the king. They use the densities to decide if the jeweler made the crown from the block the king gave him.

Cognitive Demands

Figure 13.7, page 268, summarizes the cognitive demands of the Crown items based on the items and scoring guides. Adequate performance of the item set requires that the students have some basic reading, writing, and arithmetic skills. They need to understand the context of royalty and crowns, write simple declarative sentences, and be able to add, subtract, divide and multiply. Beyond the declarative knowledge required to perform these basic skills, the cognitive demands are considerable. This is a difficult item set for grade 8 students. The research literature documents how difficult the concept of density is for students. The concept is complex and has broad application in

---

[5]The American Association for the Advancement of Science's *Atlas of Science Literacy* (AAAS 2001) illustrates the complexity of declarative knowledge contained within the National Science Education Standards and the Benchmarks. We are not suggesting here that good explanations consist of text alone. Diagrams, tables, charts, and other representations are tools of effective explanation. Responses to assessment items asking for explanations might use a diagram as the primary tool for communication. Even so, the diagram cannot stand alone and must reveal the purpose of the explanation and the reasoning to the conclusion.

Figure 13.7 TIMSS "Metal Crown" Cognitive Demands

Knowledge Type	Knowledge
Declarative	Pure Metals Metallic Mixtures Density as a Property of Substances (Metals) and Objects (Crown) Characteristics of Science Explanations Characteristics of Descriptions of Science Procedures
Procedural	Reason From Information to Conclusion Calculate Volume Measure Volume of an Irregular Object Calculate Mean Identify Sources of Experimental Error Make Multiple Measures Construct a Science Explanation Construct a Description of a Laboratory Procedure Explain One's Reasoning From Information to Conclusion
Schematic	Object, Substance Pure Substance, Mixture Density
Strategic	Planning an Approach to Solving a Complex Problem

everyday life and in science. By grade 8 students have been exposed to the concept and are developing a density schema (Smith et al. 2004). However, even the basics of the schema are difficult. Density is the ratio of mass to volume. Ratio is a difficult concept, as is volume, which has different meaning in daily life. For instance, the volume of a jar of strawberry can be construed in several ways. To the jam-maker the important volume is the volume of jam the jar holds. To the manufacturer of the glass jars, the important volume is the volume of glass. To the designer of the boxes in which the jam jars will be shipped, the important volume is the space the jar takes up in the box. In science contexts, as in daily life, students must sort out these and other ways of thinking about volume and its relationship to density.

The concept of homogeneous solid mixtures is difficult for most grade 8 students. Metals are familiar to students who have had daily contact with metals, all of which are solid mixtures of two or more metals. Even so, they have a difficult time conceptualizing how two metals such as gold and copper might be combined to form a homogeneous solid mixture. Yet it is the suspicion that the crown may be a mixture of metals that is the declarative knowledge that students need to understand the basic premise of this item set. Differences in macro-properties (density, for instance) and atomic structures of pure metals and mixtures are elements of the declarative knowledge underlying the understanding of how pure metals and mixtures of metals are different.

Declarative knowledge related to science practices is also essential to the performance of this item set. Students need to know procedures: measuring the volume of an irregular object by water displacement; using a formula and linear measurements to calculate the volume of a cube; and using a formula, mass, and volume to calculate density. The item set also asks students to engage the science practices of describing procedures, writing explanations and describing how they reasoned from data to a conclusion. Yet neither examples of descriptions and detailed descriptions of the characteristics of the explanations are contained in either the scoring guides or other parts of the framework.

While students are not required to develop a strategy for solving the king's problem, the item set is an interesting illustration of the kind of strategic knowledge students would need if they were challenged to design a plan leading to the solution of the king's problem. An item in this form would be a measure of strategic knowledge of problem solving. [Readers interested in an example of a student's response to this item will find one on page 166 of the TIMSS 2007 Frameworks available on the internet at *http://timss.bc.edu/TIMSS2007/PDF/T07_AF.pdf.*]

Concluding Statements

Students' responses to assessment items are often reviewed with a scoring mentality. The student got what the item writer intended to measure either right or wrong. However, students' responses depend on their understanding and interpretations of the item, and often the students' understanding and interpretations are not what the item writer intended. Consequently, we have illustrated ways in which teachers might review items before making judgments about the adequacy of students' responses. Our approach has been to apply our experience to identify (1) the reading and writing knowledge and skills necessary for successful performance of constructed-response items,

(2) the science knowledge and practices that constructed items are assessing, and (3) the constructed-response items' cognitive demands. Our conclusions and the only conclusions that can be drawn from such an analysis are simply hypothesized knowledge and skills required for successful performance of the items. The hypotheses can best be tested by observing students do the items while thinking aloud or by the analysis of students' responses to the items.

As the reader has undoubtedly noticed the boundaries among the categories we used for our analysis are fuzzy and even arbitrary. Separating generic reading comprehension and critical-reading skills from those that are science specific is difficult, if not impossible, as is making the distinction between science knowledge and science practices. Within science knowledge subcategories and within science practice components there is considerable overlap. Furthermore, categories of knowledge contained in the cognitive demands also overlap. A density schema contains declarative knowledge about density, including principles (such as the density of substances decreases with increasing temperature) and facts (such as water is an exception to the principle when cooled below 0°C). Considering the time necessary to engage is the process we have illustrated and the ambiguity of the results, is the struggle to sort all of this out worth the effort?

We believe that the effort helps us to better understand the challenges assessments pose to students; what we learn from the results will make us better teachers. Our analysis of constructed-response items has illuminated several matters worthy of further consideration. One is the danger of equating scoring guides with performance expectations.

Our analysis of scoring guides for constructed-response items illustrates this danger. Constructed-response items take considerable testing time and are expensive to score. They are included on assessments because the ability to communicate is an important goal of science education. The scoring of this item type is based on bulleted lists of information to appear in the response with no consideration of the literary form of the response. We believe that scoring information of this type defeats one of the essential purposes of including constructed-response items on science assessments, that is, to make effective communication a central component of the science curriculum. While standards call for students to describe science procedures, to explain natural phenomena, and to reason from data to conclusions, the field has yet to define its expectations for well-constructed descriptions, explanations, and reasoning from data to conclusions.

Another matter worthy of further consideration is the practice of placing constructed-response items in context. Placing items in context adds to the reading load of the item and to the testing time the item requires. If the context is not familiar to the student, it adds to the item difficulty and may prevent the student from demonstrating the science knowledge she does know. We have argued elsewhere (Kouba et al. 1999) that while teaching in context makes sense, we should be cautious about testing in context.

We believe that the approach we have taken is one that highly qualified teachers apply in their classrooms when they review items from external tests their students are required to take. Teachers assess the knowledge and reasoning demands of items, applying their experience to making judgments about the difficulty of the item for their students, whether the scoring guide is fair, and whether the item is an appropriate measure of the science content it claims to measure. Guiding their analyses are the teachers' expectations for how the best students in their classes will respond to the item. Based on their expectations the teachers generate hypotheses about the knowledge and skill required to meet their expectations.

Typically teachers use the information gleaned from the analysis of the items and their personal expectations to analyze their students' responses to the items. Their goal is to figure out why students did not respond as expected and to use that information to strengthen teaching practices. In more formal terms, the process is using student responses as data to test hypotheses based on cognitive demand theory.

Teachers engage in this process as individuals. Seldom do teachers share with their colleagues their performance expectations or their hypotheses about what students need to know to meet the expectations. Our experience suggests that when teachers engage in this process with their colleagues, the products of the analysis are richer and classroom practice is enhanced.

We doubt that all our readers will agree with our analyses of the PISA and TIMSS item sets. We only hope that our analyses will generate discussion to illuminate assessment practices.

References

American Association for the Advancement of Science (AAAS). 1993. *Benchmarks for science literacy*. New York: Oxford University Press.

American Association for the Advancement of Science (AAAS). 2001. *Atlas of science literacy*. Washington, DC: AAAS and NSTA.

Catley, K., R. Lehrer, and B. Reiser. 2005. *Tracing a prospective learning progression for developing understanding of evolution.* Paper commissioned by the Committee on Test Design for K–12 Science Achievement, Board on Testing and Assessment, Center for Education, National Research Council. Washington, DC: National Research Council.

Champagne, A., K. Bergin, R. Bybee, R. Duschl, and J. Gallagher. 2004. *NAEP 2009 science framework development: Issues and recommendations.* Washington, DC: National Assessment Governing Board.

Kouba, V., O. Cezikturk, S. Sherwood, and C. Ho. 1999. Setting the context for mathematics in context. *Mathematics Teaching in the Middle School* (Dec.).

Mullis, I., M. Martin, G. Ruddock, C. O'Sullivan, A. Arora, and E.Erberber. 2005. *TIMSS 2007 assessment frameworks.* The International Association for the Evaluation of Educational Assessment. Boston, MA: The International Study Center, Lynch School of Education, Boston College.

National Assessment Governing Board. 2004. *Science framework for the 2005 national assessment of educational progress.* Washington, DC: National Assessment Governing Board.

National Assessment Governing Board. 2005. *Science framework for the 2009 national assessment of educational progress.* Washington, DC: National Assessment Governing Board.

National Center for Education Statistics. 2005. What are the differences between long-term trend NAEP and main NAEP? The Nation's Report Card. Retrieved September 18, 2006, from *http://nces.ed.gov/nationsreportcard/about/ltt_main_diff.asp*

National Research Council (NRC). 1996. *National science education standards.* Washington, DC: National Academy Press.

Nohara, D. 2001. *A comparison of the National Assessment of Educational Progress (NAEP), the Third International Mathematics and Science Study Repeat (TIMSS-R), and the Programme for International Student Assessment (PISA)* (Working Paper No. 2001-07). Washington, DC: National Center for Education Statistics.

Organisation for Economic Co-operation and Development (OECD). 2003. *The PISA 2003 framework—mathematics, reading, science and problem solving knowledge and skills.* Paris, France: OECD.

Perie, M., R. Moran, and A. D. Lutkus. 2005. *NAEP 2004 trends in academic progress: Three decades of student performance in reading and mathematics* (NCES 2005–464). U.S. Department of Education, Institute of Education Sciences, National Center for Education Statistics. Washington, DC: Government Printing Office.

Smith, C., M. Wiser, C. W. Anderson, J. Krajcik, and B. Coppola. 2004. *Implications of research on children's learning for assessment: Matter and atomic molecular theory.* Paper commissioned by the Committee on Test Design for K–12 Science Achievement, Board on Testing and Assessment, Center for Education, National Research Council.

Appendix

TIMSS "Metal Crown" Item Set

A king gave a jeweler a block of pure metal. He asked the jeweler to make him a crown out of the metal.

metal crown   metal block

After the jeweler delivered the crown, the king observed it carefully. He thought that the jeweler might have substituted another pure metal or a mixture of metals to make the crown. He weighed the crown, and it had the same mass as the original block, 2400 grams. Still not satisfied, the king asked some scientists to help him find out what the crown was made of.

Questions for Metal Crown begin on the next page.

Item S032709

The scientists decided to compare the densities of the crown and a block of metal just like the original block. The density of a substance is the mass of a sample of the substance divided by its volume (density = mass/volume).

The scientists found the volume of the block and computed its density based on its known mass (2400 g). The diagram below shows the dimensions of the block of metal that the scientists measured.

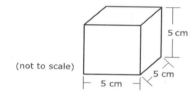

(not to scale)   5 cm   5 cm   5 cm

What is the density of the block of metal?

Answer: _____ g/cm^3

Questions for Metal Crown continue.

Scoring Guide for Item S032709

Code	Response	Item: S032709
	Correct Response	
10	19.2 g/cm^3 **Note:** Extra trailing zeroes may also be added (e.g., 19.20, 19.200)	
11	19 g/cm^3 [Rounds to nearest whole unit.]	
	Incorrect Response	
70	Shows the setup for density (mass/volume) but does not compute density or makes a computational error.	
71	125 [Computes volume but not density.]	
72	19.3 [No work shown; indicates density copied from table.]	
79	Other incorrect (including crossed out/erased, stray marks, illegible or off task)	
	Nonresponse	
99	Blank	

Item S032711

The scientists then needed to find the volume of the crown in order to determine its density. The following equipment and materials were available for them to use.

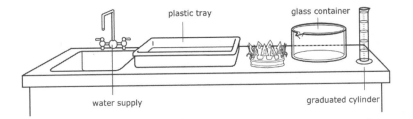

Describe a procedure that the scientists could use to find the volume of the crown using some or all of the equipment and materials shown above. You may use diagrams to help explain your procedure.

Questions for Metal Crown continue. ➤

Scoring Guide for Item S032711

Note: For full credit, responses must describe or diagram a procedure based on displacement and clearly identify how the volume of the crown is determined. Partial credit is given for procedures or diagrams that demonstrate knowledge of displacement without a complete description of the steps/measurements to be made. Responses may also implicitly refer to other materials not indicated in the diagram (e.g., ruler, marker, etc.). Because it is not totally clear from the diagram what the relative size of the crown, beaker, and tray are, credit is given for procedures that use any of these materials for displacement even if the actual procedure might not be completely successful.

Scoring Guide for Item S032711

Code	Response	Item: S032711
	Correct Response	
20	Describes or diagrams a procedure based on displacement of water using measured water level differences: i) Adding water to the beaker (sink or tray) and marking the water level. ii) Placing the crown in the beaker (sink or tray) and marking the new water level. iii) Measuring the volume difference before/after adding the crown using the graduated cylinder.	
21	Describes or diagrams a procedure based on displacement of water using measured overflow: i) Filling the beaker (or tray) with water. ii) Placing the crown in the beaker (or tray) and collecting the overflow. iii) Measuring the volume of the overflow using the graduated cylinder.	
29	Other fully correct	
	Partial Response	
10	Describes or diagrams a partial procedure that includes displacement of water but with inadequate or no description of the steps/measurements to determine the volume. *Examples: Put some water in the beaker and add the crown. Measure how much the level of water went up.* *Add the crown to the beaker filled with water. See how much overflowed.*	
19	Other partially correct	
	Incorrect Response	
70	Mentions putting the crown in the beaker (sink or tray) of water with no explicit mention that the water level will rise/overflow and no or incorrect procedure given for measuring the volume. *Example: Fill the beaker to the top with water and add the crown. You can get the volume that way.*	
79	Other incorrect (including crossed out/erased, stray marks, illegible or off task)	
	Nonresponse	
99	Blank	

Items S032712A and S032712B

The scientists measured the volume of the crown five times. They computed the density for each volume measurement. Their results are shown in the table below.

Trial	Volume of Crown (cm³)	Density of Crown (g/cm³)
1	202	11.88
2	200	12.00
3	201	11.94
4	198	12.12
5	199	12.06

A. Why did the scientists measure the volume five times?

B. The scientists reported to the king that the density of the crown was 12.0 g/cm³. Show how the scientists used their results to obtain this value for the density.

Questions for Metal Crown continue.

Scoring Guides for Items S032712A and S032712B

A: Codes for Why Scientists Repeat Measurements
Note: Priority is given to Code 10. If accuracy, precision, experimental uncertainty, measurement error, etc., is mentioned, then Code 10 should be given even if other correct codes apply.

Code	Response	Item: S032712A
	Correct Response	
10	Refers to accuracy, precision, reliability, experimental uncertainty, estimation of measurement error (or similar). *Examples: Because there is experimental error. So measuring it 5 times you can calculate the average to know how much error there is. Each time they measure the volume it is close but not exactly the same. So, it's better to measure it a few times to be sure. They want a more exact answer. To get an accurate measure of the volume. It's more reliable.*	
11	Refers only to computing an average or mean value (or median or range). *Examples: To find the average volume. To work out the mean.*	
19	Other correct	
	Incorrect Response	
70	Refers only to 'mistakes' or changes in the measurements (or similar); no explicit mention of accuracy, precision, experimental uncertainty, etc. *Examples: In case mistakes happen. To make sure it wasn't changing. To make sure the answer was right and he did not make a mistake. To make sure they did it right. To check if it was correct.*	
71	Refers only to a 'fair test' or similar; no explicit mention of computation of average, accuracy, precision, experimental uncertainty, etc. *Examples: To make sure it was a fair test. To ensure a fair test.*	
79	Other incorrect (including crossed out/erased, stray marks, illegible or off task)	
	Nonresponse	
99	Blank	

B: Codes for Obtaining Average (Median) Value

Code	Response	Item: S032712B
	Correct Response	
10	Shows (or describes) a correct method for computing the average (mean) value. *Examples:* $(11.88+12.00+11.94+12.12+12.06) = 60.$ $60/5=12.0$ $(202+200+201+198+199)/5 = 200.$ $2400/200=12.0$ *They added together all of the densities and then divided by 5 to get the average.*	
11	Shows (or describes) a correct method for determining the median value. *Examples:* *202, 201, 200, 198, 199. 200 is the median volume, so 2400/200 is the median density (12). 12 is the middle value when placed in order (12.12, 12.06, 12.00, 11.94, 11.88).*	
19	Other correct	
	Incorrect Response	
70	States that it is the average, mean, or median value with no or incorrect work shown.	
71	Shows a computation of density (mass/volume). [No determination of average or median included.] *Examples:* *They did mass divided by volume.* $2400g/200cc = 12 \text{ g/cc}$	
79	Other incorrect (including crossed out/erased, stray marks, illegible, or off task)	
	Nonresponse	
99	Blank	

Items S032713A and S032713B

The table below lists the density for different metals.

Metal	Density (g/cm³)
Platinum	21.4
Gold	19.3
Silver	10.5
Copper	8.9
Zinc	7.1
Aluminum	2.7

A. Look at the density you computed for the block of metal. What was the block of metal most likely made of?

Answer: _____

Explain your answer.

B. The density of the crown was found to be 12.0 g/cm³. What would you report to the king about what metal or mixture of metals the jeweler used to make the crown?

End of Metal Crown section.

Scoring Guide for Items S032713A and S032713B

A: Codes for Identifying Metal in Block

Note: To receive credit, responses must identify gold AND give an explanation based on density. Responses that identify gold with no or incorrect explanation are given Code 70. It is possible that a different metal or metal(s) may be identified based on an incorrect density computation in the previous question. These types of responses may be given Code 19, provided the explanation is reasonable based on the computed density.

Code	Response	Item: S032713A
	Correct Response	
10	GOLD with an explanation based on correct density computed in previous question (19.2 g/cm^3). *Examples:* Gold. Because it had the closest density. Gold. The density is the same.	
19	Other correct	
	Incorrect Response	
70	GOLD with no explanation or incorrect explanation that is NOT based on density. *Example:* Gold. Because that is what crowns are always made of.	
71	SILVER (alone or mixed). [Confuses density of crown with density of the metal block.] *Example:* It is mostly silver because the density is 12 and that's the closest one.	
79	Other incorrect (including crossed out/erased, stray marks, illegible, or off task)	
	Nonresponse	
99	Blank	

B: Codes for Reporting Composition of Crown

Note: To receive credit, responses must indicate that the crown is composed of a mixture of metals (alloy) AND identify the metals that might be included based on the density (crown density between the densities of the pure metals). Responses that indicate that the crown is made of a mixture (alloy) or is not pure gold with no further information about what other metals are included are scored as incorrect (Code 70). If responses indicate that the crown is made of Palladium (not in the table but with a density of 12 g/cm^3), they should be given a Code 19.

Code	Response	Item: S032713B
	Correct Response	
10	Reports that the crown is made of a mixture (alloy) AND names specific metal(s) that might be included (reasonable composition based on density). *Examples:* *The jeweler used some silver as well as gold.* *It might have had some copper mixed in because that would lower the density and the cost.* *The jeweler most likely used all silver except for a thin coat of gold to make it look pure gold even though it wasn't.*	
19	Other correct	
	Incorrect Response	
70	Reports only that the crown is made of a mixture or is NOT pure gold (or similar); NO specific metals are named. *Examples:* *The jeweler didn't use the block of metal that the king gave him.* *The jeweler used four more metals to make the crown.*	
71	Reports SILVER (density closest to 12 g/cm^3). *Example:* *The metal used is silver.*	
72	Reports an incorrect mixture of metals based on additive densities. *Examples:* *It's silver and aluminum (10.5 + 2.7)* *Mixture of silver and aluminum as their density adds up to 12.0 approximately.* *Copper and aluminum.*	
79	Other incorrect (including crossed out/erased, stray marks, illegible or off task)	
	Nonresponse	
99	Blank	

## CHAPTER 14

Aligning Classroom-Based Assessment With High-Stakes Tests

Marian Pasquale and Marian Grogan
Educational Development Center (EDC), Center for Science Education

Whether or not you agree that high-stakes testing is fair, appropriate, and equitable—at the state or national level—it is a reality for students. Tests vary considerably from state to state in their quality and in the depth of knowledge expected of test takers. Limitations notwithstanding, however, we at EDC's Center for Science Education have taken the position that classroom teachers can use items of high-stakes tests as models to hone their own classroom assessment and, at the same time, help prepare their students to perform well on these tests.

High-stakes science tests administered by most states consist of different kinds of questions, including those that require a written response (multiple-choice and open-ended or constructed-response questions), those that require the creation or use of a graphic in the response, and those that require a performance of some sort in the response. The best questions on high-stakes tests require students to use "higher-order thinking skills"[1] and to demonstrate their understanding of certain concepts, ideas, and skills rather than to simply recall facts.

In our work with middle-grades science teachers across the country, we hear that students have difficulty with many of the questions found on state

---

[1] By "higher-order thinking skills," we refer to Bloom's taxonomy of cognitive demand. Those thinking skills considered "higher order" are the skills needed to apply, analyze, synthesize, and evaluate information, rather than to recall facts.

tests, even if they are familiar with the science content and format of the question. Upon further investigation, we find that the questions that students encounter in science class often do not match the cognitive demand of those encountered on high-stakes tests. In addition, the kinds of science investigations and activities in which students engage in the classroom often do not provide the opportunity to learn and practice higher-order thinking.

In response to what we heard, we developed an online course for middle-grades science teachers that focused specifically on the kinds of assessments found on high-stakes tests, and how teachers could use these assessments in their classrooms. Many of the ideas presented in this chapter emerged from our work with teachers in that course.

Here, we will focus on the types of assessments found on high-stakes tests that can and should be used in middle-grades science classrooms. We will provide a model for analyzing the kinds of information these types of assessments can provide about student understanding, as well as suggestions for ways that science teachers can better align their in-class assessments with those found on high-stakes tests.

Types of Tasks Found on High-Stakes Tests[2]

Written Assessments

Multiple-choice and constructed-response assessments on high-stakes tests include a range of tasks in which students show their understanding through writing. Students are often able to pass a test constructed of questions that ask for factual information. However, when they are asked to describe, explain, or make predictions about real-world phenomena, knowing the facts doesn't help. Students can have difficulty with questions such as these because they go beyond their understanding of the factual science being assessed.

Multiple-choice questions come in several forms. One form asks students to select one correct answer. Questions like these typically require students to use higher-order thinking such as applying and analyzing. Figure 14.1, from the 2003 Massachusetts Comprehensive Assessment System

––––––––
[2]The items we have included are released items from various state science tests. They have not been selected because they are exemplary but because they are representative of the questions found on all tests.

(MCAS), is an example of this kind of question. This question requires students to apply their knowledge of dynamics, or the effects of forces on the motion of objects. To be successful, they must know something about how the shape affects the flow of air around the vehicle.

Figure 14.1 Multiple-Choice Question—Selecting One Correct Answer

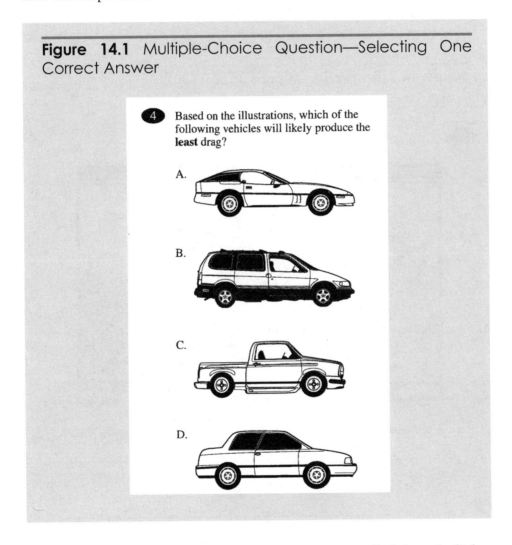

4 Based on the illustrations, which of the following vehicles will likely produce the **least** drag?

A.

B.

C.

D.

A second form of multiple-choice question is called "justified" because, after selecting an answer, students must explain their choice. By asking students to justify their answers, it becomes obvious if the answer is a guess or not. Figure 14.2, from the 2001 Grade 8 CSAP (Colorado

Department of Education), is an example of a justified multiple-choice question. To answer this question, students need to be able to read and interpret a diagram. They have to know that earthquakes occur along tectonic plate boundaries and that plate boundaries do not necessarily coincide with continental boundaries.

Figure 14.2 Justified Multiple-Choice Question

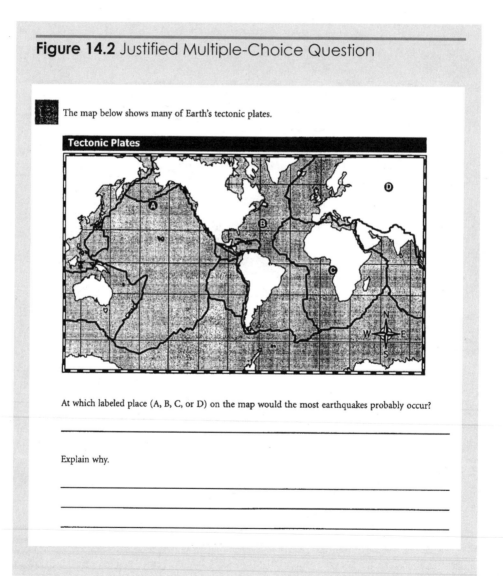

The map below shows many of Earth's tectonic plates.

At which labeled place (A, B, C, or D) on the map would the most earthquakes probably occur?

_____

Explain why.

_____

_____

_____

In **constructed-response** or **open-ended questions**, students generate their own answers. These generally take the form of a short essay. Students may be asked to apply their knowledge to a new situation or solve a problem.

Figure 14.3, from the 2004 Grade 7 Kentucky Core Content Test, is an example of a constructed-response question. Notice that the task includes Newton's first and second laws. Students need to know how to apply each law to each vehicle in question. They also need to be able to predict the sequence of reactions. To respond adequately to this task, students need to have had some experience with a situation similar to that posed in the task.

Figure 14.3 Constructed-Response Question

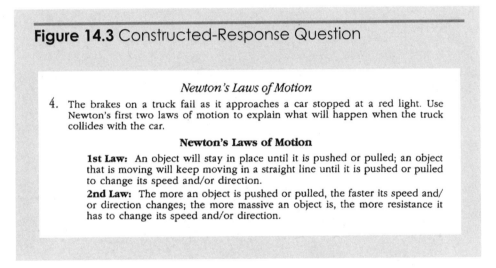

Newton's Laws of Motion

4. The brakes on a truck fail as it approaches a car stopped at a red light. Use Newton's first two laws of motion to explain what will happen when the truck collides with the car.

Newton's Laws of Motion

1st Law: An object will stay in place until it is pushed or pulled; an object that is moving will keep moving in a straight line until it is pushed or pulled to change its speed and/or direction.

2nd Law: The more an object is pushed or pulled, the faster its speed and/or direction changes; the more massive an object is, the more resistance it has to change its speed and/or direction.

In the classroom, informal written assessments are used for formative purposes. According to Fellows (1994), "Students' writing during lesson activities provides qualitative data about their understandings of the goals and concepts across the science unit." Writing then serves a dual purpose: to help students clarify their thinking and to assess *what* and *how* they are thinking. A student's writing is a representation of his or her thinking and can show how he or she is struggling with concepts.

Graphic Assessments

Graphic assessments on high-stakes tests ask students to describe, organize, represent, or analyze data. That data may be qualitative or quantitative.

Students may have to read and interpret diagrams, charts, maps, tables, and graphs or to use the information they provide to answer a question.

Figure 14.4, from the 2005 Grade 8 Massachusetts Comprehensive Assessment System (MCAS), is an example of a graphic assessment. In this task, students must know how to read a topographical map. This means that they must understand what a contour interval is, that it describes elevation, and that in this diagram the elevation reverses itself. Landforms A, a hilltop, and B, a valley, are relative to each other.

Figure 14.4 Graphic Assessment

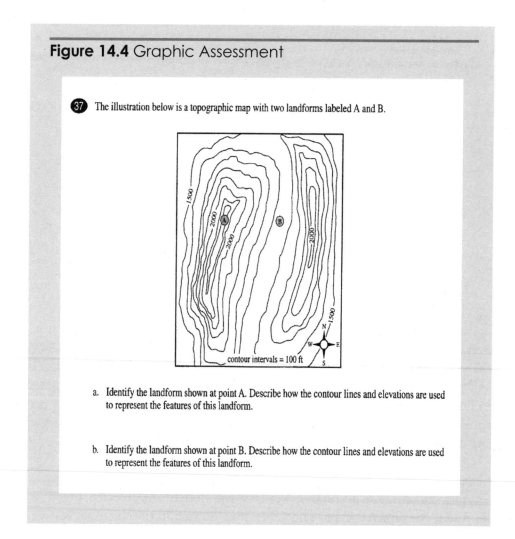

37 The illustration below is a topographic map with two landforms labeled A and B.

contour intervals = 100 ft

a. Identify the landform shown at point A. Describe how the contour lines and elevations are used to represent the features of this landform.

b. Identify the landform shown at point B. Describe how the contour lines and elevations are used to represent the features of this landform.

In the classroom, "writing through pictures" is a vital tool for sense-making. Graphics—in the form of drawings, diagrams, maps, graphs, and tables—can frequently communicate ideas in much more complete and complex ways than words alone. Students can also communicate their understanding through models—visual representations that may or may not be created with paper and pencil.

When students use graphics to represent their ideas, the process can often help them to clarify their own understanding. Drawing, in particular, can catalyze students' abilities to develop and verbally articulate new theories and explanations and to make essential connections. This is true for students from kindergarten through high school.

For middle school students, for example, in a unit on heat transfer, students might draw to explain what is happening inside an oven as a cake bakes. In an ecosystems unit, students might diagram all of the relationships and cycles within a terrarium. In a force and motion unit, students might create three different graphs that all show an object moving at a constant speed. All of these graphics provide windows into student understanding; in addition, they provide insight into what activities will best serve students' learning needs.

Performance Assessments

In a performance assessment, students craft an observable performance that often requires problem solving, inquiry, decision making, or role playing. A common characteristic of almost all performance assessments is that they are genuine activities that mirror real-life contexts. Few states include performance tasks as part of their science test because of the difficulty of administering and scoring them. Those states that do include virtual performance questions ask students to respond to multiple-step "virtual" experiments or lab situations, or to design an experiment.

Figure 14.5, page 290, from the 2001 Grade 8 CSAP (Colorado Department of Education), is an example of a virtual performance assessment. In the task, students need to analyze an investigation. For the first part of the investigation, students need to know the kinds of questions that can be investigated, how to design an investigation, and how to identify dependent and independent variables. In the second part, they need to know how to read a table and be familiar with decimals and metric units for measuring volume. In the third part, students need to know what a hypothesis is, know what a beaker is, and have some idea of how to control for experimental error by averaging results.

Figure 14.5 Virtual Performance Assessment

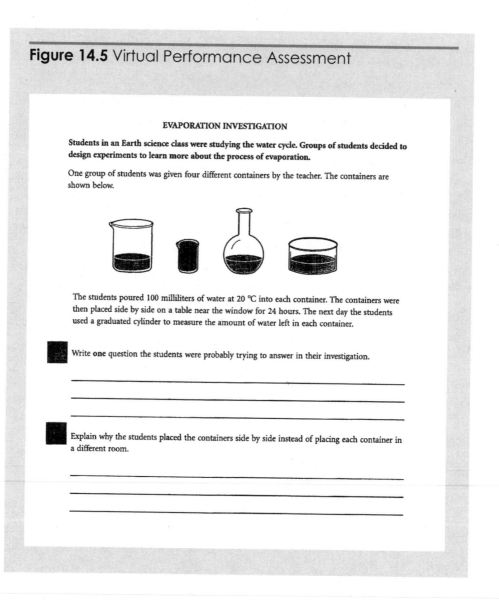

EVAPORATION INVESTIGATION

Students in an Earth science class were studying the water cycle. Groups of students decided to design experiments to learn more about the process of evaporation.

One group of students was given four different containers by the teacher. The containers are shown below.

The students poured 100 milliliters of water at 20 °C into each container. The containers were then placed side by side on a table near the window for 24 hours. The next day the students used a graduated cylinder to measure the amount of water left in each container.

■ Write **one** question the students were probably trying to answer in their investigation.

_____

_____

_____

■ Explain why the students placed the containers side by side instead of placing each container in a different room.

_____

_____

_____

A second group of students decided to see if different liquids evaporate at the same rate. The students decided to test three liquids: fresh water, salt water, and rubbing alcohol. The students poured 50 mL of each liquid into a separate beaker at room temperature, and the beakers were placed next to one another on a shelf.

After two days, the students used a graduated cylinder to measure the amount of liquid remaining in each beaker. The table below shows the results from their experiment.

Fresh Water	35.3 mL
Salt Water	38.6 mL
Rubbing Alcohol	22.7 mL

Before the experiment, the students made the hypothesis that salt water would evaporate faster than the other liquids. Do the results of their experiment support their hypothesis? Explain your answer.

_____

_____

_____

_____

In the classroom, performance assessments sometimes take place over an extended period of time. Among instructional materials, types of performance assessments vary widely. Some instructional programs use the term to refer to hands-on tasks only; others include a wider array of performances that draw on scientific knowledge and skills. Thus, performance assessments can range from presenting a proposal to simulating a town meeting to designing an experiment.

According to Jay McTighe (McBrien and Brandt 1997), "With performance tasks we look for opportunities for students to really demonstrate understanding and ability to use knowledge in meaningful ways." And different from other kinds of assessment tasks, performance tasks illustrate what students can do or the application of their knowledge in different circumstances, usually real-world contexts.

Perhaps the best thing about performance tasks is that they allow and, indeed, foster students' problem-solving ability and creativity. Since performance tasks can have more than one correct solution, they tap into different learning styles and ways of thinking and assess a student's problem-solving ability and creativity beyond specific content.

Using High-Stakes Test Items in the Classroom

How can teachers incorporate these kinds of assessments into their classroom instruction?

In the classroom, the kinds of assessments described and illustrated above can be used for both formative purposes (to inform instruction) and summative purposes (to evaluate learning at the end of a unit). Most teachers use activities in their day-to-day teaching that can be used for formative assessment purposes. When used for this purpose, the assessment becomes "embedded" in the instruction. These activities/tasks then provide data to be used as feedback to modify teaching and learning activities to better meet the needs of students. These same kinds of tasks can be used to determine how well and to what extent students have acquired knowledge and skills at the end of a unit.

Whether used for formative or summative purposes, assessment tasks such as these can provide a wealth of information about what students understand. Besides giving students an opportunity to become familiar with these kinds of tasks, it gives them the opportunity to apply their knowledge and skills in different situations. Teachers acquire a more complete picture of what their students know and can do.

If carefully crafted, all of the types of assessment mentioned above should require students to use higher-order thinking skills while demonstrating their knowledge and skills. Teachers need to be explicit about the depth of understanding they expect students to develop and match both learning opportunities and assessments experiences to that end.

How can teachers align classroom-based assessment to high-stakes tests?

Classroom assessments align with those found on high-stakes science tests when they require the same level of difficulty and cognitive demand. Our view of alignment starts with an analysis of an assessment task, framed around these questions:

1. *What type of task is this and what kind of response is called for?* Is it a written task that requires an essay or quasi-essay kind of response? Is it a graphic task that requires creation or interpretation of a graphic representation? Is it a performance task that requires a multistep performance to demonstrate understanding?
2. *What does a student have to know and be able to do adequately to respond to this task?*
3. *What is the cognitive demand of the task?* What level of thinking is required to adequately respond to the task—is it sufficient to recall facts, or does the student need to be able to explain, analyze, and/or evaluate?
4. *Have students had the opportunity to learn and practice the kinds of skills required to respond to this task?*

This four-step analysis can be applied to state test items as a way for teachers to understand what students need to be able to do to perform well on them. Teachers can also use this analysis in the design of their own assessment tasks, to bring them into closer alignment with the higher-level thinking demanded by items found on state science tests. Figure 14.6, page 294, from the 1998–99 Kentucky Core Content Test, shows an example of how this analysis can be applied to a state assessment item.

Figure 14.6 Analyzing a State Assessment Item: Fossil

Dinosaur Fossil

Features such as the bone size or structure of an animal's skeleton often help scientists determine the animal's characteristics, which include how it moves, what it eats, or even the predator/prey relationships of the animal. In 1826, scientists in Germany uncovered the fossilized bones of the dinosaur illustrated below.

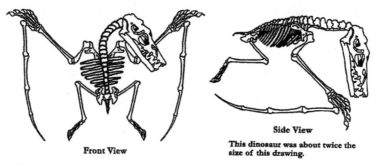

Front View

Side View

This dinosaur was about twice the size of this drawing.

a. List three features of the skeleton of this dinosaur.

b. Explain how each of the features you listed relates to a characteristic that helped the dinosaur live.

Our analysis:
- This is a *constructed-response* task that requires students to interpret and analyze a diagram (graphic).
- To respond to the task, students need to be familiar with the *structure and function of different parts of the skeletal system*.
- They must be able to *analyze* features of the skeletal form and *apply* their understanding of form to the function it provides.
- Finally, they would need to have had some *opportunity to become familiar with the skeletal systems* of various animals.

Figure 14.7 is another example in which students are required to interpret a graphic and construct a response. This example is especially interesting in that it can elicit student misconceptions about which a teacher may be unaware.

Figure 14.7 Analyzing a State Assessment Item: Water Droplets

John left the bowl of ice, pictured to the right, on the kitchen table in July. After a while, John saw water droplets on the outside of the bowl.

1. Where did the water droplets come from? [EMS25A2-1]

2. Why are there water droplets on the outside of the bowl? [EMS25A2-2]

Our analysis:
- This is a *constructed-response task* that uses a graphic to illustrate the situation.
- Students need to know that *air contains water vapor* and that *temperature has an effect on water vapor.*
- This task assesses students' *comprehension* of condensation.
- Students should be familiar with this phenomenon from everyday experience. It is common to see children tasting the liquid that forms on the outside of a drinking glass, testing to see if it is the same liquid that is inside the glass. Despite numerous experiences with this phenomenon, it usually requires a more formal classroom experience to develop an understanding of condensation. Students often think that the liquid inside the glass is seeping through and collecting on the outside.

We've discussed a variety of assessment strategies in this chapter and illustrated the strategies with examples of assessment items from state science tests. Now we'd like to share one teacher's adaptation of a test item from the TIMSS, in which he pushed the level of thinking required of students and asked them to explain their thinking. We think it is an excellent example of how an existing task can be modified to raise the level of cognitive demand. We suggest that teachers start with a released task and then modify or adapt it to meet their purposes.

Figure 14.8, from the 1999 TIMSS eighth-grade released items, shows the original task. It is a simple multiple-choice item. The teacher[3] changed it into a justified multiple-choice item by asking students to explain their answer. Then, he created a constructed-response question, asking students to evaluate the response of a hypothetical student to a related question and to explain, in writing, their reasons for agreeing or disagreeing with said student. Figure 14.9 shows his enhanced two-part question and a student's response to each part.

Figure 14.8 Original TIMSS Item

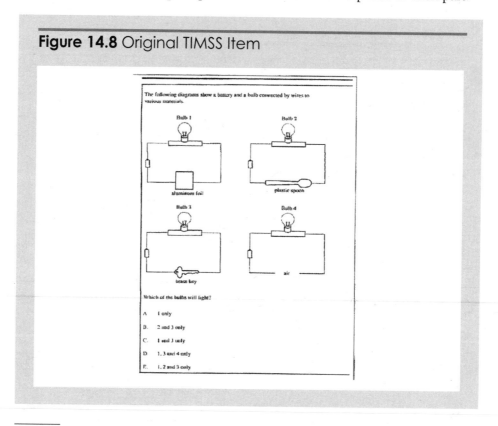

---

[3]This example was created by Richard Matthews from Pittsburgh, Pennsylvania, as part of our online course on science assessment in the middle grades, offered by EDC and facilitated by the authors in the spring of 2005.

Figure 14.9 Enhanced TIMSS Item and Student Responses

Part 1

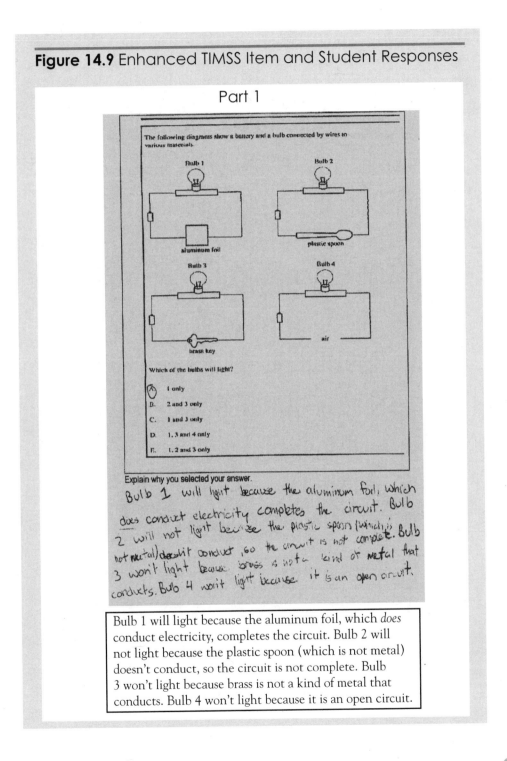

Bulb 1 will light because the aluminum foil, which *does* conduct electricity, completes the circuit. Bulb 2 will not light because the plastic spoon (which is not metal) doesn't conduct, so the circuit is not complete. Bulb 3 won't light because brass is not a kind of metal that conducts. Bulb 4 won't light because it is an open circuit.

Part 2

A student in another science classroom had the following diagram and explanation for a conductivity tester the students were asked to design.

The student wrote:
When I place the material that I want to test across the two black spots on the tester the bulb will light if the material conducts electricity. The bulb will not light if the material does not conduct electricity.

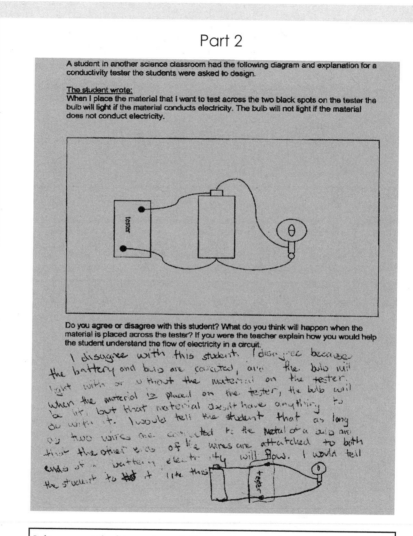

Do you agree or disagree with this student? What do you think will happen when the material is placed across the tester? If you were the teacher explain how you would help the student understand the flow of electricity in a circuit.

I disagree with this student. I disagree because the battery and bulb are connected, and the bulb will light with or without the material on the tester. When the material *is* placed on the tester, the bulb will be lit, but that material doesn't have anything to do with it. I would tell the student that as long as two wires are connected to the metal of a bulb and that the other ends of the wires are attached to both ends of a battery, electricity will flow. I would tell the student that it is like this...

We like the teacher's modification in Figure 14.9, Part 1, because it removes students' ability to take a lucky guess and requires that students articulate in writing their knowledge of the conductivity of certain materials. Figure 14.9, Part 2, requires students to articulate in writing their understanding of circuits and pushes students' thinking to evaluate, predict, and explain.

In this case, the teacher, of course, knew that his students had had opportunities to design and experiment with circuits in the classroom. His purpose, with this modified question, was to determine how much of what students had learned in the classroom had been understood and retained nearly two years after the circuit unit had been completed.

Conclusion

In closing, we do not mean to imply that high-stakes assessments are without fault: Not every item is well constructed; not every item calls for higher-order thinking skills; not every item is free of cultural or gender bias. But there are many models for good assessment, and teachers—with some analytical tools, a little creativity, and the knowledge of what they want to assess—can use these items as a resource to improve their own classroom-based assessments and, in addition, help prepare their students for high-stakes tests.

References

Fellows, N. 1994. A window into thinking: Using student writing to understand conceptual change in science learning. *Journal of Research in Science Teaching* 31(9): 985–1001.

McBrien, J. L., and R. S. Brandt. 1997. *The language of learning: A guide to education terms*. Alexandria, VA: Association for Supervision and Curriculum Development.

CHAPTER 15

Systems for State Science Assessment: Findings of the National Research Council's Committee on Test Design for K–12 Science Achievement

Meryl W. Bertenthal
Indiana University

Mark R. Wilson
University of California, Berkeley

Alexandra Beatty
National Research Council

Thomas E. Keller
National Academy of Sciences[1]

Shortly after the passage of the No Child Left Behind Act of 2000 (NCLB), the National Science Foundation asked the National Research Council (NRC) to form a committee to provide advice and guidance to states and their test development contractors on the design and implementation of quality science assessments to meet the requirements of the law. This chapter summarizes some of the findings and

---

[1]Meryl Bertenthal served as the study director and Mark Wilson was the chair of the committee that authored the report on which this chapter is based. Alexandra Beatty served as a program officer to the committee. Thomas Keller served as the chair of the state science supervisors' panel that advised the committee.

recommendations of the NRC Committee on Test Design for K–12 Science Achievement and is based on the committee's final, book-length report, *Systems for State Science Assessment* (NRC 2006).[2] While the committee's report is targeted at state policy and education decision-makers, many of the findings are of interest to teachers and others involved in the improvement of science education and assessment in their states. In this chapter we try to connect a few of the committee's findings and recommendations to issues of interest to teachers.

There are inherent difficulties in summarizing a whole book in a single chapter. One such difficulty is providing the kind of specific detail and exemplification that the reader might be looking for. Readers who seek additional examples or who are interested in learning more about the suggestions and approaches mentioned here are encouraged to consult the committee's full report, which is available from the National Academies Press.[3]

Many of the points made in the NRC report and in this chapter may apply to educational assessment in other areas. In particular, the principles that framed the committee's thinking about science assessment could guide assessment in other subject areas as well. However, there are aspects of science as a discipline—for example, the abstract nature of many of the concepts that students are expected to learn and the emphasis on scientific inquiry and investigation in many state standards—that present challenges specific to science assessment. Thus, in designing high-quality science assessment, states will need to focus on both the general precepts of sound educational measurement and the features that are unique to science assessment.

No Child Left Behind and Science Assessment

NCLB is an extension of the 1994 reauthorization of the Elementary and Secondary Education Act (ESEA). (ESEA moves beyond that law because it affects all public schools, districts, and students and because it includes science in its requirements for standards and assessments.)

NCLB rests on four pillars: (1) a set of challenging content and achievement standards for student performance; (2) assessments to measure the

---

[2]We are grateful to the committee members and to the advisory groups of science teachers, science education supervisors, and state assessment directors that assisted the committee for their contributions to the report. We acknowledge and thank them for their intellectual contribution to this chapter.

[3]The report can be viewed online at *http://books.nap.edu/catalog/11312.html*

achievement of all students relative to the state standards; (3) flexibility for states and districts to implement instruction and curriculum so that all students can achieve the standards; and (4) accountability measures that move schools toward greater effectiveness in promoting achievement. In this chapter, as in the committee's report, the focus is on the first two pillars: standards and assessment.

Under NCLB, states are required to raise the reading, mathematics, and science achievement level of all students and to narrow the achievement gap in these subjects among students of different backgrounds. To attain these goals, states are required to have a set of challenging standards, expand their standardized testing programs, analyze and report assessment results in specific ways for accountability purposes, and ensure that all students have the opportunity both to be taught by qualified teachers and to learn in well-performing schools.

NCLB puts in place strong accountability measures for schools and districts and imposes sanctions on schools when their students do not make adequate yearly progress (AYP). So that states work diligently toward these goals, there is a rigorous timetable for state compliance with NCLB regulations. Currently, science is not part of AYP formulas, but NCLB requires that results from science assessments be reported publicly along with other information on student achievement.[4] This means that science assessment under NCLB will still carry high stakes for schools and school districts. Individuals and schools will get their own results, and they, as well as other stakeholders, will be able to compare their results with those of other schools. It was the consensus of the NRC committee that teachers will tailor their instruction to the assessments and, therefore, the quality of the assessments will be of paramount importance.

Under NCLB, states must have challenging academic content and achievement standards for science in place by 2005–2006. States must begin measuring student attainment of their standards in 2007–2008 through assessments that are fully aligned with the standards. The assessments must meet accepted professional standards for technical quality for the purposes

---

[4]As this chapter was being completed, several bills were being introduced in Congress to require the inclusion of science assessment results in AYP formulas. The committee that authored the report on which this paper is based suggested that such a step should not be undertaken without careful consideration of both the positive and negative implications of including or excluding science from AYP formulas.

for which they will be used. The committee found that meeting the technical quality and alignment provisions of the law could present serious challenges for states. For example, the committee found that some states may list so many standards that tests cannot measure them all. Further, most state assessments fail to measure adequately the cognitive complexity or "depth of knowledge" described in state standards.

NCLB also specifies that states' assessment systems must incorporate multiple up-to-date measures of student achievement, including measures that assess higher-order thinking skills and understanding of challenging content. Science assessments are to be administered annually to all students, including those with disabilities and those who are not fluent in English, at least once in each of three grade bands, 3–5, 6–9, and 10–12. States are required to make reasonable accommodations for students with disabilities and limited English proficiency to allow them to participate in the assessments, and they must have in place alternate assessments for students who cannot participate in the regular assessment even with accommodations. [For more information about issues related to assessing students with disabilities or English language learners see, for example, *Keeping Score for All* (NRC 2004); *Testing English-Language Learners in U.S. Schools* (NRC 2000); and pages 136–141 of the committee's report (NRC 2006).]

Under NCLB, states' assessment systems may include many different types of assessment strategies and be comprised of a uniform set of assessments statewide or a combination of state and local assessments. Regardless of the form that a state's assessment system takes, the results must be reported publicly and be expressed in terms of its academic achievement standards. The results must be reported in the aggregate for the full group of test takers, be disaggregated for specified population groups, and provide information that is descriptive, interpretive, and diagnostic at the individual level.

Standards

Content standards are considered fundamental to science education and science assessment because they define what is to be taught and what kind of performance is expected. For standards to serve this function well, they must communicate clearly to everyone concerned—teachers, assessment developers, students, parents, and policy makers—what students are expected to know and be able to do. The NRC committee found that standards will communicate best if they are clear, detailed, and complete; reasonable in scope;

rigorous and scientifically correct; and built around a conceptual framework that reflects sound models of student learning (NRC 2006, Chapter 4).

The committee also found that many state content standards and, consequently, the curricula aligned to them contain too many disconnected topics. They found that there is not enough attention paid to how a student's understanding of a topic can be supported and enhanced from grade to grade or across subject disciplines. As a result, topics often appear to receive repeated coverage over multiple years with few opportunities for students to see connections between a topic under study and other related topics, thus giving students an incomplete foundation for further knowledge development.

In general, good state standards should be organized and elaborated in ways that clearly specify what students need to know and be able to do and how their knowledge and skills will develop over time with competent instruction. Science standards should be structured around central principles of science that represent the foundation for the concepts, theories, principles, and explanatory schemas for phenomena in a discipline. These central principles, which are sometimes referred to as the "big ideas" or the enduring understandings (see Wiggins and McTighe 1998) of science, are the foundation for the principles, theories, and explanatory schemes within a discipline. While there is no universally accepted list of "big ideas," and states may have to decide on their individual foci, possible examples might be evolution, Newton's laws, or kinetic molecular theory.

Organizing standards around big ideas is a fundamental shift from the organizational structure that many states use in which standards are grouped under subject areas or discrete topics. The committee found that a potential positive outcome of the reorganization of state standards from discrete topics to big ideas is the opportunity for a shift in emphasis from breadth of coverage to depth of coverage around a relatively small set of foundational principles and concepts. These foundational principles and concepts become the focused target of instruction and can be progressively refined, elaborated, and extended over time (NRC 2006, p. 3).

The committee identified two ideas that emerge from the science education research literature—learning progressions and learning performances—that would be useful for states to adapt in organizing, elaborating, and critiquing their standards in order to make them better able to guide curriculum, instruction, and assessment. These two ideas would also be useful for teachers, curriculum designers, and assessment developers because they provide the specificity and organizational framework that is missing in most standards documents but that is needed for coherence among curriculum, instruction, and assessment.

Learning Progressions

Learning progressions, which can be developed for lessons, units of study, yearlong courses, or an entire K–12 experience, describe in words and examples what it means to develop greater understanding of an idea or set of ideas. They are anchored on one end by students' prior knowledge and on the other by the expectations for learning over time with instruction. Learning progressions propose the intermediate understandings or benchmarks between anchor points that contribute to building a more sophisticated understanding and serve as targets for curriculum, instruction, and assessment.

The committee emphasizes that learning progressions are not developmentally inevitable; more than one path leads to competence. The pathways that individual students or groups of students follow depend on many things, including the knowledge and experience that they bring to the task, the quality and content of the instruction that supports their learning, and the nature of the specific tasks that are part of the experience. Nonetheless, some paths are followed more often than others. Organizing standards to reflect the more typical ways in which deepening understanding develops provides a structure around which states, curriculum developers, and teachers can organize learning. Also, this organization can provide clues about the types of assessment tasks that can be used to shed light on students' achievement at different points along the progression.

Ideally, learning progressions should be based on research about how competence develops in a particular domain. However, for many aspects of science learning, the research literature is incomplete and research findings may have to be supplemented with the judgments of expert teachers and others with knowledge of how students learn science and what is known more generally about cognition and science learning.

What might a learning progression look like? Figure 15.1 shows a learning progression for atomic molecular theory that was developed as an example for the NRC committee (Smith et al. 2006).

The learning progression is organized around two sets of big ideas. The first set (M1, M2) consists of big ideas about (a) the properties of matter and material kinds, (b) their constancies and changes across transformations, and (c) the role of measurement, modeling, and argument in knowing. This first set of big ideas can be introduced in the earliest grades and elaborated throughout schooling at the macroscopic level. For example, students move from describing the properties of material kinds to learning about density as the ratio of weight over volume, and from conservation of material kind and weight during melting to conservation of mass across all phase changes. In middle school and high school, the atomic-molecular theory can be introduced as a second set of big ideas. This second set of ideas (AM1, AM2, AM3, AM4) builds on the first because understanding the atomic-molecular theory depends to a great extent on the macroscopic big ideas studied earlier (e.g., the understanding that matter has weight and occupies space) and, at the same time, it provides deeper explana-

Figure 15.1 Learning Progression for Atomic Molecular Theory

Six Big Ideas of Atomic Molecular Theory That Form Two Major Clusters

M1. *Macroscopic Properties:* We can learn about the objects and materials that constitute the world through measurement, classification, and description according to their properties.

M2. *Macroscopic Conservation:* Matter can be transformed, but not created or destroyed, through physical and chemical processes.

AM1. *Atomic-molecular theory:* All matter that we encounter on Earth is made of less than 100 kinds of atoms, which are commonly bonded together in molecules and networks.

AM2. *Atomic-molecular explanation of materials:* The properties of materials are determined by the nature, arrangement, and motion of the atoms and molecules of which they are made.

AM3. *Atomic-molecular explanation of transformations:* Changes in matter involve both changes and underlying continuities in atoms and molecules.

AM4. Distinguishing data from atomic-molecular explanations: The properties of and changes in atoms and molecules have to be distinguished from the macroscopic properties and phenomena they account for.

tory accounts of macroscopic properties and phenomena. They also enable the further elaboration of macroscopic understandings. (p. 12)

Learning Performances

The committee found that most state science standards describe science content knowledge—and the general cognitive demands relative to that knowledge—without providing operational definitions of what it means to know or understand. Therefore, standards must be elaborated before they can serve as a basis for instruction or assessment. If they are not elaborated clearly, teachers and assessment developers can infer from almost any standard an array of meanings that draw on many different combinations of content knowledge and cognitive demand.

Learning performances are a way to reformulate a scientific content standard in terms of the scientific skills and practices that *use* that content, such as being able to define terms, describe phenomena, use models to explain patterns in data, construct scientific explanations, or test hypotheses (Perkins 1998; Reiser et al. 2003). By defining the most important skills and practices for which the knowledge is used, it is possible to connect the conceptual statements in the content standards with assessment performances in which students can demonstrate their understanding.

Articulated learning performances act as guides for designing assessments and can serve as the link between classroom and large-scale assessments by defining what students who meet the standard would be capable of doing with their knowledge. Assessment tasks could then be developed specifically to elicit evidence that students can use their knowledge as described in the standard. Learning performances can be targeted broadly by state and district tests and at a more fine-grained level in classroom assessments.

Assessment

Assessment is a systematic process for gathering information about student achievement. It provides important information for many different purposes that are important to the education system, including guiding instructional decision-making in the classroom, holding schools accountable for student achievement, and monitoring and evaluating educational programs. It is also the way that teachers, school administrators, and state policy-makers exemplify their goals for student learning. While assessment can do all of these things, it must be designed specifically to serve the particular purpose or purposes for which the results will be used. For example, an assessment that is designed to provide information about student achievement that is diagnostic, descriptive, and interpretive would need to test students' understanding of a few key concepts deeply and thoroughly. On the other hand, an assessment that is used to inform policy-makers about the effectiveness of the overall education system would need to cover broadly all of the topics deemed important by decision-makers. Neither of these strategies could provide results that are valid for the purposes of the other (NRC 2006, Chapter 2).

A System of Assessment

The committee concluded that a single assessment strategy would not, by itself, meet the requirements of NCLB as it could not provide results that would be valid and reliable for all of the purposes identified under the law. The committee therefore recommended that states develop a system of science assessment that collectively would meet the various purposes of NCLB and provide education decision makers with assessment-based information that is appropriate for each specific purpose for which it will be used. The system that each state develops in response to NCLB necessarily will vary according to the state's goals and priorities for science education and its uses for assessment information. For example, a state might choose to develop a single test in which students take a common core assessment that provides individual results, along with an assessment with a matrix-sampling design that provides information about the achievement of groups of students across a broad content domain. Or a state might choose to combine standardized classroom assessments that provide diagnostic, descriptive, and interpretive information with an external assessment that shows the progress that all students are making toward achieving state standards or for program evaluation purposes. Or a state may decide to give up a statewide test and instead use one of many different models in which results from local, district, or state assessments are combined, aggregated, and reported for specific purposes.[5] When multiple assessment strategies are used, they should be designed from the beginning to function as part of a coherent system of assessment.

[5]Examples of these are summarized in the committee's report (pp. 31–37) and are illustrated in detail in papers commissioned by the committee. These papers are available online at: www7.nationalacademies.org/bota/Test_Design_K-12_Science.html

A coherent system of standards-based science assessment is horizontally coherent: curriculum, instruction, and assessment are all aligned with the standards; target the same goals for learning; and work together to support a student's developing science knowledge, understandings, and skills. It is vertically coherent: All levels of the education system—classroom, school, school district, and state—are based on a shared vision of the goals for science education, of the purposes and uses of assessment, and of what constitutes competent performance. The system is also developmentally coherent: It takes into account how students' science understanding develops over time and the scientific content knowledge, abilities, and understanding that are needed for learning to progress at each stage of the process (Wilson 2005).

If a collection of tests is not coherent, the information that is produced can yield conflicting or incomplete information and send confusing messages about student achievement that are difficult to untangle. If discrepancies in achievement are evident, it is hard to determine whether the tests in question are measuring different aspects of student achievement—and are useful as different indicators of student learning—or whether the discrepancy is an artifact of assessment procedures that were not designed to work coherently together. Gaps in the information provided by the assessment system can lead to inaccurate assumptions about the quality of student learning or the effectiveness of schools and teachers and can bring about interventions that may not be necessary. In Figure 15.2 we list some of the characteristics of a high-quality, coherent assessment system. But what would a coherent system look like? Table 15.1, page 312, is a framework for an assessment system adapted from the one developed by the state of Maine in 2003 (NRC 2003). We include it here to illustrate how multiple assessments can be incorporated coherently into a single assessment system.[6]

---

[6]The system has since been modified and so the description included here does not describe the current (2007) Maine assessment system.

Figure 15.2 Characteristics of a High-Quality Science Assessment System

The following are characteristics of an assessment system that could provide valid and reliable information to the multiple levels of the education system and could support the ongoing development of students' science understanding:

1. incorporates assessments that are closely aligned to the standards that guide the system and is structured so that all elements are coherent with the goals, curriculum materials, and instructional strategies of the science education system of which it is a part;
2. includes a range of measurement approaches and multiple measures of achievement that provide a variety of evidence to support educational decision-making at different levels of the system;
3. contains measures that assess student progress over time rather than relying solely on one-time, large-scale testing opportunities;
4. is useful in the sense that the assessment results are made accessible and are reported in a timely manner to those who need them;
5. fits into a larger education system that provides the necessary resources for the development, operation, and continued improvement of both the assessment system and the education system when assessment results indicate improvement is necessary;
6. provides systematic, ongoing professional development for teachers and others on current science assessment practices, the uses and limitations of assessment results, and processes for developing and using sound assessments; and
7. is systemically valid—that is, it promotes in the education system desired curricular and instructional changes that result in increased learning and not just improvement in test scores.

Source: National Research Council (NRC). 2006. *Assessment in support of instruction and learning: Bridging the gap between large-scale and classroom assessment.* Washington, DC: National Academy Press, p. 28.

Table 15.1 Framework for an Assessment System

	Primary purpose of the assessment	Who selects or develops the assessment?	Who scores the assessment?
Classroom Assessment	Informing teaching and learning	Individual teacher	Individual teacher
School or District Assessment	Informing and monitoring	Groups of teachers and administrators	Groups of teachers (and others)
State Assessment	Monitoring and evaluating programs to ensure accountability	Groups of administrators, administrators, and/or policy-makers	Scorers outside the district
Assessment System	Informing teaching, monitoring and evaluating, certification	District assessment leadership	Both internal and external

The assessment system in Table 15.1 was designed to meet three principal goals: (1) provide high-quality information about student performance to inform teaching and learning, (2) monitor schools and administrative units and hold them accountable for their success at making sure students meet the state standards, and (3) certify that students have met the state standards.

A Developmental Approach to Assessment
In keeping with the committee's conclusion that science education and assessment should be based on a foundation of how a student's understanding of science develops over time with competent instruction, the committee called for a developmental approach to science assessment (see Masters and Forster 1996, 1997). A developmental approach means gathering evidence of the development of students' understanding over time, as opposed

to only at a specific point in time. This approach recognizes that science learning is not simply a process of acquiring more knowledge and skills, but rather a process of progressing toward greater levels of competence and understanding as new knowledge is linked to existing knowledge and as new understandings build on and replace earlier conceptions.

A hallmark of a developmental approach is the use of multiple assessments that collectively provide information about students' learning over time. Students' performances on these multiple assessments are compared against a pre-constructed progress map or learning progression that describes what students are expected to learn and how that learning is expected to unfold as the student progresses through the instructional material. Records are maintained about performance on each assessment and are combined and viewed as a whole to estimate students' level of achievement in a particular area of learning. The estimate of a student's current location on the progress map or learning progression serves as a guide to the kinds of learning experiences that would help them develop the knowledge, understanding, and skills necessary to progress.

Professional Development and Assessment

Regardless of the information it is designed to collect, assessment, on its own, cannot improve student learning—it is the way in which assessment and the results are used that can accomplish that goal. For an assessment process to function as it should, everyone who uses assessment results needs to be assessment literate. That is, they need to have a clear understanding of the purposes and limitations of assessment, goals and purposes for assessments, the ways in which different assessments function, and how to interpret and use assessment results appropriately. Those who need to develop their assessment literacy includes everyone from elected officials at the highest levels to school board members to parents to students to teachers. The teachers need to develop their knowledge of assessment the most because they are in the best position to use the results to improve student learning.

Although teachers may not be involved in designing their state's tests, they should understand the kinds of data that are produced by these tests and how the results can and will be used. Further they should have sufficient understanding of the technical properties of the assessments to put the results in context and to link them to other pieces of information they have about their students.

Assessment activities require that teachers have, in addition to a deep understanding of the content domains they are teaching, knowledge about how to develop tasks that are valid and useful in the classroom and how to use the results of these assessments in instructionally supportive ways. They must also understand the uses and limitations of external assessment, such as large-scale statewide tests, and be cognizant of the ways in which such assessment affects their teaching. Figure 15.3 is a list of some of the things that assessment-literate teachers should know and be able to do.

Figure 15.3 Assessment Competencies for Teachers

Teachers should be able to

1. choose assessment methods appropriate for instructional decisions,
2. develop assessment methods appropriate for instructional decisions,
3. administer, score, and interpret the results of both externally produced and teacher-produced assessment methods,
4. use assessment results when making decisions about individual students, planning teaching, developing curriculum, and school improvement,
5. develop valid pupil grading procedures that use pupil assessments,
6. communicate assessment results to students, parents, other lay audiences, and other educators, and
7. recognize unethical, illegal, and otherwise inappropriate assessment methods and uses of assessment information.

Source: American Federation of Teachers, National Council on Measurement in Education, and National Education Association. Standards for teacher competence in educational assessment of students. Available at www.lib.muohio.edu/edpsych/stevens_stand.pdf

The committee found that inservice teachers need to have the opportunity, and take the responsibility, to develop their assessment literacy. Taking classes, participating in professional development opportunities, and keeping up with issues surrounding assessment are some of the ways that teachers can develop their own assessment literacy. By organizing and participating in opportunities to discuss student work that results from assessment, teachers can develop a common understanding of what competence in a domain looks like and how to support student progress toward achieving it.

Conclusions

Designing high-quality science assessment is an important but difficult goal to achieve. Science assessment must target the knowledge, skills, and habits of mind that are necessary for science literacy, and it must reflect current scientific knowledge and understanding in ways that are scientifically accurate and consistent with the ways in which scientists understand the world. It must assess understanding of science as a content domain and an understanding of science as a specific way of thinking. It must also provide evidence that students can apply their knowledge appropriately and that they are building on their existing knowledge and skills in ways that will lead to more complete understanding of the key principles and big ideas of science. Adding to the challenge, competence in science is multifaceted and does not follow a singular path. Science assessment must address these complexities while also meeting professional technical standards for reliability, validity, and fairness for the purposes for which the results will be used.

The committee concluded that the goal of developing high-quality science assessment will only be achieved though the combined efforts of scientists, science educators, developmental and cognitive psychologists, experts on learning, law and policy makers, and educational measurement specialists working collaboratively rather than separately.

References

American Association for the Advancement of Science (AAAS). 1993. *Benchmarks for science literacy*. New York: Oxford University Press.

American Education Research Association (AERA). 2003. *Research points: Standards and tests, keeping them aligned*, L. B. Resnick, ed. Spring (1,1). Washington, DC: AERA.

Masters, G., and M. Forster. (1996, 1997) *Developmental assessment: Assessment resource kit*. Hawthorne, Australia: Australian Council on Educational Research Press.

National Research Council (NRC). 1996. *National science education standards*. National Committee on Science Education Standards and Assessment. Center for Science,

Mathematics, and Engineering Education. Washington, DC: National Academy Press.

National Research Council (NRC). 2000. *Testing English-language learners: Report and workshop summary*. Board on Testing and Assessment, Committee on Educational Excellence and Equity. Washington, DC: National Academy Press.

National Research Council (NRC). 2003. *Assessment in support of instruction and learning: Bridging the gap between large-scale and classroom assessment*. Board on Testing and Assessment, Committee on Science Education K–12, and the Mathematical Sciences Education Board. Washington DC: National Academy Press.

National Research Council (NRC). 2004. *Keeping score for all: The effects of inclusion and accommodation policies on large-scale educational assessments*. Committee on Participation of English Language Learners and Students with Disabilities in NAEP and Other Large-Scale Assessments. Washington, DC: National Academies Press.

National Research Council (NRC). 2006. *Systems for state science assessment*. Committee on Test Design for K–12 Science Achievement. M. R. Wilson and M. W. Bertenthal, eds. Board on Testing and Assessment, Center for Education, Division of Behavioral and Social Sciences and Education. Washington, DC: National Academies Press.

Perkins, D. 1998. What is understanding? In M. S. Wiske (Ed.), *Teaching for understanding: Linking research with practice*. San Francisco: Jossey-Bass.

Reiser, B. J., J. Krajcik, E. Moje, and R. Marx. 2003. Design strategies for developing science instructional materials. Paper presented at the National Association for Research in Science Teaching meeting, Philadelphia (March).

Smith, C., M. Wiser, A. Anderson, and J. Krajcik. 2006. (Focus Article of combined double issue of journal): Implications of research on children's learning for standards and assessment: A proposed learning progression for matter and atomic-molecular theory. *Measurement* 14 (1&2): 1–98.

Wiggins, G. P., and J. McTighe. 1998. *Understanding by design*. Alexandria, VA: Association for Supervision and Curriculum Development.

Wilson, M. 2005. *Constructing measures: An item response modeling approach*. Mahwah, NJ: Lawrence Erlbaum.

## CHAPTER 16

From Reading to Science: Assessment That Supports and Describes Student Achievement

Peter Afflerbach
University of Maryland

This chapter is based on a presentation made to the National Science Teachers Association's conference on assessment, "Science Assessment: Research and Practical Approaches for Grades 3–12 Teachers and School and District Administrators," held in November 2005. In the presentation I addressed two questions:

1. How might assessment practice from other disciplines inform high-quality science assessment?
2. What are the challenges and accomplishments of student assessment in the different content domains?

I structure the chapter in relation to these questions, focusing first on recent advances in our understanding of assessment and learning. We know much about how students learn science and how teachers effectively teach science (NRC 2005). We also know much about the materials and procedures of effective assessment (Pellegrino, Chudowsky, and Glaser 2001). Thus, we have considerable knowledge about assessment that can be used in the service of high-quality teaching and student achievement.

The second section of this chapter focuses on the challenges related to effective science assessment. I propose that there is a series of imbalances that is caused largely by high-stakes testing and that prevents teachers and students from using appropriate and productive science assessments. I then describe specific imbalances in assessment and the means by which teachers and schools can work toward creating balance. Throughout, I stress that our attention to the context in which assessment is conceptualized and conducted—and in which teachers and students, curriculum and instruction, and tradition and politics interact—is of paramount importance.

As should be quickly apparent, I am not a science educator, nor am I a researcher of science learning. I bring to this chapter experiences with reading assessment that include membership on the National Assessment of Educational Progress (NAEP) Standing Reading Committee and the NAEP 2009 Reading Framework Committee, research and writing related to high-stakes testing (Afflerbach 2002a, 2005), and a recent book on reading assessment (Afflerbach 2007). My two daughters and their ongoing school experiences are a rich source of information on the nature of assessment in their fifth- and seventh-grade classrooms. Finally, I bring my experience as a reading teacher in elementary and middle schools. I believe that these experiences with reading assessment have counterparts in assessing student learning and achievement in science. The science education community is about to experience the increased use of single, annual test scores as proxies for student achievement and teacher and school accountability, and it must be vigilant about mitigating the effects of this testing. Sadly, we in reading have considerable experience with this!

The Importance of High-Quality, Classroom-Based Assessments

Education decision-making is informed by effective assessment. For teachers, a priority is assessment that can be used to shape instruction and learning. Consider the students who populate our classrooms. In every classroom we expect that students will vary in terms of their knowledge, skills, and strategies. They will vary in their prior knowledge for particular content domains, their familiarity with processes of inquiry, and their epistemologies related to specific content domain knowledge, including science. As well, there are individual differences in students' motivations, volition, and self-esteem. These differences contribute to varied academic performances and achievements. Talented teachers use assessment to continually update their understandings of each of their students and to inform instruction.

CHAPTER
16

We teach in an era in which accountability is measured once a year with a high-stakes test. We must remember that accountability is achieved through high-quality assessment that is conducted daily in the classroom. Assessment should provide teachers with useful information about their students' development. From my perspective, Vygotsky's (1978) description of the zone of proximal development (Figure 16.1), the place in which students learn new things in relation to their existing knowledge and competencies, figures largely here. Specifically, the zone of proximal development is a theoretical space bounded on one side by a student's current level of learning and achievement and on another side by the next, expected level. For example, consider the student who is developing an understanding of a prediction strategy in reading. The student must access and use available and appropriate prior knowledge related to the topic of the text. The student must take cues from the text that hint at the content and point of the text, combine them with the prior knowledge, and produce a prediction. The student must then monitor the prediction for its accuracy, employing related metacognitive strategies. Our understanding

Figure 16.1 Assessment and the Zone of Proximal Development

Student's next level of competency and achievement:
Developing more complex prediction strategies
and using prediction strategies when reading more complex texts

Assessment and teaching

Zone of proximal development

Teaching and assessment

Student's current level of competency and achievement:
Knowledge of how and when and where to use prediction strategies

that students have called on the prediction strategy and used it appropriately results from timely assessment, which might include our observation of the student reading, teacher questions, and the request for a student think-aloud while reading. With this knowledge, we may plan a next lesson that focuses on further use of the prediction strategy, metacognition, or the reading of more complex texts. Or, careful assessment may reveal that the student needs further work on developing and successfully using prediction strategies. The zone of proximal development, or the space between the known and the new, is identified by our careful assessment.

If you believe, as I do, that teacher accountability is related to identifying students' zones of proximal development and teaching in these zones, then the centrality of classroom-based assessment is evident. Regular classroom-based assessments help us identify teachable moments for each student: They can provide the detail we need to effectively teach to individual students' needs and ascertain that immediate and long-term learning goals are met. We must know where students are—be it in terms of their reading skill and strategy development, science learning, motivation and engagement, or prior knowledge for science content—to be best positioned to help them. A robust classroom assessment program continually provides detailed information about students' current competencies and next steps: It informs our ongoing teaching in students' zones of proximal development. Throughout single lessons and on a daily basis, over marking periods and across the school year, classroom-based assessment *is* high-stakes assessment, for without it there is no progress to daily, weekly, and annual learning goals. How might assessment practice from other disciplines inform high-quality science assessment?

When we can describe in detail what it is we plan to assess, we are then in a good position to create the assessment materials and procedures that will richly describe and evaluate students' development (Pellegrino, Chudowsky, and Glaser 2001). Lessons from assessment in different disciplines suggest that thoughtful assessment uses our collected wisdom about the constructs to be measured, with the result that the inferences we make from assessment information are accurate and useful. An example would be the learning expected of students as they progress through a unit of study on the environment, reading several different texts that present divergent views on global warming. We know that students' reading and understanding of science phenomena represents important learning and reflects successful instruction (NRC 2005). We can design assessments that allow us to measure students' ability to read disparate accounts of why the Earth is warming, to take notes that help students identify authors' claims and related evidence (if any), and to make critical judgments about which articles or editorials appear most trustworthy, based in part on the ability to determine how well the author provides evidence to support claims made.

As we carefully conceptualize what it means to read and learn about global warming *and* how to assess students' learning, we can make accurate inferences about student achievement. We can examine students' ability to establish a literal understanding of texts, we can determine students' facility for identifying an author's main points or thesis, and for identifying text that supports the main points. Assessment must honor the nature of learning, with the complexity of assessment referencing the complexity of learning. We infer learning from student performance on these assessments. With confidence in our conceptualization of reading critically in science and confidence in our attempt to assess students' critical reading, we have confidence in our inferences drawn from assessment results. The results of our assessment can be used in a formative manner, to immediately shape our understanding of student development and learning and to plan instruction. We can also use the results in a summative manner, as they provide evidence that the student has (or has not) met a key learning goal.

To summarize this section: We know a lot about how students learn and develop. We also know a lot about how different assessments can provide rich measures of student learning and can reflect teacher expertise and school accountability. We must be relentless in using research findings to revise and update our understanding of effective teaching and student learning and the effective assessment of such teaching and learning.

The Influence of High-Stakes Tests and the Context of Imbalance in Assessment

Assessment must reflect a series of balances that produces information that is useful to different audiences, for their different purposes (Crooks 1988). A first-order need is useful assessment for students and teachers, and it is within classrooms that the promise of assessment must be realized. Never before have we had such detailed understanding of student learning and its development, and never before have we possessed as many potentially valid assessment options.

Although theoretical and practical knowledge of learning and assessment is rich, the implementation of useful and effective assessment may be impoverished. Our considerable knowledge may be compromised in a politically charged, often contentious environment in which school resources are used to meet the needs of audiences other than teachers and students. High-stakes tests are used to meet the legal requirement of annual assessment under the No Child Left Behind law, but these tests are "thin" (Davis 1998) when it comes to taking full measure of student achievement. A result is that the depth and breadth of learning that characterizes high-quality teaching and learning is not reflected in assessment practice.

In addition to mandated high-stakes tests, assessment habits in U.S. schools are informed by tradition, rather than by current conceptualizations of student learning and effective teaching. As an example, I am troubled that "adequate yearly progress" in reading in my daughter's elementary classroom is measured by a single test that is much like the one I took in fourth grade more than 40 years ago. Consider the two multiple-choice reading test items presented in Figure 16.2.

Figure 16.2 Sample Multiple-Choice Reading Test Items, 2006 and 1996

Example A

Read "Chinese Almond Cookies" and answer the following question: Which word could be used instead of the word creamy in Step 2?

1. lighter
2. smooth
3. solid
4. tasty

Source: Maryland State Assessment 2006. Retrieved January 24, 2007 from: www.mdk12. org/assessments/k_8/sample_grade4_reading.html

(continued)

Figure 16.2 *(continued)*

Example B

Read "The Final Report" and answer the following question:
What do you think Jamie did next?

1. hid his final project
2. sneaked away
3. scolded his mother
4. showed his final project to his parents

Source: Paraphrased from the 1966 Nelson-Denny Reading Test.

Example A is from the 2006 Maryland Student Assessment, and Example B is a paraphrased item from a test published 40 years earlier: the 1966 Nelson-Denny Reading Test.[1] While these are single items from tests that contain many items, my point here is that we know a considerable amount about the constructive nature of reading and the importance of asking students to use what they comprehend from reading in important tasks. Current high-stakes tests just don't help us with this: The advance of our knowledge of reading and assessment in the past 40 years is not reflected in the highly consequential tests that are used throughout the United States.

I recommend that educators examine (if they can) the types of items found on high-stakes tests of learning. If these items reflect most (or all) of what is currently known about how students learn and the valuable outcomes of successful teaching and learning, then the argument that tests don't represent our most current knowledge of any particular content domain and that they cannot tell us about the complexity of students' learning is moot. I am convinced that the current regimen of high-stakes tests misses important aspects of learning, cannot measure the complexity of

---

[1]Tests are secure and their contents are copyrighted. Thus, it is difficult to gain access to tests and to have public discussions about their contents. When tests are current (i.e., still in use), reproducing an item could make it available to some test takers, thereby diminishing the reliability of the item and test. It is also the case that currently used tests often defy scrutiny, and analysis of the entire array of items that make up a test (and their related benchmark performances) is next to impossible.

learning, and cannot provide a measure of how students apply what they learn. This is especially unfortunate because much school capital is invested in the practice of using single test scores to make highly consequential decisions about students and teachers, despite the fact that such practice is indefensible (AERA/APA 1999). Using single test scores to judge students' achievement and teachers' accountability also skews schools' assessment agendas and funding (Afflerbach 2005). The purchase of and other costs related to high-stakes tests each take from limited school resources and may create a poverty of assessment alternatives and the means to pursue them.

Given this situation of mandated testing, it is not surprising that the assessment of learning in American schools is marked by imbalance. In reading, a significant portion of this imbalance is attributable to the supreme attention given high-stakes testing and a resultant lack of attention to classroom-based assessments that can inform teaching and learning (Afflerbach 2002a, 2005). Working to correct these imbalances can contribute to superior teaching and learning, while ignoring them may diminish teachers' and students' achievements. The most pressing challenges in assessment relate to a lack of balance in

- meeting the needs of different audiences and purposes of assessment,
- assessments that focus on processes and products,
- formative assessment and summative assessment,
- the assessment that is done to or for students and assessment that is done with and by students,
- the assessment of cognitive and affective factors related to learning,
- the assessment of knowledge and how students use this knowledge, and
- the demands for teacher and school accountability and professional development opportunities that help teachers develop expertise in assessment.

Addressing Specific Imbalances in Assessment

Meeting the Needs of Different Audiences and Different Purposes of Assessment

Assessment is influenced by the constant but sometimes overlooked dynamic of competing audiences and purposes for assessment. Table 16.1 provides an overview of the different audiences and purposes of assessment (Afflerbach 2007). Each legitimate audience and purpose should be served by a well-planned, democratic assessment agenda. Yet, in many schools the needs of particular audiences and their specific uses for assessment are left wanting, while the needs of other groups are clearly met. Consider that as schools work to demonstrate adequate yearly progress they must attend, at a minimum, to the purchase, administration, scoring, and reporting of high-stakes tests. The opportunity cost here is substantial: The school monies, time, and effort given to this enterprise yield single test scores that may meet the needs of a powerful hierarchy of audiences outside of the classroom. Yet, test scores provide little or no information that is useful to the teacher to determine how well lesson and unit learning goals are being met or how to address students' individual differences in using reading strategies, in learning, and in science.

Table 16.1 Representative Audiences and Purposes for Assessment

Audience for Assessment	Purpose for Assessment
Students	• To report on learning and communicate progress • To motivate and encourage • To teach children about assessment and how to assess their own learning • To build student independence
Teachers	• To determine nature of student learning • To inform instruction • To evaluate students and construct grades • To diagnose student strengths and weaknesses
Parents	• To inform parents of their children's achievement • To help connect home and school efforts to support student learning
School Administrators	• To determine science program effectiveness • To prove school and teacher accountability
Politicians	• To establish accountability of schools • To inform the public of school progress
Taxpayers	• To demonstrate that tax dollars are well spent

Source: Adapted from Afflerbach, P. 2007. *Understanding and using reading assessment, K–12.* Newark, DE: International Reading Association. Reprinted with permission.

There is a clear need for balancing and coordinating assessment so that it provides needed information to particular audiences. This coordination should go beyond a simple check to ascertain that each group gets some assessment information: It should be aimed at making assessment efficient. Assessments that provide valuable information to more than one constituency deserve our attention. Consider classroom teachers and legislators, two audiences of assessment that are rarely conceived of as wanting the same sorts of information. Yet, certain assessments may provide valuable information to both groups. Performance assessments are particularly promising. They can provide detailed summative and normative information about student achievement in relation to learning outcomes because they give teachers and students formative information during the steps leading up to these outcomes. In addition, performance assessment programs can be the centerpiece of assessment systems that are useful to all school stakeholders (Valencia and Place 1994).

Balancing Assessments That Focus on Processes and Products

Using information from assessment, we make inferences about students' growth and achievement. We hypothesize about the extent of students' development. In general, process assessments focus on the nature of students' skills, strategies, and learning as lessons and learning tasks unfold. We ask a student to read instructions on how to prepare a suspended solution and we observe the student combining ingredients, mixing the solution and taking notes. At each point in our observation, we have the opportunity to examine the learning process (here combining reading and science) and to determine the appropriateness and accuracy of the student work and perhaps the need for intervention and encouragement. Our process assessment helps us determine the skills and strategies that work or do not work and the learning as it occurs.

In contrast, product assessments focus on the results of learning and performance and provide an after-the-fact account of student achievement. The information provided by product assessments can help us determine students' achievement in relation to important learning goals, ranging from benchmark reading performances to specific content area learning. Typical product assessments are quizzes, tests, and questions related to students' understanding of a lesson or unit. We can make inferences about students' achievement in relation to benchmarks, curriculum standards, and other students when we examine test scores. However, we must make large, backward inferences about the nature of student work during the process. If we want to know how our instruction contributed (or didn't contribute) to students' achievement, the inferences we make from product assessment may be built on a soft foundation. This is an important fact about product assessments: They are relatively limited in their ability to provide detail on what students can and can't do as they learn science. An apt analogy is one in which we try to determine why a soccer team won or lost a game by examining the final score. Of course the final score is important, but it tells us nothing of the means by which it was achieved. Product assessments provide little information about the learning process, thus restricting the opportunity to influence instruction to meet students' current needs.

Balancing Formative Assessment and Summative Assessment

Summative assessments are used to make highly consequential decisions. School success or failure, sanction or reward, is often determined by single summative assessment scores. The pressure to focus on such summative assessment creates an imbalance with formative assessment efforts, the very type of assessment that could help teachers and schools demonstrate accountability on a daily basis. In contrast to summative assessment, formative assessment is conducted with the goal of informing our instruction and improving student learning. At the heart of effective instruction is the classroom teacher's detailed knowledge of each student. This knowledge is constructed through ongoing, formative assessments, conducted during the school day and the school year.

Formative assessment, such as teacher questioning, may be tailored so that it provides immediate and useful information. Asking questions during instruction helps the teacher develop a detailed sense of how well students are "getting" the lesson. Teachers' questions can focus on both skills and strategies, the cognitive growth toward lesson goals, and the affective outcomes of learning. The teacher uses students' responses to questions to gauge how students are progressing toward lesson goals and to inform the ongoing instructional focus. These questions are used to gather information about student learning and thinking at different levels of complexity; they also describe how students use the knowledge they gain from reading. Consider the range of questions that a teacher can ask as students read a chapter about weather in a science textbook: "What are the characteristics of cumulonimbus clouds?" "What types of clouds are associated with different weather systems and different altitudes?" "How do clouds form?" "Can you explain your reasoning?" "Where do you get the information contained in your explanation?" "How can your ability to identify clouds influence your planning for outdoor activities?" "Does the author provide evidence that supports the claims made in the text?"

Questions such as these, based on Bloom's taxonomy of educational objectives (1956), evoke varied student responses that demonstrate the nature of their understanding. From students' responses, the teacher constructs an understanding of students' achievement. And from this understanding comes action: a decision to move ahead or to re-teach, a decision to slow the pace of instruction or speed it up, a decision, perhaps, to have more class discussion around the key concept of clouds. Formative assessment is conducted *in situ*, as the process of teaching and learning unfolds. Creating balance will result in formative assessment describing students' ongoing growth as it occurs and summative assessment providing summary statements about students' achievement.

Balancing the Assessment of Knowledge, Skills, and Strategies With the Assessment of How Students Use This Knowledge

Throughout our lives, and sometimes in classrooms, we use what we have learned to accomplish things. Too often, the goal of classroom instruction and assessment appears focused on what we might call learning-for-learning's-sake: Students are asked to learn content and the endgame of assessment is to see if they can give this content back to us. In reading, the classic example is reading text and responding to literal and inferential comprehension questions. Although we certainly need to determine whether students have comprehended the texts we ask them to read in class, this should not be the sole focus of our assessment. We need to know what students can do with what they learn, whether it involves the application of scientific method, generalization from single instances to trends, and the like. Indeed, the National Assessment of Educational Progress 2009 Reading Framework is based on the following conceptualization of reading:

CHAPTER
16

Reading is an active and complex process that involves
- *Understanding written text,*
- *Developing and interpreting meaning, and*
- *Using meaning as appropriate to type of text, purpose and situation* (NCES 2005, p. 2)

It is the "using meaning" concept that is most radical (at least for tests), and this idea should be used to analyze our contemporary assessments. In science classrooms, students should learn content and then use the information they learn from science inquiry to perform meaningful, related tasks. Many current assessments focus on the former: The bulk of assessment seeks to describe the student's learning of science content. We can assess students' ability to determine classes of elements on the periodic table, and we can ask students to locate and identify the parts of a cell. Each of these assessments focuses on understanding science content as the final goal of science learning. Such a perspective is important, but it does not fully represent what students might do with what they learn.

We must remember that learning science to answer literal and inferential understanding questions, while common school practice, is less common outside of school and a fair distance from students being able to ask their own questions. Our assessment should also focus on students using what they understand. When students read guidelines for conducting hands-on experiments to help guide their science inquiry, reading involves these two goals: to learn science content (in this case procedural knowledge of scientific inquiry) and to use what is learned in a related task or performance. This is authentic assessment.

When students are assessed only to determine what they have learned and remembered in science class, we are but halfway to our goal. We must complement this type of assessment with information about how students use what is learned. Performance assessment, a form of authentic assessment, can be focused on the science learning that students do and the types of things we expect them to do with knowledge gained from learning (Baxter and Glaser 1998). For example, fifth-grade students read instructions and guidelines for conducting a hands-on science experiment. Of course, we focus on their comprehension and resultant learning of how to do things, but we are also very interested in their application of what is learned in conducting the science experiment, which can include identification and use of laboratory tools and following procedural steps accurately.

Balancing the Assessment of Cognitive and Affective Learning Outcomes and Characteristics

Existing assessments in areas outside of science can inform the effort to assess students' growth and accomplishment in science beyond the cognitive. There are assessments that focus on students' motivations for reading and valuing reading (Motivation to Read Profile; Gambrell et al. 1996), students' self-concepts as readers (Reading Self-Concept Scale; Chapman and Tunmer 1995), and students' attitudes toward reading (Elementary Reading Attitude Survey; McKenna and Kear 1990). These assessments can provide ideas for developing similar assessments within science. Assessment that charts science learning in relation to students' levels of motivation, their sense of agency, their self-esteem, and their understanding of the uses and value of science will provide more full measure of science instruction success. Particular motivations and self-concepts can support science learning, and they can be outcomes of science learning.

The vast majority of classroom assessments focus on the measurement of students' cognitive development. In reading, as in science, most school and standards-based goals are set in cognitive frames. Unfortunately, there is little or no attention paid, from the assessment perspective, to the factors that can support and enhance learning and development. Experienced classroom teachers and parents know that having content domain knowledge and related skills and strategies is essential to students' success but does not guarantee this success. Successful students are engaged (Guthrie and Wigfield 1997): They identify themselves as successful learners, they persevere in the face of learning challenges, and they consider learning to be an important part of their daily lives.

When we think of our teaching successes, do we think only of students who scored well on tests? Or do we also think of students who went from being reluctant learners to enthusiastic learners? Do we think of students who evolved from easily discouraged learners to learners whose motivation helped them persevere through challenges? Do we remember students who avoided engagement in all aspects of the classroom routine evolving into students who loved learning? Certainly, we can count such students and our positive influence on them among our most worthy teaching accomplishments.

If we are serious about accountability, we need to have balance in the assessment that demonstrates that high-quality teaching and effective instructional programs change students' lives. To achieve balance we need assessments that are capable of measuring and describing student growth that is complementary to the acquisition of reading and science knowledge. This growth may involve self-esteem, motivation, and perseverance in the face of difficulty. Measuring this growth can move us toward a richer understanding of the accomplishments of students and their teachers. As well, our determination that students lack motivation and self-esteem for learning is valuable information. Achieving balance would result in more of these assessments being built into the routines of classrooms and schools and attention and respect being given to the information that these assessments provide.

The goal of balancing the assessment of students' cognitive and affective development is certainly attainable. For example, current funding from the National Science Foundation supports research that examines the assessment of student motivation in science and math (MSP-Motivation Assessment Program 2006). The development of such measures can help us gauge student science achievement in relation to motivation to learn, as well as the interactions of students' science performance with motivation. Such assessments will help tell a more complete story of the effects of science instruction (Stefanou and Parkes 2003). The importance of determining students' motivation in relation to science learning and achievement is also reflected in aspects of the National Assessment of Educational Progress (2004), the "Nation's Report Card." For example, a subgroup of all students taking the science NAEP responds to items that probe their beliefs and feelings about science, including the statements "I like science," "I am good at science," "Science is boring," and "Science is good for solving everyday problems." Such assessment allows us to examine complementary aspects of students' learning and take a fuller measure of students' growth and accomplishments.

Balancing the Assessment That Is Done to or for Students With Assessment That Is Done With and by Students

A majority of students move through school with assessment done to them or for them. Students remain outside the culture of assessment. A result is that assessment appears to be a "black box" (Black and Wiliam 1998). Typically, students learn course content and then take a quiz or test that is graded and returned to the student. The student earns a score, but gains little or no understanding of how assessment works. Students do not learn to do assessment for themselves. Even with the questions we ask in class, without our explanation of why we ask these questions or how we arrive at our evaluations of student responses, students will not understand how the evaluation of their learning is done. There may be lost opportunities for students to learn to conduct assessment on their own. Our classroom-based assessment must provide students with the means to eventually assume responsibility for their own learning at the same time that it provides valuable information that can inform our understanding of students and our instruction.

To create balance, we can provide opportunities in which students learn both the value of self-assessment and the means to conduct assessment for themselves. Thus, as we use scaffolded instruction to help our students learn the details of scientific inquiry, we must focus on a concurrent and complementary goal: that of modeling and explaining assessment so that students can learn what assessment is, why and when we use it, and how valuable it is. This can be a long and challenging process, so a good start is modeling straightforward assessment routines and helping students learn to begin and successfully complete the routines independently. Consider a checklist (below) for use by fourth-grade teachers (the checklist combines elements I have created and have seen used in different fourth-grade classrooms). This checklist may be used as a common instructional tool in reading class, and it can be used in anticipation of its value in science. As students read their science textbook, the teacher regularly asks the students to refer to the checklist and engage in the assessment thinking that it requires. The teacher models using the checklist and expects that the students will learn to use it as they read independently.

_____ I check to see if what I read makes sense.
_____ I remind myself why I am reading.
_____ I focus on the goal of my reading while I read.

_____ I check to see if I can summarize sentences and paragraphs. If reading gets hard,

_____ I ask myself if there are any problems.

_____ I try to identify the problem.

_____ I try to fix the problem.

_____When the problem is fixed, I get back to my reading, making sure I understand what I've read so far.

Such checklists can be developed to focus on science learning or on science and reading. For example, a checklist that reminds students of the important steps (and sequence of steps) in a specific scientific inquiry can serve initially as an external prompt to remembering how to do inquiry well. Over time, a student can internalize the contents of the checklist, in the form of a procedural schema, and then call on it when future assignments involve scientific inquiry.

The teacher also models the use of the checklist by asking related questions during science class and thinks aloud about why she asks the questions and her answers to the questions. The predictable presentation of self-assessment strategies can help students on the path to self-assessment.

I should note that checklists like the one above are also scalable: They can reflect specific instructional goals and different levels of complexity. If we are interested in helping students learn to assess their own critical reading abilities in science, we may complement the above list with the following items:

_____ I check the text to see if the author provides evidence to support the claim that one consequence of automobile emissions is global warming.

_____I compare the information in the text with what I already know about air pollution and global warming.

We do not give up our responsibility to conduct classroom assessments when we promote student self-assessment. Rather, we look for opportunities when using our assessments to help students learn assessment themselves (Afflerbach 2002b). Creating balance is imperative, for if in all our teaching students do not begin to learn how to do self-assessment, how will they ever become truly independent?

Balancing the Demands for Teacher and School Accountability With Professional Development Opportunities to Develop Expertise in Assessment

Each of the balances that I advocate for in this chapter is dependent on teachers' professional development in assessment. Professional development opportunities help teachers become practicing experts in classroom-based assessment. Specifically, professional development can help teachers learn and use effective assessment materials and procedures that best influence the daily teaching and learning in the classroom (Johnston 1987; Stiggins 1999). The support provided by professional development helps teachers use and refine the formative, process-oriented assessments that are so critical to the daily successes of the classroom. These assessments provide information that helps teachers recognize and capitalize on the teachable moment, and these daily successes add up to the accomplished teaching and learning that is reflected in scores on accountability tests. But accountability is not achieved through testing—it is achieved through the hard work that surrounds successful classroom assessment and instruction.

Successful classroom-based assessment demands teacher expertise, and professional development helps teachers develop this expertise. Unfortunately, the school funds spent on high-stakes tests are taken from school budgets that are otherwise limited. This means that money spent on tests cannot be spent on initiatives that would actually help teachers become better at classroom-based assessment. Thus, there needs to be a better balance between the call for teacher and school accountability and the means to help teachers and schools establish and maintain that accountability.

Summary

As teachers of reading or science (or of reading and science), we are challenged to provide effective instruction for each and every student. Effective instruction is dependent on assessment that helps teachers and students move toward and attain daily and annual teaching and learning goals. This chapter describes the imbalances that must be addressed if assessment is to reflect our best efforts to measure students' achievement. We are not wanting for description and detail of how classroom-based assessment can help our teaching and how our teaching helps students develop. There must be a concerted effort to bring classroom-based assessment into the spotlight and when the time arrives, to be ready to deliver on its promise.

The imbalances identified in this chapter need our attention. Righting these should lead to assessment programs that are more integral to the daily life of teachers, students, and classrooms, providing evidence that assessment can contribute to student learning and achievement (Crooks 1988). When we focus on process assessment, we can accurately determine what aspect of a science inquiry strategy students do and don't understand. When we assess and determine how a student's motivation grows as the result of gaining control of scientific procedure, we are describing a compelling success story. And when we share our assessment knowledge with our students, we are preparing them for a balanced approach to the assessment of their own science learning, fostering independence.

We have the means to develop assessment that is central to the identification and accomplishment of teachable moments and assessment that reflects student achievement in relation to our most recent understanding of science and science learning. This must be complemented by teachers' professional development and the public commitment to examine our new conceptualizations of science learning and science assessment and to support those assessments that best describe students' related achievement.

References

Afflerbach, P. 2002a. The road to folly and redemption: Perspectives on the legitimacy of high stakes testing. *Reading Research Quarterly* 37: 348–360.

Afflerbach, P. 2002b. Teaching reading self-assessment strategies. In C. Block and M. Pressley (Eds.), *Comprehension instruction: Research-based best practices* (pp. 96–111). New York: Guilford Press.

Afflerbach, P. 2005. High stakes testing and reading assessment. *Journal of Literacy Research* 37: 1–12.

Afflerbach, P. 2007. *Understanding and using reading assessment, K–12*. Newark, DE: International Reading Association.

American Educational Research Association (AERA)/American Psychological Association (APA). 1999. *Standards for educational and psychological testing*. Washington, DC: AERA/APA.

Baxter, G., and R. Glaser. 1998. Investigating the cognitive complexity of science assessments. *Educational Measurement: Issues and Practice* 17: 37–45.

Black, P., and D. Wiliam. 1998. Inside the black box. *Phi Delta Kappan* 79: 139–148.

Black, P., and D. Wiliam. 2005. Assessment for learning in the classroom. In J. Gardner (Ed.), *Assessment and learning*. London: Sage.

Bloom, B. 1956. *Taxonomy of educational objectives, Handbook I: The cognitive domain.* New York: David McKay.

Chapman, J., and W. Tunmer. 1995. Development of young children's reading self-concepts: An examination of emerging subcomponents and their relationship with reading achievement. *Journal of Educational Psychology* 87: 154–167.

Crooks, T. 1988. The impact of classroom evaluation on students. *Review of Educational Research* 58: 438–481.

Davis, A. 1998. *The limits of educational assessment.* Oxford, UK: Blackwell.

Gambrell, L., B. Palmer, R. Codling, and S. Mazzoni. 1996. *Motivation to read profile (MRP).* Athens, GA: National Reading Research Center.

Guthrie, J., and A. Wigfield. 1997. *Reading engagement: Motivating readers through integrated instruction.* Newark, DE: International Reading Association.

Johnston, P. 1987. Teachers as evaluation experts. *The Reading Teacher* 40: 744–748.

McKenna, M., and D. Kear. 1990. Measuring attitude towards Reading: A new tool for teachers. *The Reading Teacher* 43: 626–639.

MSP-Motivation Assessment Program. 2006. Tools for the evaluation of motivation-related outcomes of math and science instruction. Retrieved from: *www.mspmap.org*

National Assessment of Educational Progress (NAEP). 2004. *Student background questionnaire: Science.* Washington, DC: National Center for Education Statistics.

National Center for Education Statistics (NCES). 2005. *2009 NAEP reading framework.* Washington, DC: NCES.

National Research Council (NRC). 1996. *National science education standards.* Washington, DC: National Academy Press.

National Research Council (NRC). 2005. *How students learn: Science in the classroom.* Washington, DC: National Academy Press.

Pellegrino, J., N. Chudowsky, and R. Glaser. 2001. *Knowing what students know: The science and design of educational assessment.* Washington, DC: National Academy Press.

Stefanou, C., and J. Parkes. 2003. Effects of classroom assessment on student motivation in fifth-grade science. *Journal of Educational Research* 96: 152–162.

Stiggins, R. 1999. Evaluating classroom assessment training in teacher education. *Educational Measurement: Issues and Practices* 18: 23–27.

Valencia, S., and N. Place. 1994. Literacy portfolios for teaching, learning, and accountability. In S. Valencia, E. Hiebert, and P. Afflerbach (Eds.), *Authentic reading assessment: Practices and possibilities* (pp. 134–156). Newark, DE: International Reading Association.

Vygotsky, L. 1978. *Mind in society: The development of higher psychological processes.* Cambridge, MA: Harvard University Press.

SECTION 4

Professional Development: Helping Teachers Link Assessment, Teaching, and Learning

S ection 4 describes professional development that builds teachers' assessment expertise and the impact of these programs on teaching and student learning.

Discussion Questions
- What are key components of effective assessment-centered professional development programs?
- What do these programs recommend about balancing the challenges of classroom-based and high-stakes assessments?
- What evidence supports the claim that more effective classroom-based assessment leads to improved student scores on high-stakes tests?
- What are some strategies for improving teachers' skills in designing and analyzing assessments and providing student feedback?

Chapter Summaries
Researchers Okhee Lee from University of Miami and Kathryn LeRoy of the Duval County (Jacksonville, Florida) Public Schools describe their Miami-based study with two populations of non-native speakers. Their project provided in-

tense professional development organized around materials and ongoing assessments developed for the project. The authors present compelling data showing the improved performance on the Florida statewide standardized achievement test (F-CAT) by students whose teachers are part of this program.

The chapter by Elaine Woo and Kathryn Show from the Seattle, Washington, Public Schools describes an elementary science professional development program in Seattle that emphasizes writing in science. Teachers receive instruction in the basics of integrating science and writing, assessing students' science notebook entries, and providing meaningful feedback to students. Independent researchers studying the Seattle program examined the high-stakes test performance of students from participating teachers' classroom. The authors present recent evidence of students' improvement on the Washington Assessment of Student Learning (WASL) relative to comparison groups from other communities.

Janet Struble, Mark Templin, and Charlene Czerniak, researchers at the University of Toledo, initiated a professional development program for elementary science teachers in Toledo, Ohio, that focuses on enhancing teachers' formative assessment skills. The authors' present data from their research that analyzed student performance on high-stakes tests and found significant correlations between students' scores and the amount of professional development their teachers received.

Paul Kuerbis of Colorado College and Linda Mooney from the Colorado Springs Public Schools describe a professional development model in which teachers collaborated with science content experts, assessment experts, and researchers to create a series of formative assessments to accompany each of their science units. As part of this program, which serves several communities in Colorado, all teachers received professional development in the effective use of the assessments. The authors report on research that monitors students' performance on statewide tests administered by the Colorado Student Assessment Program.

Another approach to professional development is described by Paul Hickman, a science education consultant, and his teacher colleagues, Drew Isola from Allegan, Michigan, and Marc Reif from Ruamrudee International School in Bangkok. These authors advocate use of the Reformed Teaching Observation Protocol (RTOP), a tool originally developed for research on secondary physics teachers' classroom practices, by science teachers for self-evaluation of own assessment practices.

Nancy Love, at Research for Better Teaching, studies the impact of school communities' analysis of multiple sources of student assessment data on their instructional decisions and organizational responses. Her chapter poses simple questions that school groups can use as they engage in collaborative inquiry. Teams of teachers and administrators use evidence of student achievement to analyze what is and isn't working and then recommend program changes that support successes and eliminate ineffective programs. Love's recommendations build from her ongoing research with large school districts around the country.

## CHAPTER 17

What Research Says About Science Assessment With English Language Learners

Kathryn LeRoy
Duval County, Florida, Public Schools

Okhee Lee
University of Miami

As high-stakes testing and accountability in science looms on the horizon, science teachers are faced with the dilemma of identifying key science concepts and effective instructional practices to maximize achievement for all students. Education reform and specifically the No Child Left Behind (NCLB) law of 2001 require that *all* students achieve high academic standards in core subject areas. Assessment of science achievement necessitates consideration of fairness to different student groups. Fairness in this context means "the likelihood of any assessment allowing students to show what they understand about the construct being tested" (Lawrenz, Huffman, and Welch 2001, p. 280).

How do we ensure that assessments are valid and equitable for all students? With English language learners (or ELL students), assessments should distinguish among academic achievement, English language proficiency, and general literacy. Although large-scale assessments may include various accommodation strategies, they are rarely administered in languages other than English. Even when assessments are administered in students' home language in addition to English, ensuring the comparability of assessment instruments between two languages is complicated.

This chapter addresses what research says about science assessment with ELL students. Specifically, we draw from our ongoing research and development efforts to promote science and literacy achievement of ELL students in a large urban school district (visit our project website at *www.education.miami.edu/psell*). First, we describe our intervention with a focus on inquiry-based science teaching and learning. Second, we describe how we design science and literacy assessment instruments for ELL students in our research. Third, we describe our efforts to align classroom assessments in science and literacy with high-stakes assessments for ELL students. Finally, based on our work and other research findings, we offer suggestions for what teachers can do to assess science and literacy achievement of ELL students.

Science Instructional Intervention With ELL Students

What science instructional practices will support increased science and literacy achievement of ELL students? Some empirical evidence has demonstrated success with inquiry-based science; however, there is a huge variation in what that looks like in today's science classrooms. With content standards in place and inquiry at the cornerstone of what science curriculum should include, how can this be translated into quality teaching in today's classrooms that meets the needs of all students?

Through university and school district collaborative research, we implement an instructional intervention—called "Promoting Science among English Language Learners (P-SELL) in a High-Stakes Testing Policy Context"—to promote ELL students' science and literacy achievement in the context of high-stakes testing and accountability. Our intervention involves teachers and students at grades 3 through 5 located in 15 elementary schools in a large urban district. These schools enroll greater-than-district proportions of ELL students and students from low socioeconomic status backgrounds and have performed poorly according to the state's accountability plan. The research tests two questions: (1) Can ELL students learn academic subjects, such as science, while also developing English proficiency? and (2) Can ELL students, who learn to think and reason scientifically also perform well on high-stakes assessments?

Research on science instruction with ELL students highlights hands-on, inquiry-based science that enables these students to develop scientific understanding and acquire English language proficiency simultaneously

(Amaral, Garrison, and Klentschy 2002; Lee et al. 2005; Rosebery, Warren, and Conant 1992). First, hands-on activities are less dependent on formal mastery of the language of instruction, thus reducing the linguistic burden on ELL students. Second, hands-on activities through collaborative inquiry foster language acquisition in the context of authentic communication about science knowledge and practice. Third, inquiry-based science promotes communication of students' understanding in a variety of formats, including written, oral, gestural, and graphic. Fourth, by engaging in the multiple components of science inquiry, ELL students develop their grammar and vocabulary as well as their familiarity with scientific genres of speaking and writing (Lee et al. 2006). Finally, language functions (e.g., describing, hypothesizing, explaining, predicting, and reflecting) can develop simultaneously with science inquiry and process skills (e.g., observing, describing, explaining, predicting, estimating, representing, inferring) (Casteel and Isom 1994). By engaging in inquiry-based science, ELL students learn to think and reason as members of a science learning community.

Essential to our intervention are the project-developed curriculum units and instructional practices that promote science inquiry and English language development of ELL students. Our intervention emphasizes the use of the Science Inquiry Framework (see Figure 17.1, p. 344) that supports science inquiry with diverse student groups. Although some advocates of more open-ended, student-centered inquiry would argue against a framework for organizing and planning inquiry, our practical experience as urban science educators with ELL students has demonstrated the importance of such a framework as an initial step for teachers and students. While making the inquiry process explicit, the framework also allows for flexibility to foster students' initiative and responsibility for their own learning. The icons in the framework serve as points of reference for assisting students in thinking about and organizing their own inquiry. The icons also encourage the use of graphic representations in communicating science, especially for ELL students and students with limited literacy development (Lee et al. 2005). In using the framework, it is important to recognize that the process is not to be followed as a lock-step fashion, but to be considered as a guide. The way students engage in inquiry will vary depending on their prior experience with science, their level of literacy development, and the kinds of questions that are promoted through interaction and discussion.

Figure 17.1 Science Inquiry Framework

1. Questioning 	*State the problem* • What do I want to find out? (written in the form of a question) *Make a hypothesis* • What do I think will happen?
2. Planning 	*Make a plan by asking these questions (think, talk, write)* a. What materials will I need? b. What procedures or steps will I take to collect information? c. How will I observe and record results?
3. Implementing  	*Gather the materials* • What materials do I need to implement my plan? *Follow the procedures* • What steps do I need to take to implement my plan? *Observe and record the results* • What happens after I implement my plan? • What do I observe? • How do I display my results? (using a graph, chart, table)
4. Concluding 	*Draw a conclusion* • What did I find out? • Was my hypothesis correct or incorrect?
5. Reporting 	*Share my results (informal)* • What do I want to tell others about the activity? *Produce a report (formal)* • Record what I did so others can learn. • Consider different ways to express my information.

Our intervention also scaffolds student initiative and responsibility in conducting inquiry as teachers gradually reduce their level of guidance. The National Research Council (2000) presents a continuum from "teacher direction" to "learner self-direction" as it relates to essential features of science inquiry. Initially, students may need a great deal of assistance to engage in inquiry. As they develop inquiry skills, they will need less and less assistance. Eventually, they can explore and do inquiry on their own. As students engage in multiple components of inquiry, they learn to engage in some areas more easily (e.g., implementing activities and reporting results), while they require more assistance and experience in other areas (e.g., questioning and applying findings). The Science Inquiry Matrix (see Figure 17.2) illustrates this continuum, as teachers gradually relinquish authority and encourage students to assume responsibility for inquiry.

Figure 17.2 Science Inquiry Matrix

Inquiry Levels	Questioning	Planning	Implementing	Concluding	Reporting
0	Teacher	Teacher	Teacher	Teacher	Teacher
1	Teacher	Teacher	*Students*	Teacher	*Students*
2	Teacher	Teacher	*Students*	*Students/ Teacher*	*Students*
3	Teacher	*Students/ Teacher*	*Students*	*Students*	*Students*
4	*Students/ Teacher*	*Students*	*Students*	*Students*	*Students*
5	*Students*	*Students*	*Students*	*Students*	*Students*

Science inquiry should also involve such conventions as control of variables, use of multiple trials, and accuracy of measurement. Controlling variables will allow students to establish the cause-and-effect relationship between the variable being tested (i.e., independent variable) and the variable that is being measured (i.e., dependent variable). Multiple trials enhance reliability of the results, and this can be achieved by having groups of students or individual students do an experiment and then compare the results. Furthermore, students should strive for accuracy and precision when they use the tools of science for measurement (e.g., metric rulers, balances, graduated cylinders, thermometers). These conventions are integral to the science inquiry framework (see Figure 17.1), as teachers gradually withdraw assistance and students learn to take initiative and assume responsibility for conducting inquiry (see Figure 17.2).

How to Design Science and Literacy Assessment Instruments for ELL Students

In our research, we use multiple approaches of science assessment with ELL students that include (a) a project-developed pre/post science test directly aligned to the curriculum, (b) a writing prompt that addresses a key science concept covered within the curriculum, and (c) a reasoning interview protocol that engages students in a performance task. Although these instruments are developed in English to maintain continuity between the language of instruction and the language of assessment, teachers are instructed to use students' home language if needed and students are allowed to use their home language.

Science Test

We developed a pre/post science test for each grade level. The test measures students' knowledge of key concepts and big ideas of the science topics in the curriculum that is taught during the school year. The test also measures students' understanding of science inquiry by asking them to construct graphs and tables using the data provided, offer explanations for the data, and draw conclusions. The test consists of project-developed items, public-release items from the state science assessment, public-release National Assessment of Educational Progress (NAEP) items, and public-release Trends in International Mathematics and Science Study (TIMSS) items. Item formats include multiple-choice items and short and extended written response items, with many of the items having multiple components. For the project-developed items, we developed a scoring rubric. For public-release items from the state science assessment, NAEP, and TIMSS, we use the available scoring rubrics. The maximum points for each specific item or item subcomponents vary, depending on the level of cognitive or conceptual difficulties.

Writing Prompt

We developed a pre/post expository writing prompt to assess both literacy (writing) and science content. The writing prompt on the water cycle for third-grade students and the scoring rubric is presented in the Appendix, page 352. Students receive two scores on the writing prompt: (a) *form* measures students' use of conventions, organization, style, and voice; and (b) *content* measures students' ability to give scientific explanations and to use scientific vocabulary. The distinction between form and content allows teachers to assess science learning and English language proficiency separately.

Reasoning Task

We developed a "reasoning interview" protocol for each grade level. Reasoning interviews are conducted individually and videotaped. The topics include measurement with third-grade students, force and motion with fourth-grade students, and Earth systems with fifth-grade students. In addition to measuring students' conceptions of each science topic, reasoning interviews measure students' ability to apply measurement concepts using scientific tools, to conduct a scientific experiment on a force and motion task, and to engage in scientific discourse on an Earth systems task. The three reasoning tasks represent critical elements of students' ability to engage in science inquiry (NRC 2000). The performance-based reasoning interviews with individual students complement the larger-scale assessment of the science test and the writing prompt being used with all students in our research.

Assessment Results

Together, these three assessments enable us to examine the first research question—Can ELL students learn academic subjects, such as science, while also developing English proficiency? Using hierarchical linear modeling (HLM) analysis, the results of our first-year intervention with third graders revealed that students in the treatment group displayed a statistically significant increase in science achievement test scores at the end of the school year (see Table 17.1, p. 348; detailed results are reported in Lee et al. 2008). Furthermore, the results of the HLM analysis showed that the students who were enrolled in English for Speakers of Other Languages (ESOL) programs made achievement gains comparable to the students who had exited from ESOL or never been in ESOL. Over the five-year period of our research, we will continue to examine the research question using longitudinal data.

Table 17.1 Descriptive Statistics for Science Test Scores
(total possible score = 24 points)

Variable	Test	Subgroup	N	M	SD
All Students	Pre		818	7.40	3.36
	Post		818	14.34	4.30
	Gain		818	6.95	4.18
ESOL	Pre	ESOL levels 1 to 4	118	6.55	3.40
		ESOL exited or non-ESOL	698	7.53	3.34
	Post	ESOL levels 1 to 4	118	12.39	4.62
		ESOL exited or non-ESOL	698	14.67	4.16
	Gain	ESOL levels 1 to 4	118	5.84	4.08
		ESOL exited or non-ESOL	698	7.14	4.17

How to Align Classroom Assessment in Science and Literacy With High-Stakes Assessments for ELL Students

The research on science assessments, especially in the context of high-stakes assessments with diverse student groups including ELL students, is currently limited. Until very recently, science has not been part of high-stakes assessments in most states. Even when science is tested, it usually does not count toward accountability. As a result, research on science assessment *accommodations* with ELL students is even more limited. As science becomes a part of accountability measures under No Child Left Behind, more research is expected.

A primary motivation for our research involves ongoing concerns about low science achievement of ELL students, especially given the national context of the impending high-stakes testing policy in science under No Child Left Behind. When the third-grade students in our research advanced into fifth grade in 2006, they were be the first cohort of students for whom the statewide science assessment was factored into school accountability.

The state assessments are administered in reading, mathematics, writing, and science. Reading and mathematics are assessed in grades 3–10 and writing and science are assessed once at the elementary, middle, and high school levels. The state science assessment includes multiple-choice and performance task items in the physical and chemical sciences, the Earth sciences, the life and environmental sciences, and scientific thinking. Items on the state science assessment encourage student reasoning, planning, analyzing, and using scientific thinking. Students should be proficient in determining the logical steps or outcome in an experiment; comparing or contrasting structures or functions of different organisms or systems; applying and using concepts from a scientific model or theory; drawing conclusions; analyzing an experiment to identify a flaw and propose a method for improving it; and predicting outcomes as a result of a change within a system.

Our ongoing research is enabling us to examine the second research question: Can ELL students, who learn to think and reason scientifically, also perform well on high-stakes assessments? Following the first year of our intervention with third graders, the treatment group students showed higher scores on the state mathematics test, particularly on the measurement strand emphasized in the intervention, than the comparison group students (detailed results are reported in Lee et al., in press). Of the 8-point maximum score for the measurement strand, the mean of the treatment group was 5.00 ($SD = 1.91$) compared to the mean of 4.39 ($SD = 2.02$) in the comparison group. We will continue examining this question during the final year of our research (through 2009) when the state science assessment will be in place.

What Teachers Can Do to Assess Science and Literacy Achievement of ELL Students

An important aspect of classroom assessment includes the use of meaningful and relevant topics, tasks, and activities. Teachers can employ certain assessment practices for ELL students that will benefit all students in a classroom:

1. Teachers may use two separate scoring criteria for writing prompts or short- and extended-response tasks to assess science learning and English language proficiency separately (see Appendix). This assessment practice enables teachers to identify strengths and weaknesses of ELL students in each area (Lee et al. 2005).

2. Teachers may assess ELL students in their home languages as well as in English. Allowing students to communicate their science knowledge and abilities in their home languages promotes both general literacy and academic learning, which, in turn, promotes English language proficiency. Achievement in these three areas develops simultaneously (Lee and Fradd 1998).

3. Teachers should promote the use of multiple representational formats, keeping in mind that the goal is to move ELL students toward established literacy standards. Students who cannot write in either their home languages or English can express ideas in drawings, graphs, and tables, as well as in oral communication. When students realize that they are expected to produce meaningful representations of their knowledge in assessment settings, they engage in science learning activities and tasks in more meaningful ways.

4. Teachers who are aware of linguistic and cultural influences on ELL students' responses support the assessment process. Whereas efforts have traditionally focused on eliminating these influences, an emerging approach advocates understanding how home language and culture can be incorporated to guide the entire assessment process (Solano-Flores and Trumbull 2003).

Conclusions

Since instruction and assessment complement and reinforce each other, it is essential to provide high-quality instruction for ELL students and to assess their achievement outcomes in a manner that can guide subsequent instruction. Science assessment with ELL students presents both promises and challenges for their science learning and English language development. Assessment in ELL students' home language in addition to English enables them to express their science knowledge and abilities; however, this does not ensure valid and equitable assessment if the language of instruction is in English. As science becomes a part of accountability measures across the states under No Child Left Behind, accommodations for ELL students should be addressed. Unfortunately, research on science assessment accommodations with ELL students is insufficient to guide large-scale assessments.

In this chapter, we describe our own ongoing research and development on science instructional intervention and assessment with ELL students.

Preliminary results indicate that (a) ELL students can learn science while developing English proficiency and that (b) ELL students can engage in science inquiry and reasoning while performing well on high-stakes assessments. The emerging literature in this field can offer insights for valid and equitable science assessment of ELL students, which, in turn, can be used to further enhance science instruction for all.

Author Note

This work is supported by the National Science Foundation (NSF Grant #ESI–0353331). Any opinions, findings, conclusions, or recommendations expressed in this publication are those of the authors and do not necessarily reflect the position, policy, or endorsement of the funding agencies.

References

Amaral, O. M., L. Garrison, and M. Klentschy. 2002. Helping English learners increase achievement through inquiry-based science instruction. *Bilingual Research Journal* 26: 213–239.

Casteel, C. P., and B. A. Isom. 1994. Reciprocal processes in science and literacy learning. *The Reading Teacher* 47: 538–545.

Lawrenz, F., D. Huffman, and W. Welch. 2001. The science achievement of various subgroups of alternative assessment formats. *Science Education* 85(3): 279–290.

Lee, O., C. Buxton, S. Lewis, and K. LeRoy. 2006. Science inquiry and student diversity: Enhanced abilities and continuing difficulties after an instructional intervention. *Journal of Research in Science Teaching* 43(7): 607–636.

Lee, O., R. A. Deaktor, J. E. Hart, P. Cuevas, and C. Enders. 2005. An instructional intervention's impact on the science and literacy achievement of culturally and linguistically diverse elementary students. *Journal of Research in Science Teaching* 42(8): 857–887.

Lee, O., and S. H. Fradd. 1998. Science for all, including students from non-English language backgrounds. *Educational Researcher* 27(3): 12–21.

Lee, O., J. Maerten-Rivera, R. Penfield, K. LeRoy, and W. Secada. 2008. Science achievement of English language learners in urban elementary schools: Results of a first-year professional development intervention. *Journal of Research in Science Teaching* 45(1): 31–52.

National Research Council (NRC). 2000. *Inquiry and the national science education standards: A guide for teaching and learning.* Washington, DC: National Academy Press.

Rosebery, A. S., B. Warren, and F. R. Conant. 1992. Appropriating scientific discourse: Findings from language minority classrooms. *Journal of the Learning Sciences* 21(1): 61–94.

Solano-Flores, G., and E. Trumbull. 2003. Examining language in context: The need for new research and practice paradigms in the testing of English-language learners. *Educational Researcher* 32(2): 3–13.

Appendix

The Water Cycle

Writing Prompt:
Pretend you are a drop of water. Before you begin writing, think about how water changes form in the water cycle. Explain to the reader how you are changed as you go through the water cycle.

Writing Scoring Rubric for *Form*

The writing scoring rubric for <u>form</u> considers the following components:

Convention
- Spelling
- Correct plurals and comparisons
- Capitalization and punctuation
- Subject/verb agreement and verb and noun forms

Organization
- Indentation for new paragraphs
- Idea development

Style/Voice
- Sentence structures to communicate ideas
- Coherence from sentence to sentence

Ratings:
4 – Complete/Comprehensive
- Spells all high frequency and most irregular words correctly with up to 3 errors
- Uses correct plurals and comparisons (e.g., good, better, best)
- No errors in capitalization and punctuation
- Subject/verb agreement and verb and noun forms are generally correct with up to 2 errors
- Uses accurate indentation for each new paragraph
- Ideas are presented logically
- Uses a variety of sentence structures to communicate ideas
- Writing is highly coherent

3 – Adequate
- Spells most high frequency (up to 3 errors) and many irregular words (with up to 5 errors) correctly
- Generally uses correct plurals and comparisons (e.g., good, better, best) (with up to 1 error)
- Few errors (up to 3) in capitalization and punctuation
- Occasional errors in subject/verb agreement, which do not impede communication (with up to 3 errors)
- Generally uses accurate indentation for the majority of each new paragraph
- Ideas are developed fairly logically
- Uses a variety of sentence structures to communicate ideas, although most (more than half of all sentences) are simple constructions
- Writing is fairly coherent

2 – Emerging/Expanding
- Some errors in spelling (up to 8 errors)
- Some errors in plurals and comparisons (e.g., good, better, best) (up to 3 errors)
- Some errors in capitalization and punctuation (up to 5)
- Errors in subject/verb agreement may somewhat impede communication
- Writing may not be organized into paragraphs
- Ideas are partially developed
- Sentence structures are limited to simple constructions with little attempt at variety
- Writing is partially coherent

1 – Minimal/Inaccurate
- Frequent errors in spelling (9 or more errors)
- Frequent errors in plurals and comparisons (e.g., good, better, best) (4 or more errors)
- Frequent errors in capitalization and punctuation (more than 5 errors)
- Lacks subject/verb agreement; verb and noun forms are incorrect and impeded communication
- No attempt to organize writing into paragraphs
- Ideas are not developed
- Sentence structure is fragmented/incomplete; lack of proper sentence structure may impede communication
- Writing lacks coherence

0 – No response/Unscorable
- The response is simply a rewording of the prompt
- The response is simply a copy of published work
- The student refused to write
- The response is illegible

- The response contains an insufficient amount of writing to determine if the student was addressing the prompt

Writing Scoring Rubric for *Science Content*

Elements of Writing:

Since the prompts require expository writing, the scoring rubric considers science vocabulary and explanation. Use of science vocabulary and a comprehensive explanation include the following components:
- scientific vocabulary
- three states of water (solid, liquid, gas, ice water, water vapor)
- three processes of change during the water cycle (evaporation, condensation, and precipitation)
- heating and cooling related to the water cycle
- concept of a cycle (continuous, ongoing, repetitive)
- sequence of events

Ratings:

4 – Complete/Comprehensive
- accurate use of science vocabulary with adequate explanations
- correctly names all three states of water as related to the water cycle
- accurately describes all three processes of change during the water cycle
- mentions process of both heating and cooling
- mentions concept of a cycle and the ongoing, repetitive nature of the process
- does not demonstrate sequencing errors

3 – Adequate
- accurate use of science vocabulary with adequate explanations
- correctly names all three states of water as related to the water cycle
- accurately describes all three processes of change during the water cycle
- may mention process of heating or cooling but does not include both elements
- mentions the concept of a cycle
- does not demonstrate sequencing errors

2 – Emerging/Expanding
- expanding accurate use of scientific vocabulary with some explanation
- correctly names two of three states of water as related to the water cycle
- accurately describes two of three processes of change during the water cycle
- does not mention process of heating or cooling
- does not mention the concept of a cycle
- may demonstrate sequencing errors

1 – Minimal
- inaccurate use of scientific vocabulary or minimal accurate use of scientific vocabulary without explanation
- accurately names only one of three states of water as related to the water cycle
- accurately describes one of three processes of change during the water cycle
- does not mention process of heating or cooling
- does not mention the concept of a cycle
- inaccurate description of sequence of events

0 – No response, unidentifiable, and/or irrelevant content

Washington State's Science Assessment System: One District's Approach to Preparing Teachers and Students

Elaine Woo and Kathryn Show
Seattle Public Schools

Responding to the reauthorization of the Elementary and Secondary Education Act of 2001 (No Child Left Behind), Washington State's Office of the Superintendent of Public Instruction published *Science K–10 Grade Level Expectations: A New Level of Specificity* in 2005 as a guide for Washington educators. In addition, state science and assessment leaders, science teachers, and scientists developed a high-stakes science assessment, the Washington Assessment of Student Learning (WASL), which is now fully implemented at the fifth, eighth, and tenth grades. This chapter will describe how the elementary science staff in the Seattle Public Schools extended the district's infrastructure and support system to help teachers prepare their students for the state science WASL. The original resources for developing this infrastructure came from a National Science Foundation (NSF) Local Systemic Change (LSC) grant.

For well over a decade, students and teachers in our state have benefited from various NSF, state-level Math Science Partnership, Washington State LASER (Leadership and Assistance for Science Education Reform), and regional foundation grants for pursuing science education reform. Seattle Public Schools received one of the earlier LSC grants from NSF in August 1996 for the purpose of bringing about systemic change in elementary science education. In addition, the Puget Sound area has many scientific research institutions and industries whose leaders and employees care deeply about the success of student achievement in K–12 science. The support provided by individuals and outreach groups from these institutions has complemented the support from NSF and has allowed concentrated reform work for 12 years. Seattle's elementary science consultants have focused on offering high-quality professional development and curricular support for teachers so that students can acquire a foundation for science literacy as well as the skills and knowledge necessary to be successful on the state science WASL.

The State Science Standards and Assessment

There are three state science standards, called Essential Academic Learning Requirements (EALRs): Systems, Inquiry, and Application. In the *Science K–10 Grade Level Expectations* document, the three standards are displayed in a science "symbol" (Figure 18.1) with the following explanation:

> *The Washington State science symbol describes the relationship among the systems of the natural world, how those systems are investigated through inquiry, and how the knowledge and process of science are applied to solve human problems using scientific design. Inquiry contributes to new knowledge about systems. The application of our knowledge of systems, time and again, results in new tools for science (e.g., computers, telescopes, DNA sequencers). It is just such tools and the creativity of the human spirit that lead to greater understanding of systems. In this way the science symbol reflects the structure of the modern scientific endeavor and the transformation of modern society.* (p. 6)

The concept of systems was new for Washington State educators. Before 2005, we had referred to this first standard as "content." The idea of systems comes from the National Science Education Standards (NRC 1996), which note that viewing content through the organization of systems mirrors the language and thinking of scientists. The grade-level expectation document defines systems as "an assemblage of interrelated parts or conditions through which matter, energy, and information flow" (OSPI 2005, p. 54). We needed to reinforce the idea that everything in the world is part of a system, and we wanted to help teachers and students

Figure 18.1 Washington State Science Symbol

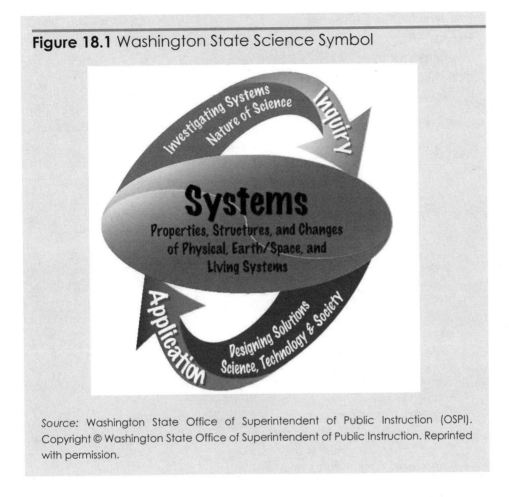

Source: Washington State Office of Superintendent of Public Instruction (OSPI). Copyright © Washington State Office of Superintendent of Public Instruction. Reprinted with permission.

understand that everything they were learning was interrelated and part of a larger system. Systems thinking is not explicit in our adopted elementary curricular units, so we had to extend the concepts in those units to include envisioning the world in terms of systems. Seattle's science consultants have included this information in the Supplementary Curriculum Guides used in professional development so that teachers would apply systems thinking to their science units.

The Grade Level Expectations are the essential content and processes to be learned. Figure 18.2, page 360, shows the first two standards and examples of Grade Level Expectations listed under those standards.

Figure 18.2 Examples of Two Essential Academic Learning Requirements (EALRs), Grade Level Expectations, and Evidence of Learning

EALR (Standard) One—SYSTEMS: The student knows and applies scientific concepts and principles to understand the properties, structures, and changes in physical, earth/space, and living systems.
Grade Level Expectation: Understand how to use properties to sort natural and manufactured materials and objects.
Evidence of Learning: Identify which states of matter (solid, liquid, or gas) can change shape and which can expand to fill a container. (p. 16)

EALR (Standard) Two—INQUIRY: The student knows and applies the skills, processes, and nature of scientific inquiry.
Grade Level Expectation: Understand how to plan and conduct simple investigations following all safety rules.
Evidence of Learning: Generate a logical plan for, and conduct, a simple controlled investigation with the following attributes:

- prediction
- appropriate materials, tools, and available computer technology
- variables kept the same (controlled)
- one changed variable (manipulated)
- measured (responding) variable
- gather, record, and organize data using appropriate units, charts, and/or graphs
- multiple trials

Source: Washington State Office of Superintendent of Public Instruction (OSPI). 2005. *Science K–10 Grade Level Expectations: A New Level of Specificity.* Olympia, WA: Washington State Office of Public Instruction, p. 38. Copyright © Washington State Office of Superintendent of Public Instruction. Reprinted with permission.

The language of the state science assessment (WASL) is taken directly from the Grade Level Expectations and the bulleted lists of Evidence of Learning for each Grade Level Expectation. Consequently, teachers need a thorough knowledge of the Grade Level Expectations to prepare their students to be successful on the assessment. One critical task our science consultants had to complete was to adjust the language of our current science units to match the language of the Grade Level Expectations and the Evidence of Learning lists. We will discuss this effort later in the chapter.

The state assessment is made up of three types of scenarios: Systems Scenarios, Inquiry Scenarios, and Application Scenarios. A scenario is a short, hypothetical example of what students might encounter in a school science investigation or in the world outside of school. The scenarios are based on the language of the Grade Level Expectations and the Evidence of Learning lists. The science assessment is made up of 40% Systems points, 40% Inquiry points, and 20% Application points. The 40% focus on inquiry is driving the implementation of inquiry throughout the K–12 classrooms in the state. The test items include multiple-choice, short-answer, and extended-response question items. (See Appendix A for the Washington Assessment of Student Learning released fifth-grade scenario "Hold That Soil." This is an Inquiry Scenario based on an investigation into an Earth system with six multiple-choice and two short-answer items. To see a Powerful Classroom Assessment (PCA) based on this scenario, go to *www.k12.wa.us/assessment/wasl/science.)*

Achievement of Seattle's Fifth Graders on the Science WASL

How did Seattle's fifth graders do on the state science assessment after their teachers participated in a well-developed professional development system supported with NSF and other funds? Washington's fifth-grade students started piloting the state science assessment in 2001. The fifth-grade exam became operational in 2004 and mandatory in 2005. The bar graph in Figure 18.3, page 362, shows that Seattle's students had a higher average percent proficient in the first four years of the science WASL than did fifth graders statewide. This is demographically unexpected given the diverse demographics of the student population within the Seattle Public Schools. The district is made up of almost 46,000 K–12 students; about 40% receive free and reduced-price lunch and approximately 58% are students of color. These students speak 129 different languages.

In addition, the program's researchers found the difference between Seattle's fifth graders' scores and the scores of fifth graders statewide to be statistically significant:

> When comparing the district's performance on the Washington Assessment of Student Learning (WASL) to the performance of other districts with similar demographics, the achievement is even more significant. Researchers from the National Center for Research on Evaluation, Standards, and Student Testing (CRESST) made that comparison and concluded that the district performed considerably better than would be expected: 6.3 percent more Seattle students met the standard than did students in districts of similar size and demographics. (Soholt 2005, p. 1)

Figure 18.3 Impact of Reform on Seattle's Fifth-Grade Science WASL Scores

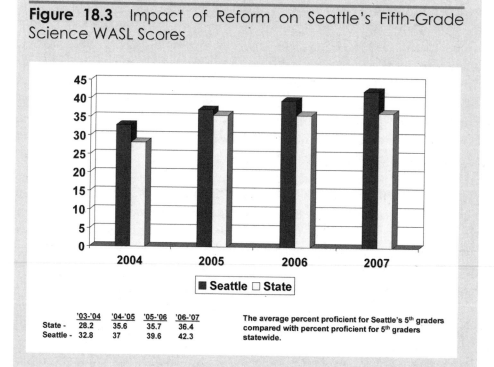

■ Seattle □ State

	'03-'04	'04-'05	'05-'06	'06-'07
State -	28.2	35.6	35.7	36.4
Seattle -	32.8	37	39.6	42.3

The average percent proficient for Seattle's 5th graders compared with percent proficient for 5th graders statewide.

CHAPTER
18

The level of, and steady improvement in, Seattle's fifth graders' scores is not an accident. The ongoing professional development and teachers' efforts have had a positive impact. The professional development has evolved into an effective support system, in part because of continual feedback from teacher participants. The district science staff was able to make substantial adjustments to the adopted curriculum materials; develop expertise among teachers, particularly among the teachers in the two leadership groups (one focusing on Initial Use classes for the science units and the other focusing on science writing); provide a basic level of professional development for all new teachers; and develop a nationally recognized integrated science-writing program. The science staff made this progress as the result of having

- a focused, common vision among stakeholders,
- strong collaboration among and support from community partners, including scientists,
- continuity of highly qualified staff,
- intensive, ongoing professional development for the science consultants and other staff,
- a credible feedback loop for teacher input,
- use of current data, research, and best practices, and
- a belief in taking time to fully develop cutting-edge program components, and an effective outside evaluator.

Defining the Professional Development System

Figure 18.4, page 364, shows the legacy of the NSF Local Systemic Change grant work. District science consultants developed courses with the overall goal of providing teachers with support so that all students would acquire the foundation to become scientifically literate by the end of the 12th grade. The consultants have been released from the classroom for several years and have had a lot of professional development of their own on how to design and implement high-quality professional development as well as how to coach and mentor teacher leaders and all teachers.

Figure 18.4 K–5 Inquiry-Based Science Program in the Seattle, Washington, Public Schools

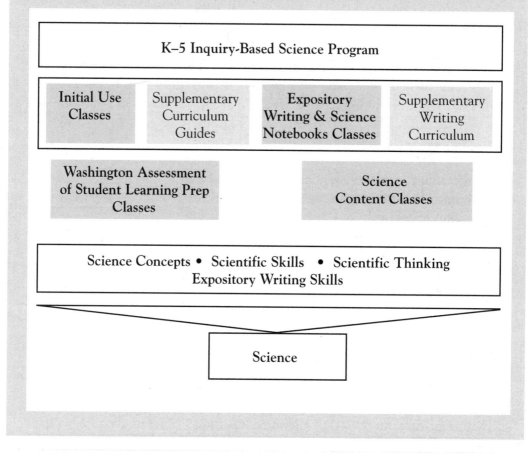

The Initial Use and Expository Writing and Science Notebook courses shown in Figure 18.4 are the two most basic courses classroom teachers attend. Teachers are required to take Initial Use classes in order to receive science units for their grade level. During the Initial Use class, teachers receive an extensive Supplementary Curriculum Guide, and later the kit of materials including a teacher manual. The Expository Writing and Science Notebooks, WASL Prep, and Science Content courses are optional.

In its evaluation report at the end of the LSC project, Inverness Research Associates of Inverness, California, noted the following:

- *LSC staff have designed and delivered a high-quality, developmentally appropriate sequence of professional development experiences that support improved science instruction on a districtwide basis.*

- *Teachers, principals, and students express a high degree of satisfaction with and ongoing demand for the services of the program.*

- *Classroom observations indicate that full implementation of the program leads to higher quality science instruction.*

- *The LSC science project has been both creative and successful in cultivating strands of curricular integration with other key subject areas — particularly writing and mathematics.* (St. John, Fuller, and Tambe 2002, pp. 7–12)

In their feedback to the program staff, classroom teachers indicate satisfaction with and appreciation for the professional development support, which includes districtwide classes as well as on-site support. Below are comments from teachers who have participated in the professional development.

*What I like best is the long-term, ongoing training that allows me to go to class, apply the concepts in my classroom, return to class, etc. [There is] constant consultation and support and pay for our overtime. All kits are inclusive. This is the best training I've ever had. —*Kindergarten teacher, Gatzert School, 2000

*Immersing myself in the Inquiry-Based Science Program transformed the science instruction in my classroom. It [has] significantly improved student learning among the students in my classes since 1995. —*Fifth-grade teacher, Bryant School, 2001

I would like to thank you and your staff for the excellent direction of the science curriculum that has been assembled. I have been following the lesson plans verbatim and am astounded by the results that my students have achieved. —Fifth-grade teacher, T. Marshall School, 2004

Initial Use Classes and Supplementary Curriculum Guides

To sustain the Initial Use classes once the NSF grant was completed, our consultants started developing a team of lead teachers, which now includes about 40 classroom teachers who come after school to teach Initial Use classes. At least two, and sometimes three, lead teachers work together to teach the class for one science unit. Before the lead teachers take over the instruction of a unit, the consultants work with them and model several sessions. It usually takes two years of modeling before the lead teachers instruct independently. Before the class, the lead teachers meet with a consultant to plan the class. They meet again after the session to discuss what went well and what might be changed. Most of the Initial Use classes are six hours and take place during two three-hour sessions after school. (Because of the complexity of two intermediate units, the Initial Use classes for those two units are nine hours or three three-hour sessions. Program staff increased the length of these two classes because of teacher demand.)

In an Initial Use class, teachers

- become familiar with the scientific content in the unit,
- become familiar with all the lessons and how they build to develop conceptual understanding,
- learn how to use the Learning Cycle when planning and implementing inquiry-based science lessons,
- deepen understanding of inquiry by actively participating in key lessons,
- become familiar with the Supplementary Curriculum Guide, which has been developed specifically for each science unit and provides modifications, extensions, and additional lessons in order to address the state standards and Grade Level Expectations,
- learn about classroom-based assessment strategies, and
- become familiar with materials and management strategies to ensure success with the unit.

The district's science consultants have partnered with scientists to develop the Supplementary Curriculum Guides mentioned in the fifth bullet 'on the previous page. These guides are meant to support teachers in the classroom as they implement the science unit. The Supplementary Curriculum Guides include

- instructional modifications meant to focus learning on the standards, such as strategies for introducing systems thinking, emphasizing the planning of controlled investigations, and introducing specific vocabulary from the state Grade Level Expectations;
- clarifications and extensions of the teacher manual for each unit;
- tips for each lesson in the unit that provide ideas for use of materials, introduction of extended concepts, and use of strategies to promote skill development. Consultants have developed these tips by observing students doing investigations in classrooms, working with scientists, and incorporating national-level research on best practices; and
- additional lessons, developed by the consultants and scientists, that provide teachers with a lesson plan in which the Learning Cycle is explicit. An example is a lesson supporting teachers in addressing the second standard (inquiry), which focuses on fair tests/controlled investigations. Planning fair tests begins on a very basic level in kindergarten and grows in complexity in successive grades. For example, in a kindergarten unit, an investigative question is "What do land snails eat?" In a fifth-grade unit, an example of an investigative question is "What effect does vegetation have on soil erosion?"

Figure 18.5, page 368, provides a brief outline included in the Supplementary Curriculum Guides that shows teachers how to guide students in planning and conducting fair tests/controlled investigations and develop skills needed for proficiency on the state science assessment. This outline helps teachers see how this modification fits into the flow of the student science unit.

Figure 18.5 Steps to Help Students Prepare for the Washington State Science Assessment

An Instructional Strategy Based on What Scientists Do

- Students engage in explorations and investigations in order to gain ideas about a science concept as they participate in the science unit.
- Teacher charts student ideas about the concept students have been investigating.
- Teacher models how to turn student ideas into investigative questions.
- One question is chosen as the investigative question.
- Students plan an investigation with guidance from teacher.
- Students conduct investigation, collecting and interpreting data.
- Teacher elicits conclusions.

Consultants analyze the WASL-released items each year, leading to the modification or addition of lessons to support teachers in addressing the Grade Level Expectations more thoroughly. The consultants do not support teaching only to the state assessment; rather, they encourage teachers to address the Grade Level Expectations more specifically in their instruction so that students are likely to show proficiency on the assessment.

Expository Writing and Science Notebooks Classes and Supplementary Writing Curriculum

In the Expository Writing and Science Notebooks classes and Supplementary Writing Curriculum, the developer/consultant explains how to teach expository writing in the context of teaching science and how to develop and deepen students' scientific thinking and understanding in the process. This instructor also models how science notebooks should be used for formative assessment. These classes and curriculum are part of the Expository Writing and Science Notebooks Program, which the consultant has been developing over the last nine years. The program has three components:

1. *Professional development* available to all elementary teachers in the district. This component consists of a series of four workshops per grade level that introduce the overall approach and apply it to each specific unit following suggestions in the Supplementary Writing Curriculum.
2. A *Supplementary Writing Curriculum* for each of the 18 elementary science units. Each curriculum includes suggestions for integrating science and expository writing instruction in every lesson. Like the workshops, the curriculum explains how to use word banks, graphic organizers, and scaffolding for teaching expository writing and scientific thinking.
3. *Teacher leadership development* for four to nine teachers per grade level. These Lead Science Writing Teachers assist in developing and field-testing supplementary curriculum strands and materials for other elementary teachers, provide writing samples to be used in professional development and the Supplementary Writing Curriculum, and improve their own instructional practices by planning lessons and analyzing student notebooks with their colleagues using protocols developed in the program.

When students begin a science lesson, they take out their notebooks, write the date at the top of the next page, and when appropriate, write the focus or investigation question for the lesson. These questions keep both the students and teachers focused on the conceptual storyline of the unit. After an introductory class discussion, students may write their prediction about the outcomes of the session's investigation. Then, as they conduct their investigation, students collect and record qualitative and/or quantitative data. Students then discuss and analyze their data during reflective class discussions about their investigation.

In a separate instructional period, the teacher helps students learn how to write scientifically about their investigation; then, the teacher may provide scaffolding such as sentence starters or graphic organizers to guide students as they write in their science notebooks. Several times during the unit, teachers collect the notebooks to analyze certain entries and provide constructive feedback that focuses on the students' strengths and addresses weaknesses by asking questions that scientists might have about the entries.

After four years of intensive research on the science writing program, Inverness Research Associates came to the following conclusions:

- *Independent experts judge that the student work in science notebooks is, on the whole, more sophisticated in quality, and reflective of greater rigor and a higher level of learning of both science and writing, than is typical in science programs in other schools and districts that use similar science units.*
- *The Writing Program thus enhances to a significant degree the district's elementary science program, and it helps bolster the district's literacy program, including the extent to which those programs help students meet state standards.*
- *Participants in the Writing Program spend more time teaching science, teach more writing in science, have higher expectations for students with special needs, and follow the district's science curriculum more consistently than teachers who have little or no experience with the Expository Writing Program.* (Stokes, Hirabayashi, and Ramage 2003, p. ii)

The National Center for Research on Evaluation, Standards, and Student Testing (CRESST) at the University of California in Los Angeles (UCLA) has completed additional research on the impact of the science writing program on students' science learning. The center completed a detailed analysis of Seattle's fifth-grade students' state science assessment scores to determine if there is a significant impact when students' teachers have 7.5 hours or more of professional development in the science writing program. The existing data suggest the following:

> [The] elementary science program and particularly the Expository Writing and Science Notebook Program ... are showing benefits for students' fifth-grade science learning, as judged by WASL science scores; ... the program may be especially beneficial for low SES classrooms and schools; ... the findings reinforce the value of extended professional development; ... the findings for low SES classrooms, i.e., classes with relatively high concentrations of students who qualify for free and reduced lunch, reinforce the value of consistent and strong curriculum across grade levels, particularly for students

whose instructional resources outside the classroom may be relatively meager. (Choi and Herman 2005, pp. 24, 25)

Finally, the researchers found that students taught by teachers who had had three days of inquiry-science professional development for the grade level taught and over seven and a half hours of classes in the science writing outperformed other students with significantly higher scores (Figure 18.6). Ways that teachers can implement this successful approach is the subject of the book *Writing in Science: How to Scaffold Instruction to Support Learning* by Betsy Rupp Fulwiler (2007), the developer/implementer of the science writing program.

Figure 18.6 Impact of Teacher Participation in Science Writing Program on Student Science WASL Scores

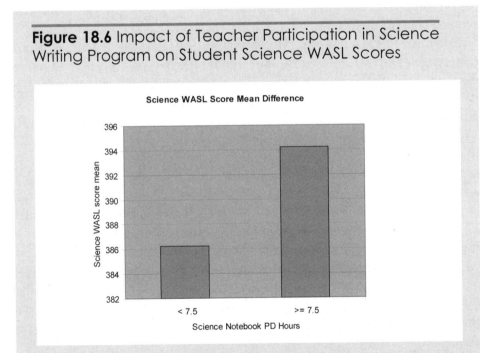

Source: Choi, K., and J. Herman. 2005. *Seattle School District Expository Writing and Science Notebooks Program: Using Existing Data to Explore Program Effects on Students Science Learning.* Los Angeles: National Center for Research on Evaluation, Standards, and Student Testing, UCLA Graduate School of Education and Information Sciences. Copyright © 2005 Choi and Herman, CRESST. Reprinted with permission.

WASL Prep Classes

The WASL Prep classes have focused on Grade Level Expectations for the first and second standards (systems and inquiry) because this is where scores are lowest. For the first standard, Properties, Structure, and Changes in Systems, teachers are shown how to make systems more explicit as students conduct investigations in their science units. To start, teachers are introduced to the state definition of a system: "an assemblage of interrelated parts or conditions through which matter, energy, and information flow" (p. 54 in the Grade Level Expectations document). This definition is broken down into manageable parts and teachers discuss the meaning of the terms *input, output, matter,* and *energy.* This language and the underlying concepts are new to many of the teachers, so the consultants model how to introduce the language in meaningful and relevant ways within the science unit lessons.

For students to understand the vocabulary, it is imperative to move from concrete to abstract, so a consultant models a three-part strategy to make this shift. The first step is to focus on the concrete system with which students are familiar. The next step is to focus on an illustration or diagram of the concrete system, which leads to using an abstract graphic organizer. In a WASL Prep class for fifth-grade teachers, for example, the consultant would model the three-part strategy with a stream table used in the Earth science unit. Teachers are shown how to introduce the stream table as a system. Teachers are next directed to have students identify the parts of this stream table system: the cup with a hole at the bottom, the plastic tub with a hole at one end, the soil (sand, gravel, humus, and clay), the water, and the bucket that receives the runoff. The consultant then leads a discussion to show teachers how to elicit from students how the parts of the system are interconnected and what the function is of each of these parts. Finally, the teachers are shown how to guide students in thinking about what keeps the stream table system functioning—the inputs and outputs of matter and energy.

After this discussion, the consultants show the teachers an illustration of the system that can be given to students. The teachers are shown how to guide the students in labeling the parts of this illustration as well as the inputs and outputs (see Figure 18.7). Then, the consultant models how to move the discussion from the illustration to the abstract graphic organizer (Figure 18.8, p. 374). After practicing with several systems using this three-part strategy, students can use the graphic organizer in Figure 18.9, page 374, to represent any system in the world around them. Teachers have said their students are excited about the holistic systems approach and are suddenly noticing interconnected systems all around them in the real world.

Figure 18.7 Stream Table System

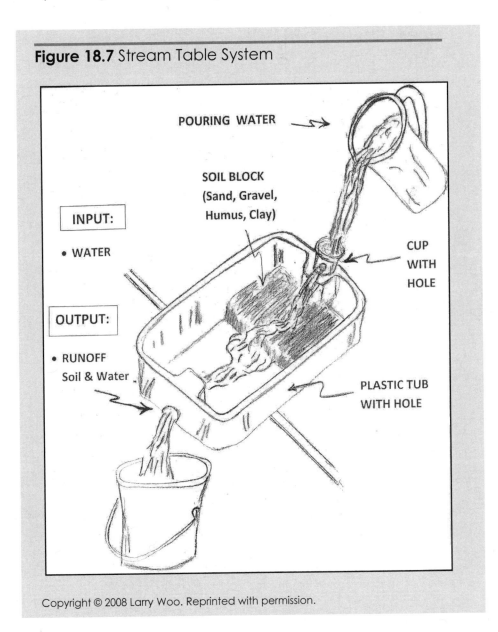

INPUT:

• WATER

OUTPUT:

• RUNOFF
Soil & Water

POURING WATER

SOIL BLOCK
(Sand, Gravel,
Humus, Clay)

CUP
WITH
HOLE

PLASTIC TUB
WITH HOLE

Figure 18.8 Graphic Organizer to Accompany "Stream Table System"

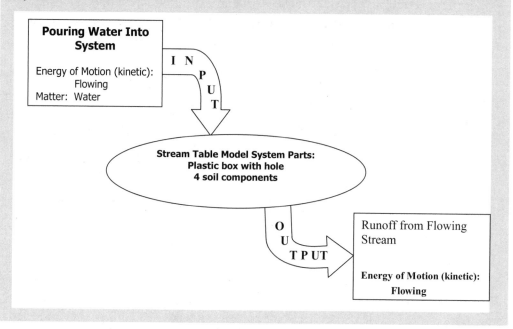

Figure 18.9 Graphic Organizer to Represent Any System

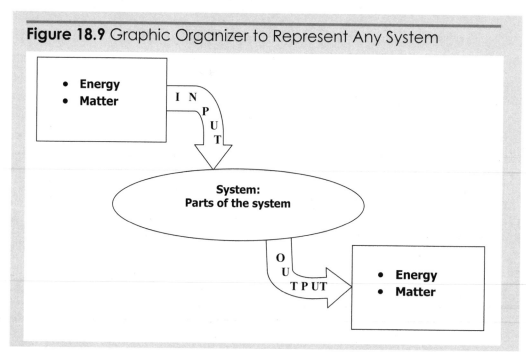

The other part of the WASL class addresses the second standard, or inquiry, by guiding students in planning a fair test/controlled investigation. The science consultant explicitly models for the teachers how to guide students in coming up with testable questions and then how to plan and conduct fair tests/controlled investigations where variables are identified. Finally, the science-writing consultant presents instructional writing strategies that support success on the science WASL. The consultant explicitly models, step-by-step, how to guide students in interpreting generated data and in writing a conclusion that includes the components required by the state assessment. The consultant also gives the teachers a suggested template that supports this modeling process. The template includes components that are not assessed on the state science assessment but that also promote higher-level thinking and more complex expository writing skills.

The program offers WASL Prep classes separately for third-, fourth-, and fifth-grade teachers so that the strategies can be learned specifically in the context of their own science units. Each of these classes is three hours. The classes are offered both after school and during the school day with substitute release time. One two-hour class is offered after school for teachers in the K–2 grade band. Primary teachers are shown how they lay the foundation for student understanding in the intermediate grades. In all the WASL Prep classes, the consultants emphasize how each grade level contributes to preparing the students for the fifth-grade state assessment. Teachers have reported that the components of this class helped them immensely because the content of the class is tied specifically to the concepts and skills of the science units they are teaching.

Science Content Courses

For seven years, the elementary science program offered summer content courses taught by scientists with assistance from consultants and/or lead teachers for each grade level so that the teachers would understand the content at a higher level than that at which the students are taught. Consultants and teacher leaders worked with the scientists to make sure that the course aligned with our curriculum and instructional approach and that the scientists were aware of the realities of the classroom. The courses were usually offered in August and sometimes in June right after school was let out. They were usually from 12 to 20 hours in length.

In recent years, the program has not been able to offer many science content courses. In most of these recent summers, a science partner collaborated with us and offered a content course for a particular grade level. In summer 2007, for example, the Seattle Aquarium, funded through its Ocean Science program (a NOAA [National Oceanic and Atmospheric Administration] Ocean Literacy Grant), offered a content course that addressed complex life science concepts underpinning both fourth- and fifth-grade life science units. Our staff values these courses and intends to include them when we can access the needed funds and time. Meanwhile, the consultants infuse as much content as possible into the Initial Use, the Expository Writing and Science Notebooks, and the WASL Prep classes. We continually work with scientists as we refine instructional strategies and develop supplementary curriculum for each science unit.

Conclusion

The elementary science consultants in Seattle Public Schools have been adamant about supporting teachers in learning how to address specific state Grade Level Expectations in each science unit. This focus is in contrast to giving teachers general ideas that they then have to apply on their own. Evidence of the success of this approach has been shown over the past five years in students' science notebooks and steady improvement in state science assessment scores. The expanded infrastructure developed with the National Science Foundation and other regional funds clearly has had a dramatic impact on how teachers prepare their students and how Seattle's fifth-grade students perform on the state science assessment.

Acknowledgments

This chapter is based on work supported by the National Science Foundation (Grants No. ESI-9554605 and EIS-0554651), the Stuart Foundation (Grants No. 2004-250, 2003-261, 2002-257, 2001-182, and 2000-175), and the Charles Simonyi Fund for Arts and Sciences. In addition, the work has been supported by Seattle's Alliance for Education; Amgen; The Boeing Company; the City of Seattle: the Seattle Aquarium and Seattle Public Utilities; Commonweal Foundation; George Rathmann Foundation; Institute for Systems Biology; Medtronic; Mental Wellness Foundation; Nesholm Family Foundation; Seattle Pacific University, Department of Physics; University of Washington, Department of Biostatistics, Department of Medicine, Division of Medical Genetics, and the Physics Department; Washington State LASER; and ZymoGenetics. An early phase of the science writing work was funded by Social Venture Partners. Any opinions, findings, and conclusions or recommendations expressed in this chapter are those of the authors and do not necessarily reflect those of the foundations or institutions.

References

Choi, K., and J. Herman. 2005. Seattle School District Expository Writing and Science Notebooks Program: Using existing data to explore program effects on students' science learning. Los Angeles: National Center for Research on Evaluation, Standards, and Student Testing (CRESST), UCLA Graduate School of Education and Information Sciences.

Fulwiler, B. R. 2007. *Writing in science: How to scaffold instruction to support learning.* Portsmouth, NH: Heinemann.

Office of Superintendent of Public Instruction (OSPI). 2005. Science K–10 Grade Level Expectations: A new level of specificity, Washington State's Essential Academic Learning Requirements, OSPI Document Number 04-0051.

Soholt, S. 2005. *Seattle science expository writing and notebooks.* San Francisco: KSA Plus Communications for the Stuart Foundation.

St. John, M., K. A. Fuller, and P. Tambe. 2002. Seattle partnership for inquiry-based science: A local systemic change initiative. End-of-project report. Inverness, CA: Inverness Research Associates. *www.inverness-research.org/reports/ab2002-05_Rpt_SeattleLSC_EndRpt.htm*

Stokes, L., J. Hirabayashi, and K. Ramage. 2003. Writing for science and science for writing: A study of the Seattle Elementary Science Expository Writing and Science Notebooks Program as a model for classrooms and districts. *www.inverness-research.org/reports_proj.html#Projects_S-U*

Washington State Office of the Superintendent of Public Instruction's website: *www.k12.wa.us.*

Appendix

"Hold That Soil"
Scenario Summary

Title: Hold That Soil						Grade: 5			
Description: A student investigated how adding grass sod to dirt affected the amount of dirt washed away.									
Item Descriptor	**EALR* Strand, Learning Target, and Item Characteristic**					**Item Type**			
	Properties of Systems	Structure of Systems	Changes in Systems	Inquiry in Science	Designing Solutions	Multiple Choice	Short Answer	Extended Response	Cognitive Level
1 Identify the volume of water poured into each plastic container as the variable kept the same (controlled).				IN02 2.1.2 b		C			I
2 Identify the variable changed (manipulated) as the amount of grass in each plastic box.				IN02 2.1.2 c		C			I
3 Describe how humans removing plants from the edge of a stream affects the health of stream ecosystem.					DE07 3.2.4 c	B			II
4 Generate a scientific conclusion including supporting data from the investigation.				IN03 2.1.3 a			SA		II
5 Describe and sort soil based on the physical property of particle size.	PR05 1.1.5 b					A			I

Source: Washington State Office of Superintendent of Public Instruction. Released Fifth-Grade Scenario, "Hold That Soil." Copyright © Washington State Office of Superintendent of Public Instruction. Reprinted with permission.

"Hold That Soil" Scenario (continued)

6	Describe the force acting on the soil as water flowing.			CH02 1.3.1 b				C		I
7	Identify the source of energy in the system as energy of motion.	PR04 1.2.2 a						B		I
8	Identify and describe how erosion changes the surface of the earth.			CH04 1.3.4 a					SA	II

Total	6	2	0	I: 6 pts II: 4 pts
Ideal Totals	3-6	1-2	0-1	I: 70% II:30%

* EALR = Essential Academic Learning Requirements (for the state of Washington science standards)

Directions: Use the following information to answer numbers 1 through 9.

After a rain storm, Margo saw a lot of soil on the sidewalks. Sidewalks next to grassy areas stayed much cleaner. Sidewalks next to areas without grass were covered with soil. Margo did the following investigation to see if grass would help protect soil from being washed away. Margo used grass sod (a layer of grass with its roots in soil) in her investigation.

Prediction:
Water will wash away less soil from the plastic box with *grass sod* compared to the plastic box with only soil.

Materials:
2 plastic boxes with drain holes
2 clear round pans
2 pieces of *grass sod*
2 water pitchers
water
2 metersticks
6 wooden support blocks

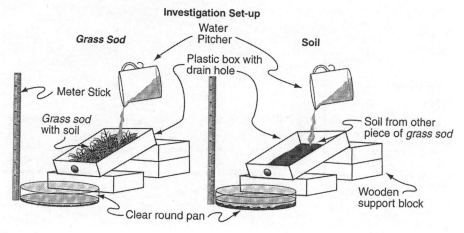

Investigation Set-up

Procedure:

1. Place one piece of *grass sod* into a plastic box with a drain hole as shown in the Investigation Setup diagram.
2. Remove the soil from the other piece of *grass sod* and place the removed soil into the second plastic box as shown in the Investigative Setup diagram.
3. From a height of 75 cm, pour 1 liter of water into each plastic box.
4. Let the soil that collects in each round pan settle for 15 minutes.
5. Measure the depth of the soil at the bottom of each clear round pan. Record this data in the table.
6. Repeat steps 3–5 two more times.

Data:

Material in Plastic Boxes vs. Depth of Soil in Clear Round Pans

Material in Plastic Boxes	Depth of Soil in Clear Round Pans After Pouring Water (cm)		
	1st Pour	2nd Pour	3rd Pour
Grass Sod	1.0	1.8	2.5
Soil Only	2.0	3.7	5.0

Item 1 What variable was kept the same (controlled) in this investigation?
  A. Amount of grass in each plastic box
  B. Depth of soil in each clear round pan
  C. Volume of water poured into each plastic box

Item 1 Information

Correct Response:	C
EALR Strand:	IN: Inquiry in Science
Grade Level Expectation:	IN02 (2.1.2) Planning and Conducting Safe Investigations Understand how to plan and conduct simple investigations following all safety rules.
Evidence of Learning:	(b) Given a description of a scientific investigation, items may ask students to identify the question being answered in an investigation.

Performance Data

Use the space below to fill in student performance information for your school and district.

Item 1 Responses * correct response	Item 1 Percent Distribution of Responses		
	School	District	State
A			9.9
B			11.0
*C			79.0
NR (No Response)			0.2

Item 2 Which variable was the changed (manipulated) variable in this investigation?
  A. Amount of soil in each plastic box
  B. Amount of grass in each plastic box
  C. Amount of water poured into each plastic box

Item 2 Information

Correct Response: B
EALR Strand: IN: Inquiry in Science
Grade Level Expectation: IN02 (2.1.2) Planning and Conducting Safe Investigations
                           Understand how to plan and conduct simple investigations
                           following all safety rules.
Evidence of Learning: (c) Given a description of a scientific investigation, items may
                           ask students to identify one variable changed (manipulated)
                           in the given investigation.

Performance Data

Use the space below to fill in student performance information for your school and
district.

Item 2 Responses * correct response	Item 2 Percent Distribution of Responses		
	School	District	State
A			24.2
*B			53.0
C			22.3
NR (No Response)			0.5

Item 3   Based on the results of Margo's investigation, what might happen to a stream
ecosystem if people removed all the plants from a hillside next to the stream?

    A. More water would flow into the stream ecosystem.

    B. More soil would be washed into the stream ecosystem.

    C. More oxygen would be carried into the stream ecosystem.

Item 3 Information

Correct Response:	B
EALR Strand:	DE: Designing Solutions
Grade Level Expectation:	DE07 (3.2.4) Environmental Resources and Issues Understand how humans depend on the natural environment and can cause changes in their environment that affect their ability to survive.
Evidence of Learning:	(c) Given an adequate description of an appropriate system, items may ask students to identify or describe the effects of humans on the health of an ecosystem.

Performance Data

Use the space below to fill in student performance information for your school and district.

Item 3 Responses * correct response	Item 3 Percent Distribution of Responses		
	School	District	State
A			14.9
*B			76.1
C			8.2
NR (No Response)			0.8

Item 4 Write a conclusion for this investigation.

In your conclusion, be sure to:
- Describe whether the prediction was correct.
- Include **supporting** data from the Material in Plastic Boxes vs. Depth of Soil in Clear Round Pans table.
- Explain how these data **support** your conclusion.

Prediction: Water will wash away less soil from the plastic box with grass sod compared to the plastic box with only soil.

_____

_____

_____

_____

_____

Item 4 Information

Score Points:	2
EALR Strand:	IN: Inquiry in Science
Grade Level Expectation:	IN03 (2.1.3) Explaining
	Understand how to construct a reasonable explanation using evidence.
Evidence of Learning:	a) Given a description of a scientific investigation, items may ask students to identify or write a conclusion that answers the investigative question or explains whether the prediction was correct including supporting data from the investigation (e.g., *Margo's prediction is correct. In the box with grass sod, only 2.5 cm of soil ended up in the clear pan. In the box with soil and no grass 5.0 cm of soil ended up in the clear pan. Less soil washed away with the grass sod*).

Performance Data

Use the space below to fill in student performance information for your school and district.

Item 4 Score Points	Item 4 Percent Distribution of Score Points		
	School	District	State
0			57.7
1			16.2
2			24.3
NR (No Response)			1.9
Mean			0.7

Linking Assessment to Student Achievement in a Professional Development Model

Janet L. Struble, Mark A. Templin, and Charlene M. Czerniak
University of Toledo

This chapter reviews assessment as a key component in a professional development model implemented in a National Science Foundation–funded program entitled Toledo Area Partnership in Education: Support Teachers as Resources to Improve Elementary Science (TAPESTRIES). We illustrate how multiple aspects of the professional development model were linked to assessment and how assessment became the driving force in participating teachers' professional development, in their classroom practice, and in the leadership of the program and the school districts.

We begin this chapter by providing background information about the project and participating school districts. Next, we describe the role that assessment plays in each of the four phases that make up the Haney-Lampe professional development model (Haney and Lampe 1995), which provided the project's framework. In the *planning stage*, the TAPESTRIES directors used the 5-E Learning Model (Bybee 1997). We used nontraditional assessments to gauge teachers' understanding at each stage of that model. A description of a typical content-based session provides an example of the teachers' Summer Institute experiences, which initiated the *training phase*. In the *application phase* of the Haney-Lumpe model, which started the school year following the Summer Institute, participating teachers taught the curriculum that they had experienced in the summer. Support Teachers—teachers who were released full time from teaching to serve as leaders in the district—helped participating teachers implement the newly adopted science curriculum. We illustrate one school year session pertaining to "research lessons" (commonly known as the lesson study approach) in which each classroom teacher participated.

Lastly, we describe how assessment was deeply rooted in the professional development of the teachers to answer the question, "How do we know they know?" (*follow-up phase*). We depict the process of working with the support teachers to develop assessments aligned to the Ohio Proficiency Tests that classroom teachers field-tested and later implemented in their classrooms. We discuss the obstacles that were encountered in implementing an assessment-focused professional development design.

Background

TAPESTRIES was a collaborative partnership conducted from 1998 to 2004, among the fourth largest urban school district in Ohio—Toledo Public Schools (TPS); a suburban district—Springfield Local Schools (SLS); and the Colleges of Education and Arts and Sciences at two universities— University of Toledo (UT) and Bowling Green State University (BGSU). TAPESTRIES was designed to achieve a comprehensive, systemwide transformation of K–6 science education and to improve science teaching and learning through sustained professional development of all K–6 teachers.

TAPESTRIES used the Haney-Lumpe (1995) science teacher professional development model, which is made up of four phases: planning, training, application, and follow-up. Because a district's strong commitment to assessment-based reform efforts can profoundly affect its effectiveness (Aschbacher 1993), UT and BGSU science education faculty and scientists, along with TPS and SLS district administrators, principals, and teachers, were involved in all four phases of the project.

Planning Phase

Project Staff Retreat

To establish a cohesive project staff with shared philosophies, expectations, and true collaborative decision making, the entire project staff (science educators, scientists, elementary Support Teachers, and graduate assistants) attended a two-day retreat each spring from 1998 to 2003. Retreat leaders prepared staff for the upcoming Summer Institute by informing them of the latest research on science teaching and learning, by reflecting on comments made by teachers' evaluations from previous years, and by developing a plan of action in content and pedagogy for the Summer Institute and the following academic year.

Support-Teacher Professional Development

Support Teachers (16 elementary teachers who were given full-time release from teaching responsibilities) helped classroom teachers to implement science inquiry and to deal with state assessments. Support Teachers designed and executed district action plans for improving science literacy. They also honed their leadership skills by participating in the two-week Summer Institutes, two three-semester-hour courses, a staff retreat, and a spring symposium.

TAPESTRIES Support Teachers served as lead teachers for their districts. They acted as liaisons between the project faculty and staff, the district, and the classroom teachers; modeled and assisted in classroom teaching and assessment; assisted with the district plan for building material distribution and replacement; and acted as peer coaches for teachers. The Support Teachers aligned the K–6 science curriculum with Ohio's proficiency standards because both the TPS and SLS faced severe challenges in raising student achievement in science according to those standards. In 1997, in fact, TPS had been ranked as an "academic emergency" school district, having attained only six of the 27 performance standards. Once the Support Teachers discovered that some topics covered on the state tests were not being taught in the TPS curriculum, the Support Teachers determined the appropriate grade level(s) for teaching the missing science topics and then created lessons to fill the gaps and assessment items similar to items on the proficiency tests.

The classroom teachers then piloted the assessments in their classrooms, analyzed the assessments' effectiveness in measuring student learning of the science topic, and submitted to the Support Teachers an analysis of the assessment items along with student work samples. The Support Teachers analyzed the teachers' critiques of the assessment items and revised the items accordingly. The revised assessments were piloted and further revised. The final products were placed on the TAPESTRIES website so all teachers could have access to them and so that they could be used in future professional development.

An example of how this procedure worked was the teacher use of the FOSS "Earth Materials" kit. The module, which was taught in grade 3 in the Toledo Public Schools, dealt with the following grade 4 proficiency outcomes: (1) create and/or use categories to organize a set of objects, organisms, or phenomena; (2) select instruments, make observations, and/or organize observations of an event, object, or organism; and (3) evaluate a simple procedure to carry out an exploration. After identifying the proficiency outcomes addressed by "Earth Materials," the Support Teachers decided on the science concept or process skill to be assessed. The sample assessment in Figure 19.1 was used to assess the student's understanding after the completion of the "Earth Materials" kit. Next, Support Teachers identified a learner outcome as well as assessments that evaluated student knowledge on each of the three levels of difficulty measured on the Ohio tests: acquiring information, processing information, and extending knowledge.

The learner outcome for the example shown in Figure 19.1 is "The student will use his or her prior knowledge about Earth materials and apply it to a new situation." The assessment item included an answer key or rubric. Classroom teachers critiqued the assessment and the associated rubric and submitted suggestions for improvement on both by providing the Support Teachers with student samples (samples that demonstrated the flaws in the assessment and rubric). Notice in the sample in Figure 19.1 that the Support Teacher did not receive any student samples for a score of 4.

We used the same method to develop assessment items using Thinking Works strategies (Carr, Aldinger, and Patberg 2000). Figure 19.2, page 392, is an Anticipation Guide (Carr, Aldinger, and Patberg 2000) for "Earth Materials." Before and after the activity, the Anticipation Guide gave the teacher information about the student's understanding of "scratch test."

The Support Teachers attended national conferences to meet with lead teachers from other funded grant projects to compare district successes and challenges. They were especially surprised to learn that they faced many challenges in common with other projects. The TAPESTRIES grant directors routinely included Support Teachers in their research presentations and encouraged them to present at local, state, and national meetings. TAPESTRIES Support Teachers served as lead teachers, and their contributions were so valued that the Toledo Public Schools found additional funding to keep the services of half of these individuals beyond the original NSF funding.

Figure 19.1 Extending Knowledge Assessment Item for FOSS "Earth Materials"

You are given a mineral that came from the Moon. What could you find out about it?

Describe two tests that you could perform on the mineral and list the properties of the mineral that you can identify.

ANSWER KEY:
TESTS: vinegar or acid; hardness; observation

RUBRIC:

4 points	Student explains 2 tests and the properties of the mineral.
3 points	4 POINT ANSWER with minor errors.
OR	
	Student only explains the 2 tests.
2 points	Student explains 1 test andthe property of the mineral.
1 point	Student explains 1 test.

**STUDENT SAMPLES
4 POINTS**

none available

3 POINTS

scratch test - how hard it is
vinegar
look for minerals

Student does not mention specific minerals for the vinegar test.

smell test , penny scratch
How it small | How hard it is

Student does not mention specific minerals for the smell test.

You are given a mineral from the moon. Name two tests that you could perform on it to list some of its properties. What could you find out about it?

scratched it it will have stones and conarocks. it fizing

Student explains 2 tests, but does not tell the properties.

Figure 19.2 Anticipation Guide for "Earth Materials"

Grade Level: 3
FOSS KIT : EARTH MATERIALS
Activity 2, "Scratch Test"
Anticipation Guide

Name_____

Before Activity   *After Activity*

☐ **Agree**   ☐ **Disagree**   1. Rocks are made of materials.   ☐ **Agree**   ☐ **Disagree**

☐ **Agree**   ☐ **Disagree**   2. Minerals can be broken down into other earth   ☐ **Agree**   ☐ **Disagree**
                                  materials.

☐ **Agree**   ☐ **Disagree**   3. You can scratch all minerals.   ☐ **Agree**   ☐ **Disagree**

☐ **Agree**   ☐ **Disagree**   4. Minerals are put in order by how hard they are. ☐ **Agree**   ☐ **Disagree**

Created by Lori Seubert.

School Administrators and Principals Development

The building principals and other district administrators attended yearly administrator meetings and retreats. The administrator meetings focused on inquiry-based science, cooperative learning, and assessment. A few principals participated in the entire TAPESTRIES program, including the Summer Institute and academic year monthly meetings. They were encouraged to develop special projects for use back in their own schools. Over half of the principals from the Toledo and Springfield public schools also participated in a one-day retreat and a follow-up session during the academic year.

CHAPTER 19

Training Phase

The training phase of the professional development model started with six, two-week-long Summer Institutes that occurred annually from 1998 to 2003 on the UT and BGSU campuses. The sessions were aligned with the National Science Education Standards (NRC 1996) that focus on science content knowledge, science process, inquiry-based instruction, and assessment. The institutes, which were grouped according to grade levels, were co-taught by scientists from UT and BGSU, science educators, and Support Teachers from TPS and SLS and focused on assessment as the driving theme of professional development.

We defined *assessment* in our professional development model as an ongoing process of gathering information so that the teacher can make decisions to further student understanding of a science concept. These assessments, therefore, needed to be embedded not only in the curriculum of TPS and SLS but also in the learning model used within the professional development effort. The National Research Council (2001b) advocates such an approach as an essential design consideration in science education reform.

Teachers participated in two types of sessions: (1) general informational sessions presenting topics such as inquiry, constructivism, cooperative learning, graphic organizers, and nature of science and (2) content-based sessions in which teachers worked through activities of the districts' K–6 adopted curriculum (FOSS, STC [Science and Technology for Children], and Scholastic kits) and discussed the activities with scientists. Both types of sessions used the constructivist 5-E Learning Model (Bybee 1997) as the structure for planning and delivering the lessons. (The five Es of the model are *engagement, exploration, explanation, elaboration,* and *evaluation.*)

The following is a description of a typical morning or afternoon content-based session with assessment as a focus and following the 5-E model format. The science educator/scientist team started the session with an "engagement." The engagement stage served to assess teachers' preconceptions and science content understanding and set the stage for the lesson. "Engagement tools"—some type of graphic organizer, such as a brainstorming wheel, concept map, or KWL chart, or a Thinking Works[1] Strategy (Carr,

---

[1]Thinking Works (2000), developed by Dr. Eileen Carr and associates at the University of Toledo, provides strategies that students can use to decode information in reading text and to develop their writing skills. More information can be found on their website: *www.getthinkingworks.com.*

Aldinger, and Patberg 2000) such as an anticipation guide or concept of a definition—were used to determine the teachers' prior knowledge and assess their understanding of the science concept. Using the FOSS "Variables" kit as an example of a grade 5 session, teachers filled out a KWL chart, first writing "What I **K**now" and then "What I **W**ant to Find Out" about pendulums for the first investigation, entitled "Swingers." The science educator/scientist team used what the teacher wrote to determine the course of the next stage, "exploration." See Appendix A, page 402, for a list of the Classroom Tools that were used at each stage of the 5-E Learning Model to assess teachers' understanding.

During the "exploration" stage, teachers completed an inquiry-based activity from their adopted curriculum (FOSS, STC, and Scholastic kits) for their grade levels. Using our example, grade 5 teachers worked through the "Swingers" lesson. Each pair of teachers constructed a "swinger" of a specific length using string, a paper clip, and a penny. They hung the pendulum from a pencil and counted the number of swings for 15 seconds. Data from the investigation were collected and graphed. In addition to the worksheets provided in the kit, teachers completed a graphic organizer, "Science Activity Report," by filling in information about the activity under the following headings: Problem/Question to Explore, Methods/Procedures, Data Collection, and Conclusions/Discussion. The scientist and science educator used the graphic organizer to assess the teachers' learning that took place in the activity and made decisions for how to proceed. By working through each activity of the adopted curriculum as if they were students, the teachers came to realize the joys, frustrations, and challenges of lessons that their students would experience in the coming school year.

Based on the embedded assessment from the "exploration" phase, the scientist and teacher educator decided what needed to transpire in the "explanation" phase. Scientists helped the teachers through that phase by presenting the scientific information necessary to understand the concepts in the lesson. The scientist used the misconceptions identified in the "engagement" phase to clarify the teachers' understandings. Discussions revolved around the principles of a pendulum—for example, the weight of the bob on the pendulum does not affect the number of swings that occurred. The teachers learned how to identify misconceptions their students might hold and ways to help students to understand the science concept. The session leaders linked the "student" information discussion to a discussion of proficiency outcomes. The "Swingers" investigation was aligned to two grade 6 Ohio Science Proficiency Outcomes: (1) Student makes inferences from observations of phenomena and/or events and (2) Student recognizes the advantages and/or disadvantages to the user in the operation of simple technological devices.

The science inquiry lesson, as well as the science content, was extended in the "elaboration" stage by applying the principles learned to other situations, including real-life examples. For the "Swingers" investigation, teachers investigated double-decker pendulums and the history of pendulums by researching clocks as well as the biographies of scientists such as Galileo, Huygens, or Foucault. The "Column Note Taking" graphic organizer (a chart where the topics or main ideas are listed on the left side and supporting details for the subtopic are listed on the right side) was used to assess teachers' understanding at the "explanation" and "elaboration" stages. Scientists and science educators helped teachers assess their understanding of the science content and pedagogy needed to teach the lesson.

In the "evaluation" stage, kit-based or/and teacher-designed assessments and "closure" were featured. In addition to completing a classroom tool for the final stage, teachers completed assessments derived from the curriculum kits, the state of Ohio Science standards, other NSF projects, and teacher-developed assessments.

Closure is defined as a process of reviewing and clarifying the key points of a lesson, tying them into a coherent whole, and making sure the students understand the concepts that were taught. The assessment is embedded in the closure of the lesson. This last assessment informs the teacher about student learning, and the teacher decides on the next actions in the classroom. In our example, teachers filled in the "L" in the KWL chart ("What I Learned") and answered questions from FOSS "Variables." To provide closure to the session, the scientist and science educator led the teachers through a reflective process of looking at what had transpired throughout the lesson. Examples of questions posed by the scientist and science educator were "What did we do as an engagement?" "How did I know that you understood …?" "What does your assessment tell me about your learning?"

By engaging in this "think out loud" process, along with filling in a 5-E lesson plan graphic organizer (a lesson plan organizer we developed that lists each phase of the 5-E model with a description of what the teacher and students are doing during the inquiry-based lesson), teachers planned and reflected on assessment as they implemented lessons in the classroom. This key component of the TAPESTRIES Summer Institute is captured in the National Research Council's *Classroom Assessment and the National Science Education Standards* (NRC 2001): "Reflection and inquiry into teaching, and the local and practical knowledge that results, is a start towards improved assessment in the classroom" (p. 81).

Application Phase

The "application" phase of the Haney-Lumpe professional development model occurred during the academic year. Its focus was on the implementation of the curriculum, teaching strategies, and assessments that teachers learned about during the Summer Institutes. The National Science Education Standards (NRC 1996) state that professional development must be continuous and must provide teachers with opportunities to apply, examine, and reflect on their practices. Thus, a critical piece of this phase was the Support Teachers. These teacher leaders were in the classrooms of the teachers who had participated in the Summer Institute. The Support Teachers helped their peers in implementing the curriculum, modeling the teaching strategies, providing the extra pair of hands often needed in a science classroom, and encouraging teachers to continue implementing inquiry teaching even when strategies seemed to fail.

The Support Teachers were also heavily involved with the planning of the professional development sessions that took place each month during the academic year. This planning and implementation of the monthly sessions was a collaborative effort among university (UT and BGSU) science education faculty and scientists, TAPESTRIES staff, and district support teachers. The Support Teachers visited an assigned cohort of teachers biweekly to help facilitate the science reform in the classroom. These teachers had to be very flexible and adaptable. Their roles included the following:

- Providing assistance with science curriculum preparation
- Giving strategies for teaching science
- Supplying science content background information (if necessary, with the help of the university scientists)
- Assisting with classroom and district science performance-based assessments
- Modeling science lessons
- Offering peer coaching for the classroom teacher.

We stated the theme of assessment for the professional development during the school year by asking, "How do we know they know?" We used the 5-E Learning Model lesson organizer to plan each of these academic school year sessions. The sessions occurred once a month from September to April and lasted two hours. Appendix B lists these professional development sessions. During the academic year, teachers completed an assignment, which included the following components: planning and teaching a science lesson using the 5-E Learning Model lesson organizer; developing an assessment and providing student work samples; writing a reflective essay focusing on the appropriateness of their assessments and student learning; and reading research articles pertaining to their lesson.

The professional development plan for January's class (to assess teacher's skills in implementing and assessing inquiry) is used to illustrate a typical monthly session. In the "engagement phase," classroom teachers were asked in groups to discuss what they remembered about the NSF–Horizon Research's "Classroom Observation Protocol" (*www. horizon-research.com/LSC/*) presented at the Summer Institute by filling in the "Brainstorming Wheel." This graphic organizer contains a circle with spokes, on each of which the teacher writes something the students have stated knowing about the concept being taught. The Horizon protocol evaluates a lesson in four categories: design, implementation, content, and culture. In the "exploration" stage, teachers in groups evaluated a videotaped lesson according to the protocol. The teachers gave the video a low quality rating of "ineffective instruction." For the "explanation stage," the teachers gave evidence for their rating and small group discussions took place. Teachers discussed the ineffectiveness of the lesson and offered suggestions for improvement. For the "elaboration" stage, the teachers viewed and rated another video. This video received a very high rating of "exemplary instruction"—that is, highly likely to enhance most students' understanding of the science concept being taught. The teachers then compared both videos using a "Compare and Contrast" chart (a chart listing two concepts/items with similarities and differences according to several attributes), noting the similarities and differences of the teaching featured in the videos. This activity was used to assess the teachers.

For January's assignment, the classroom teacher conducted a "research lesson" (a Japanese-style lesson study). The teacher wrote a lesson in the format of the 5-E Learning Model lesson organizer. The teacher's assigned Support Teacher viewed the lesson, critiqued its effectiveness using the Horizon "Classroom Observation Protocol," and provided written feedback to the teacher. Subsequently, the teacher wrote a two-page reflective analysis of the lesson identifying specific strengths and weaknesses of the lesson, giving suggestions for improvement in the teaching and assessment of the lesson. The research lesson assignment gave each teacher an opportunity to analyze his or her teaching and receive constructive feedback from a peer in a nurturing environment.

The Relationship Between TAPESTRIES and Student Achievement

We conducted two research studies to test the effectiveness of TAPESTRIES with regard to student achievement: (1) an analysis of grades 4 and 6 lessons that were written by teachers at the end of the Summer Institute and at the school year and (2) an examination of the program's relationship to Ohio science proficiency scores. A summary of each study follows.

The first study sought to determine whether lesson plans written by classroom teachers improved after the yearlong professional development. The research question was, "Did the quality of teacher's lesson plans improve after being enrolled in a professional development program?" Effective lesson planning was defined as using the criteria established by Horizon Research (1998). For this study, we used only the design category for analysis. We designed two qualitative rubrics to triangulate the Horizon ratings, one to measure lesson improvement and the other to measure teachers' perceptions of student learning. Qualitative rubrics measured lesson improvement and teachers' perceptions of student learning. We selected for analysis a random sample of 22 fourth- and 22 sixth-grade teachers' lesson plans. These lessons were developed by teachers in TAPESTRIES from 1998 through 2003. The sample included only those lessons that had pre-lesson plans (lessons written by the teachers at the beginning of the program), post-lesson plans (lessons written during the academic year following summer training and academic year training), and reflective analyses (teacher's self-reflections after teaching the lesson). We analyzed the lessons quantitatively and qualitatively.

The TAPESTRIES program appeared to have an impact on the quality of instructional design. The lesson improvement rubric, which assessed the effective aspects of the science lesson by looking at its design, appeared to be a good indicator of lesson plan quality for both grade levels as defined by the Horizon instrument. The student learning rubric, which analyzed teachers' reflective essays pertaining to lesson improvement and student learning, was a good indicator of teachers' attention to student learning at the fourth-grade level, but less so at the sixth-grade level.

The second study tested the effectiveness of the professional development model with relationship to science achievement scores. This study is discussed at length in Czerniak et al. (2005). In summary, the findings were as follows:

- Science proficiency scores improved after the implementation of the TAPESTRIES program in Toledo Public Schools.
- Sixth-grade proficiency scores improved at schools highly involved in TAPESTRIES as compared to scores at schools minimally involved in the program.
- Student achievement (fourth and sixth grade) differed significantly between the schools with the highest percentage of teachers' professional development (PD) hours and those with the lowest.
- The cumulative effect of TAPESTRIES-trained teachers is associated with increased student achievement.
- When comparing the percent pass rate of fourth- and sixth-grade students whose teachers participated in the TAPESTRIES program, the TPS schools outranked all other large urban school districts in Ohio (Toledo is the fourth largest city) on the 2002 science proficiency tests.

Some of the obstacles we faced in the TAPESTRIES program are the same ones that other large systemic change programs face. School reform is a slow process, and school/university partners must promise long-term commitments to sustain systemic change. Not all of Toledo and Springfield administrators and teachers joined us in the science curriculum reform efforts. To improve both science education and student proficiency scores, schools and teachers could follow the lead of the Toledo and Springfield teachers and administrators that participated in TAPESTRIES. District boards and administrators need to actively support TAPESTRIES (and/or similar) programs after funding ceases to help maintain and further refine these effective educational innovations.

Conclusion

The TAPESTRIES project was successful in developing a comprehensive school science program through the sustained professional development of K–6 teachers in the Toledo Public Schools and Springfield Local Schools. The project helped prepare scientifically literate students who can comprehend and use science while being successful on high-stakes statewide science assessments in Ohio. We believe that the focus on assessment throughout the program helped us focus the program's activities and reach targeted outcomes.

Acknowledgments

Aspects of this chapter were adapted from Czerniak, C. M., S. Beltyukova, J. Struble, J. J. Haney, and A. T. Lumpe. 2005. Do you see what I see? The relationship between a professional development model and student achievement. In R. E. Yager (Ed.), *Exemplary Science in Grades 5–8: Standards-Based Success Stories*, pp. 13–43. Arlington, VA: NSTA Press. This research was supported in part by funding from the National Science Foundation (NSF), project no. 9731306. The views expressed here are not necessarily those of NSF.

References

Aschbacher, P. R. 1993. *Issues in innovative assessment for classroom practice: Barriers and facilitators.* (CSE Technical Report 359.) Los Angeles, CA: National Center for Research on Evaluation, Standards and Student Testing.

Bybee, R. 1997. *Achieving scientific literacy: From purposes to practices.* Portsmouth, NH: Heinemann.

Bybee, R., and N. M. Landes. 1990. Science for life and living: An elementary school science program from Biological Sciences Curriculum Study (BSCS). *The American Biology Teacher* 52(2): 92–98.

Carr, E., L. Aldinger, and J. Patberg. 2000. *Thinking Works for early and middle childhood teachers.* Toledo, OH: The University of Toledo.

Czerniak, C. M., S. Beltyukova, J. Struble, J. J. Haney, and A. T. Lumpe. 2005. Do you see what I see? The relationship between a professional development model and student achievement. In R. E. Yager (Ed.), *Exemplary Science in Grades 5–8: Standards-Based Success Stories* (pp. 13–43). Arlington, VA: NSTA Press.

Czerniak, C. M., J. L. Struble, M. A. Templin, and L. Ballone. 2005. An examination of lesson quality using the 5 E model in a professional development program. Paper presented at the Association for Science Teacher Education National Conference, Colorado Springs, CO.

Haney, J. J., and A. T. Lumpe. 1995. A teacher professional development framework guided by science education reform policies, teachers' needs, and research. *Journal of Science Teacher Education* 6(4): 187–196.

National Research Council (NRC). 1996. *National science education standards.* Washington, DC: National Academy Press.

National Research Council (NRC). 2000. *Inquiry and the national science education standards.* Washington, DC: National Academy Press.

National Research Council (NRC). 2001a. *Classroom assessment and the national science education standards.* Washington, DC: National Academy Press.

National Research Council (NRC). 2001b. *Knowing what students know: The science and design of educational assessment.* Washington, DC: National Academy Press.

Appendix A

"Classroom Tools" for Each 5-E Learning Model Stage to Assess Understanding

Stage of 5-E Learning Model	Classroom Tools		Classroom Tools	
	Graphic Organizers	Purpose of the Assessment	Thinking Works™	Purpose of the Assessment
Student focus				
Engagement				
Exposes prior knowledge	Brainstorming Wheel (a circle with spokes; on each spoke, teacher enters something that students state they know about the concept)	To learn students' prior knowledge through a spontaneous response	Anticipation Guide (four to seven statements that present the "big ideas" of the science concepts that will be taught; the student is asked to agree or disagree with the statements)	To learn students' prior knowledge about the "big ideas" of the science concepts that will be taught
Focuses thinking on the concept to be learned	Concept Map (a visual representation of a concept with underlying ideas written in a circle connected by lines and descriptions to depict relationships between the ideas)	To help students to look at the relationships between topics within a concept		
Provides questions that one wants to find answers to	KWL Chart (K & W part) (a three-column chart in which the student answers the following questions: "What do I Know?" "What do I Want to find out?" and "What have I learned?")	To activate background knowledge to create questions about the science concept		

Exploration

Benchmark	Activity	Purpose
Plans and carries out a scientific investigation	Deciding (a chart in which the student finishes the following prompts: "The problem is," "I think that," "I find out by," "I found out that," and "Conclusions") Science Report Activity (a sheet that the student fills in with information under the following headings: "Problem/Question to Explore," "Methods/Procedures," "Data Collection," and "Conclusions/Discussion")	To act as a guide for the student to carry out an investigation in which the student identifies the problem, writes a plan to solve it, supports his or her thinking by gathering evidence, makes observations, and then draws a conclusion
	Exploration Chart (student fills in information in the following boxes: "Exploration Question," "Prediction," "Observations and Resource Information," "Think Pair Share: Conclusions," and "New Questions")	To guide the student as he or she completes a science investigation by having the student write predictions, record observations and data, and summarize major findings
Plans and carries out a scientific investigation that contains more than one solution	My Best Guess (a chart that the student finishes using the following categories: "Title/Topic," "What I Know," "Actions: Option 1, Option 2, Option 3," "Results 1, 2, 3," "Best Choice," and "Reflection")	To guide student thinking by first exposing prior knowledge and then designing an investigation in which there are several ways to test one's idea, state the results, and predict what might happen if something is changed

Records observations	Observation/Inference T-Chart (a chart in form of a "T" with left side headed "Exploration Observations," right side headed "Exploration Inferences," and lower center box labeled "Exploration Conclusions")	To guide students to record observations and infer what is taking place		
Explanation				
Explains concepts learned	Central Idea Chart (a chart that contains a circle where the topic is written, with connecting boxes to list related items, issues, events, or categories)	To illustrate students' understanding of how items, issues, events, or categories are clustered around a central idea	Concept of a Definition (student defines a word by answering the following questions: What is it? What is it like? What are some examples?)	To help the student demonstrate understanding of science vocabulary
Uses the experiences of the "Exploration" phase to explain the newly learned concepts	Storyboard (a sequence of boxes like a cartoon strip to depict the order of ideas or events) Time Line/Sequence Chart (a vertical line with horizontal divisions to list the sequence of ideas or events)	To guide students to identify main ideas or events and then sequence them in a logical order	About Point (student writes the topic with three supporting phrases, then composes a paragraph about the topic)	To help the student construct sentences and paragraphs in a systematic way to explain science concept(s)

	Assessment Tool	Purpose	Assessment Tool	Purpose
Listens and tries to comprehend the explanation the teacher offers	Column Note Taking (a chart in which the student lists the subtopics or main ideas on the left side and provides supporting details for the subtopics on the right side)	To help students think about, organize, and record information		
Elaboration				
Applies newly learned concepts or skills in a new context	Noting What I've Learned (a chart in which the student fills in four "Main Ideas/Key Words/Questions/Drawings" on the left side and writes three supporting facts for each on the right side under the heading "What I've Learned")	To have a student illustrate his or her thinking, together with a written explanation of the science concept(s)	Question, Clues, Response (student writes a question across the top of the paper, lists clues or supporting evidence to answer the question, synthesizes the information under the "About Point Guide," and summarizes the information by writing a paragraph)	To help the student formulate questions, provide evidence to answer the questions, and write a general conclusion
	Concept Map (student adds newly learned information to original concept map completed before the lesson)	To have student link newly learned information to the concept map drawn at the beginning of the lesson	About Point Writing Response (tool combines three to five "About Points" so that a student can write a summary made up of three to five paragraphs with a concluding paragraph)	To help the student construct sentences and paragraphs with details to explain science concept(s)

Evaluation		
Provides evidence of achieving the learner outcomes	Compare and Contrast (a chart in which the student lists two concepts and tells how the concepts are alike and different. For the differences, the student lists the attribute on the center line and the differences on opposite sides)	To illustrate students' understanding of how issues or themes can have similar and different attributes or characteristics
	KWL Chart (L part) (complete the third column in the KWL chart: "What have I Learned?")	To provide evidence of learning under the heading "What have I Learned?"
	Venn Diagram (a chart with two or three overlapping circles; the commonalities of concepts or objects are listed on the overlapping areas and the differences are written on the area of the circle outside the overlapping area)	To provide a visual representation of students' knowledge of important relationships between similar objects or events by listing categories that are overlapping
	Classifying Chart (a chart with columns; general categories are written at the top of each column and attributes are listed under each category)	To guide students to classify information by categories
	Anticipation Guide (student fills in the same anticipation guide that he or she completed before the activity/lesson after the lesson is finished)	To assess what was learned by responding to statements at the end of the lesson

Appendix B

School-Year Professional Development

Month	Topic	TAPESTRIES Objectives for the Session
September	Assessment: What is it?	To inform the teachers about the National Assessment Standards and to provide examples of how to translate these standards into their classroom To help teachers understand the difference between formative and summative assessment
October	Questioning Strategies	To use Bloom's taxonomy of cognitive development to examine the levels of questioning being asked in the classroom and write questions for all levels To examine effective questioning strategies
November	Teacher Observation: Checklists and Anecdotal Notes	To show teachers how to use checklists and anecdotal notes as assessments in an inquiry-based lesson To provide teachers with guidance on how to use these kit-based assessments To help teachers create their own checklists
December	Type 1 Assessments (Proficiency-like Questions)	To have teachers learn how to write good multiple-choice questions To teach teachers how to write questions like those that appear on the Ohio proficiency tests
January	Effective Teaching "Research Lessons"	To revisit the "Classroom Observation Protocol" developed by Horizon Research, Inc. To engage teachers in "research lessons" to analyze their teaching
February	Developing Rubrics for Short Answer and Extended-Response Questions (Proficiency-like Questions)	To inform teachers about rubrics, their types, and how to develop them To examine questions being asked on the Ohio Proficiency Tests To have teachers develop their short-answer and extended-response questions with rubrics and use them in their classrooms
March	Achieving Closure	To present the recent brain-based research on how a student learns To implement "closure" for each lesson being taught
April	Type II and III Assessments—Performance Assessments	To provide a complete picture of types of assessments To encourage teachers to use performance-based assessments when assessing students in their classrooms

## CHAPTER 20

Using Assessment Design as a Model of Professional Development

Paul J. Kuerbis
Colorado College

Linda B. Mooney
Colorado Springs Public Schools

I f you were challenged to refine and revise embedded assessments within units of study from FOSS (Full Option Science System), STC (Science and Technology for Children), and Insights kit-based science programs, how would you proceed? The leaders of the Science Teacher Enhancement Project-unifying the Pikes Peak region (STEP-uP) met this task through the design and implementation of high-quality professional development. In doing so, they discovered that the process of developing the required products was at least as valuable as the products themselves. The assessment project STEP-uP embarked on in 2000 provides a model for how districts, teams of teachers, assessment experts, and science faculty can work together to become more familiar with curriculum analysis, formative and summative assessments, and real application of science content standards.

Developing a comprehensive high-quality assessment system is no easy task, yet one we found worth tackling. Furthermore the general profession-development process we describe in this chapter is validated by emerging research showing increased student achievement as measured by standardized tests that are part of the Colorado Student Assessment Program (CSAP) (Revak et al. 2007).

Introduction

With supplemental funding from the National Science Foundation, STEP-uP agreed to revise the assessments within 20 of the kits selected from FOSS, STC, and Insights. Across five participating districts in the Pikes Peak region of Colorado, including 80 elementary schools, no single science curriculum prevailed. Instead, in the mid-1990s, the five districts examined three kit-based programs in light of state and district model science standards and selected a mixture of 42 kits/guides from the three programs.

These kit-based programs had been developed prior to the release of the National Science Education Standards (NRC 1996). Although developers of the programs had made some revisions, no major assessment work had been undertaken at the time the STEP-uP project began in 2000. Furthermore, the original kit assessments had been created just as science educators began to explore active and authentic assessments (Hein and Price 1994). Many kit guides did contain an embedded summative performance assessment, an initial step in assessment reform. A few guides noted that a mid-point lesson could be used as a form of embedded assessment. A few others included observation sheets on which teachers could make notations about student learning. These represented significant initial steps toward assessment reform in the early 1990s.

The strengths of the embedded assessments were their tight alignment to the science curriculum and instruction and the fact that student learning continued as understandings were assessed. The weakness of the embedded assessments was that even highly skilled teachers thought of them as just another "activity" within the unit and did not recognize the value of assessment being embedded within the context of learning. STEP-uP sought to use and highlight the value of embedded assessment while restructuring the activities based on current best practices in performance assessment and adding scoring rubrics and exemplars of student work.

As the 21st century began, teachers faced increased calls for accountability, including implementation of state-developed standards and assessments. The need for reexamining the assessments in the kits was clear. The STEP-uP assessment project undertook to make revisions in 20 of the adopted kits. The agreed-upon revisions included the following:

- Aligning embedded assessments to Colorado science standards, which are based on the national standards

- Developing scoring guides (rubrics) for embedded assessments, including summative performance assessments
- Collecting exemplars of work reflecting various levels of student performance: unsatisfactory, partially proficient, proficient, and advanced
- Adding a student science notebook component
- Piloting and field-testing the scoring guides, rubrics, and exemplars
- Developing a training program for teachers in the use of the assessments
- Developing a reporting system for documenting student outcomes
- Disseminating the products as well as the process

Knowing the power of assessment to drive curriculum and instruction, project leaders made sure that the goal of supporting inquiry through assessment was paramount.

Initial Planning

The leadership of STEP-uP (two co–principal investigators, project science resource teachers, and school district administrators) realized that achieving these outcomes presented an opportunity to implement a quality professional development plan. This process should not be separated from the proposed products. Sound professional development is characterized by approaches that mimic what we know about active student learning (see, e.g., NRC 1996; Sparks and Hirsh 1997; Loucks-Horsley et al. 1992; Loucks-Horsley et al. 2003). Professional development means "providing occasions for teachers to reflect critically on their practice and to fashion new knowledge and beliefs about content, pedagogy and learners" (Darling Hammond and McLaughlin 1995, p. 597). How could STEP-uP invent a professional development approach that would yield stronger, more robust formative and summative assessments and develop a new group of teacher-leaders fully invested in quality active assessment and with increased understanding of science content and pedagogy? Could a plan be designed so other districts might adopt and adapt a similar approach with local, low-cost resources?

The assessment leadership team (two principal investigators, science resource teachers, an assessment expert, and several scientists from Colorado College) for the STEP-uP project came together for a day of brainstorming and reflection on the question "What activities should a leadership course on assessment be made up of?" Several additional shorter meetings took place as the team took several weeks to outline a professional development plan with assessment design as the centerpiece. The plan continued to evolve over four years as the group implemented continuous improvement of the original design. Although we had substantial human and financial resources, we think others can adopt and adapt key elements to their particular context and fine-tune the approach depending on the extent of local resources. Questions that others might ask themselves in preparation for a similar project include the following: Who might we call on to lead the project? Will we want to have meetings after school? On weekends? During the week? What is a reasonable time frame for our work? What science and assessment specialists from the community might we involve? Is there a university or college with whom we can collaborate?

Below, we describe the key STEP-uP components that emerged from 2001 through 2006.

A Professional Development Plan

Given the extensive work necessary, STEP-uP decided to embed the assessment development work within a three-semester-hour course for teacher-leaders that was offered over a six-month time frame. The course was easily established because of an ongoing partnership between Colorado College and the STEP-uP project. The three course instructors had combined expertise in teacher development, science education, and assessment, with the assessment expert taking the lead in most of the class sessions. The broader logistics work of the project—including pilot testing, gathering of exemplars, and design of the final distribution package—fell to the project leadership group, primarily the project's science resource teachers.

Assessment Leadership Course: Phase One

This phase began with a concept and lesson content analysis of a chosen kit's teacher guide. The aim was to develop what the project calls a "conceptual storyline"—a flowchart of standards, concepts, and linked lessons that visually display how lessons support student learning of the concepts and subconcepts.

Over a period of three to four months, a group of science faculty, graduate students, and specialists developed the conceptual storyline. Each member of the group reviewed the draft storyline independently to judge the validity of the content. A separate assessment design team, composed of project leadership, assessment experts, classroom teachers, and science resource teachers, then reviewed and made modifications to the storyline as the assessment development proceeded. With time and experience, the leadership learned it was more beneficial to include science experts as members of the assessment design teams rather than as independent external reviewers. The conversations—frequently punctuated by in-depth conversations about concepts, developmental appropriateness, misconceptions, and reading/discussion of such resources as the American Association for the Advancement of Science's *Benchmarks* and *Atlas of Science Literacy* and *Science for All Americans* (1993, 2001, 1990); *Science Matters: Achieving Scientific Literacy* (Hazen and Trefil 1992); and *Making Sense of Secondary Science: Research Into Children's Ideas* (Driver et al. 1993)—led teachers to a fuller understanding of science concepts. Keeley's *Science Curriculum Topic Study* (2005), published after this project, simplifies the process of locating resources related to particular science concepts and is a resource other projects will want to use.

In the final year of the project, course leaders compressed the conceptual storyline work into two intense days of team-based work interspersed with multiple study-group meetings in order to develop a comprehensive draft conceptual storyline. Each year (2001–2004) we chose and analyzed 6–7 kit guides.

Each kit-based team first determined the Big Idea (K–12 conceptual idea) for the kit on which they were working. Next, teams identified the grade-level concepts of the kit that it claimed were part of the student learning. If the kit guide provided a list of concepts (as most do), then the team had to determine if the concept list was accurate and valid by checking the concepts against the lessons. Finally, in order to make the storyline manageable, the team collapsed the concepts into 3 to 5 subconcepts under a single concept for a given kit.

Following this, teams matched the kit concepts to Colorado Science Model Content Standards and Benchmarks. The Colorado standards are based on the National Science Education Standards, so many of the benchmarks in STEP-uP conceptual storylines will be familiar to science teachers. Matches had to be significant in order for the standards and benchmarks to be included in the conceptual storyline. Deep familiarity with the science kit and the concepts and subconcepts covered was essential to the process.

The team also studied each lesson to make sure that benchmarks were in fact sufficiently developed, not just alluded to or introduced. The matching allowed teacher-participants to see how activities in a kit module related to what they knew would be part of a forthcoming fifth-grade science achievement test. That test was developed using the same standards and benchmarks, so the inclusion of standards and benchmarks in the storyline, and in the scoring rubrics for the performance assessment, was essential to help teachers see how a given kit "taught to the test" in the best sense of that phrase.

The process was iterative, with science faculty providing an important validation of what the teachers were determining in their analyses. More important, inclusion of science experts as team members and multiple conversations based on resources such as the National Science Education Standards and the AAAS Benchmarks were crucial to this phase of the leadership course. Indeed, the process was as important as the emerging product. Expert classroom teachers found that their conceptual understandings deepened through the process.

Kit lessons were then placed on the conceptual storyline matched to the 3 to 5 subconcepts covered in the kit. Lessons often fit under more than one subconcept because multiple content benchmarks might be touched on in a given lesson. Included with each lesson title on the storyline was a brief description of the student activity and process skills within that particular lesson or investigation. The conceptual storyline helps the teacher see the importance of completing all lessons rather than only part of a kit. It also points out to the teacher how a given subconcept needs more than a single lesson to cement students' understanding and attainment of standards and benchmarks.

As a result of the conceptual storyline, even first-year teachers, once thought to be able to manage only kit materials, were able to focus on the concepts of the kit and to facilitate minds-on discussions of concepts. The storyline is organized by concepts and subconcepts; however, the lessons in the kits should be taught sequentially as determined by the science kit developer.

Figure 20.1 displays a sample conceptual storyline from the STC "Electric Circuits" kit. Figure 20.2, page 417, displays a sample conceptual storyline from the FOSS "Insects" kit. Note the three tiers of a storyline: standards, concepts/subconcepts, and lesson boxes.

Figure 20.1 Electric Circuits Conceptual Storyline

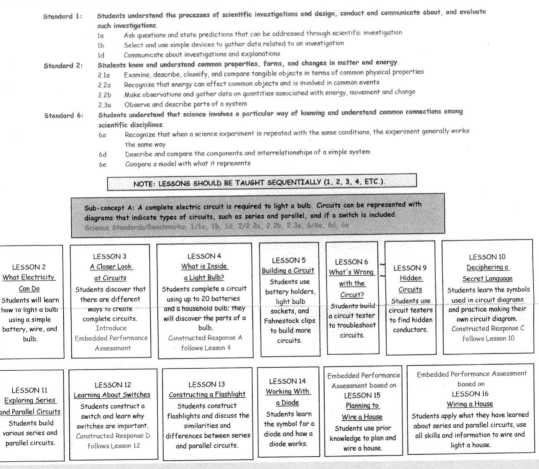

ELECTRIC CIRCUITS CONCEPTUAL STORYLINE
GRADE 3/4 - Science and Technology for Children™ (STC)

K-12 Conceptual Idea(s): Energy comes in many forms. It can be converted from one form to another. All forms of energy involve a system that is capable of exerting a force.

Grade Level Concept: Electricity is energy that flows through circuits to power devices.

Colorado Science Model Content Standards and K-4 Benchmarks

Standard 1:   Students understand the processes of scientific investigations and design, conduct and communicate about, and evaluate such investigations.
 1a   Ask questions and state predictions that can be addressed through scientific investigation
 1b   Select and use simple devices to gather data related to an investigation
 1d   Communicate about investigations and explanations

Standard 2:   Students know and understand common properties, forms, and changes in matter and energy.
 2.1a   Examine, describe, classify, and compare tangible objects in terms of common physical properties
 2.2a   Recognize that energy can affect common objects and is involved in common events
 2.2b   Make observations and gather data on quantities associated with energy, movement and change
 2.3a   Observe and describe parts of a system

Standard 6:   Students understand that science involves a particular way of knowing and understand common connections among scientific disciplines.
 6a   Recognize that when a science experiment is repeated with the same conditions, the experiment generally works the same way
 6d   Describe and compare the components and interrelationships of a simple system
 6e   Compare a model with what it represents

NOTE: LESSONS SHOULD BE TAUGHT SEQUENTIALLY (1, 2, 3, 4, ETC.).

Sub-concept A: A complete electric circuit is required to light a bulb. Circuits can be represented with diagrams that indicate types of circuits, such as series and parallel, and if a switch is included.
Science Standards/Benchmarks: 1/1a, 1b, 1d, 2/2.2a, 2.2b, 2.3a, 6/6a, 6d, 6e

| LESSON 2 What Electricity Can Do Students will learn how to light a bulb using a simple battery, wire, and bulb. | LESSON 3 A Closer Look at Circuits Students discover that there are different ways to create complete circuits. Introduce Embedded Performance Assessment | LESSON 4 What is Inside a Light Bulb? Students complete a circuit using up to 20 batteries and a household bulb; they will discover the parts of a bulb. Constructed Response A follows Lesson 4 | LESSON 5 Building a Circuit Students use battery holders, light bulb sockets, and Fahnestock clips to build more circuits. | LESSON 6 What's Wrong with the Circuit? Students build a circuit tester to troubleshoot circuits. | LESSON 9 Hidden Circuits Students use circuit testers to find hidden conductors. | LESSON 10 Deciphering a Secret Language Students learn the symbols used in circuit diagrams and practice making their own circuit diagram. Constructed Response C follows Lesson 10 |

| LESSON 11 Exploring Series and Parallel Circuits Students build various series and parallel circuits. | LESSON 12 Learning About Switches Students construct a switch and learn why switches are important. Constructed Response D follows Lesson 12 | LESSON 13 Constructing a Flashlight Students construct flashlights and discuss the similarities and differences between series and parallel circuits. | LESSON 14 Working With a Diode Students learn the symbol for a diode and how a diode works. | Embedded Performance Assessment based on LESSON 15 Planning to Wire a House Students use prior knowledge to plan and wire a house. | Embedded Performance Assessment based on LESSON 16 Wiring a House Students apply what they have learned about series and parallel circuits, use all skills and information to wire and light a house. |

(continued)

Figure 20.1 *(continued)*

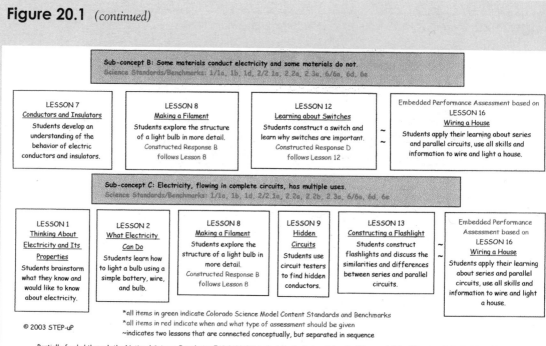

Sub-concept B: Some materials conduct electricity and some materials do not.
Science Standards/Benchmarks: 1/1a, 1b, 1d, 2/2.1a, 2.2a, 2.3a, 6/6a, 6d, 6e

LESSON 7	LESSON 8	LESSON 12	Embedded Performance Assessment based on
Conductors and Insulators	Making a Filament	Learning about Switches	LESSON 16
Students develop an understanding of the behavior of electric conductors and insulators.	Students explore the structure of a light bulb in more detail. Constructed Response B follows Lesson 8	Students construct a switch and learn why switches are important. Constructed Response D follows Lesson 12	Wiring a House Students apply their learning about series and parallel circuits, use all skills and information to wire and light a house.

Sub-concept C: Electricity, flowing in complete circuits, has multiple uses.
Science Standards/Benchmarks: 1/1a, 1b, 1d, 2/2.1a, 2.2a, 2.2b, 2.3a, 6/6a, 6d, 6e

LESSON 1	LESSON 2	LESSON 8	LESSON 9	LESSON 13	Embedded Performance
Thinking About Electricity and Its Properties	What Electricity Can Do	Making a Filament	Hidden Circuits	Constructing a Flashlight	Assessment based on LESSON 16
Students brainstorm what they know and would like to know about electricity.	Students learn how to light a bulb using a simple battery, wire, and bulb.	Students explore the structure of a light bulb in more detail. Constructed Response B follows Lesson 8	Students use circuit testers to find hidden conductors.	Students construct flashlights and discuss the similarities and differences between series and parallel circuits.	Wiring a House Students apply their learning about series and parallel circuits, use all skills and information to wire and light a house.

© 2003 STEP-uP

*all items in green indicate Colorado Science Model Content Standards and Benchmarks
*all items in red indicate when and what type of assessment should be given
~indicates two lessons that are connected conceptually, but separated in sequence

Partially funded through the National Science Foundation ESI 0196127, Agilent Technologies, Colorado College and School Districts 2, 11, 12, 20, and 38

Leadership Course: Phase Two

This part of the leadership course focused on revision of a kit's final performance assessment and development and validation of formative assessments to supplement the guide. The initial conceptual storyline allowed the kit-based assessment development teams within the leadership course to consider the guide's assessments and options for revision.

Guiding conditions/assumptions of the assessment work were developed by the STEP-uP leadership team and adhered to throughout the assessment project. Those guiding principles were as follows:

- The assessments should be driven by the content standards and key concepts taught in the kit(s).
- The "heart" of the assessment program should be the embedded assessments within the kits (performance assessments, structured observations

Figure 20.2 Insects Conceptual Storyline

INSECTS CONCEPTUAL STORYLINE
GRADE 2 - Full Option Science System ™ (FOSS)

K-12 Conceptual Idea(s): Living organisms have physical characteristics and structure, undergo changes within life cycles and interact with each other and their environment.

Grade Level Concept: Insects have specific needs. As the lifecycle of insects proceed, their body structures change and their needs change.

Colorado Science Model Content Standards and K-4 Benchmarks

Standard 1:   Students understand the processes of scientific investigations and design, conduct, communicate about, and evaluate such investigations.
    1a   Ask questions and state predictions that can be addressed through scientific investigation
    1b   Select and use simple devices to gather data related to an investigation
    1c   Use data based on observations to construct a reasonable explanation
    1d   Communicate about investigations and explanations
Standard 3:   Students know and understand the characteristics and structure of living things, the processes of life, and how living things interact with each other and their environment.
    3.1b   Classify a variety of organisms according to selected characteristics
    3.1c   Describe the basic needs of an organism
    3.3c   Describe life cycles of selected organisms
    3.4a   Identify characteristics that are common to all individuals of a species
    3.4b   Recognize that there are differences in appearance among individuals of the same population or group

Sub-concept A: Needs of insects include food, water, air, and space. Insects interact with their environment to meet their needs.
Science Standard/Benchmarks: 1/1a, 1b, 1c, 3/3.1c, 3.4/4a

ACTIVITY 1 (and 2,3,4,5,6)*
PART 1
What are the needs of living organisms?
Students discuss the needs of insects.

ACTIVITY 1 (and 2,3,4,5,6)*
PART 1
Habitat Set-Up
Students construct habitats that provide for the needs of each type of insect.

Do Constructed Response A
after mealworm habitat has been set up.

Sub-concept B: Insects have specific body structures and characteristics including 3 different body regions. The insects studied in this kit have 2 antennae, 4 wings, 6 legs, and a skeleton on the outside that doesn't grow and needs to be shed.
Science Standard/Benchmarks: 1/1a, 1b, 1c, 1d, 3/3.1b, 3.4a, 3.4b

(continued)

Figure 20.2 *(continued)*

```
┌─────────────────────────────┐        ┌─────────────────────────────────┐
│  ACTIVITY 1 (and 2,3,4,5,6)* │        │    ACTIVITY 1 (and 2,3,4,5,6)*   │
│           PART 2             │        │             PART 2               │
│  Discovering Insect Structures │      │    Discovering Insect Structures  │
│  Students observe insect     │        │ Students observe insect structures │
│  structures in the larval stage. │    │       in the adult stage.        │
│                             │        │                                 │
│                             │        │   Do Constructed Response B      │
│                             │        │   after second insect studied.   │
└─────────────────────────────┘        └─────────────────────────────────┘
```

Sub-concept C: Insects undergo simple or complete metamorphosis within their life cycles.
Science Standard/Benchmarks: 1/1a, 1b, 1c, 1d, 3/3.3c, 3.4a, 3.4b

```
┌─────────────────────────────┐        ┌─────────────────────────────────┐
│  ACTIVITY 1 (and 2,3,4,5,6)* │        │                                 │
│           PART 3             │        │  Introduce Embedded Performance  │
│ Students observe the stages of │      │  Assessment after first insect studied. │
│  complete or                │        │                                 │
│  simple metamorphosis.      │        │  Do Embedded Performance Assessment │
│                             │        │  after all insects have been studied. │
│  Do Constructed Response C   │        │                                 │
│  after third insect studied. │        │                                 │
└─────────────────────────────┘        └─────────────────────────────────┘
```

© 2003 STEP-uP

* Each activity addresses a specific insect's needs and habitat (Part 1), body structure (Part 2), and lifecycle (Part 3).
*all items in green indicate Colorado Model Science Standards and K-4 Benchmarks
*all items in red indicate when and what type of assessment should be given

Partially funded through the National Science Foundation ESI 0196127, Agilent Technologies, Colorado College and School Districts 2, 11, 12, 20 and 38

of habits of mind, and academic prompts within science notebooks), making the assessment as unobtrusive as possible.

- The administration, scoring, and recording of results of the science assessments need to be highly manageable and efficient for classroom teachers.

- The compilation of student data and program data should be addressed with district technology experts, at the front end of the project, ensuring efficient data recording and access.

- Science resource teachers and classroom teachers should be the primary designers and decision makers in the assessment design process, with consultants available for guidance, feedback, and review.

- This work is an evolving process, requiring experimentation, piloting, and review.
- Assessment data will be compiled and used in such a way, as to inform instruction and increase student learning.

Teachers rallied to the words of Grant Wiggins (1989): "Reform begins by recognizing that the test is central to instruction. Any tests and final exams inevitably cast their shadows on all prior work. Thus, they not only monitor standards, but also set them" (p. 704). Project staff and teachers were dedicated to producing assessments that supported students' learning, inquiry instruction and learning, the integration of standards (a hallmark of STEP-uP professional development), *and* prepared students for the upcoming fifth-grade Colorado Student Assessment Program in Science. The staff and teachers asked themselves the following questions: How would a group of teachers who were committed to the STEP-uP approach to assessment successfully work with teachers who were advocates of pencil-and-paper tests? Would performance assessments and notebook constructed responses adequately assess student learning and ensure high performance on state testing?

Using McTighe and Wiggins's (1999) backward design process, teams examined the final performance assessment in the kit, usually part of the final lesson(s) in the sequence. In the case of "Electric Circuits," for example, students are called on to apply all they have learned about circuits in the design (lesson 15) and actual wiring (lesson 16) of a small house—in the form of a four-compartment cardboard box. They also present and explain their work to classmates and the teacher. The teams examined the performance task and then decided whether the task actually assessed for all subconcepts in the kit and for all of the standards and benchmarks the team had identified during the storyline development phase. Design teams usually found that the kit's intended embedded final performance assessment needed to be restructured, using language and processes currently identified as best practices in science assessment and with a tighter focus on standard assessment.

Phase two of the leadership course included sessions on best practices in performance assessment in science, rubric development, and development of a user's guide. Design teams rewrote embedded performance tasks based on "Characteristics of Authentic Tasks," developed by

O'Rourke and O'Rourke (2004). Using an adapted Critical Friends protocol, "Descriptive Consultancy Protocol," each team received feedback from another team on a troubling aspect of the performance task. (For information on Critical Friends, see, for example, Appleby 1998.) Design teams met between course sessions to refine the task. Initially, these tasks were submitted to the STEP-uP leadership for critique, with feedback for modifications given back to the team. In much the same way as the conceptual storyline development changed over time, these feedback sessions became "fishbowls," with the leadership discussing strengths and weaknesses of the task and rubrics while design members listened and observed. Although team members described the fishbowl experience as "grueling," they also highly valued the feedback and felt that products were improved and that they had learned a great deal by observing how the leadership processed the critique.

Once the summative performance tasks were formalized, teachers learned best practices in writing rubrics using "Characteristics of Well-Designed Rubrics" (O'Rourke and O'Rourke 2004). Knowing that rubrics are always revised based on student work, the teams set about designing rubrics for teachers that would allow them to know what students knew, understood, and could do, as well as rubrics for students so they would know what their learning targets were. (See Figures 20.3 and 20.4, p. 422, for examples.) Dimensions to be assessed through rubrics were conceptual understandings/standards, procedures, use/application of evidence, and communication. Critical Friends protocols and "fishbowl" techniques were used to refine the rubrics.

We cannot overemphasize the importance of mapping the standards onto the conceptual storyline, into the embedded assessments, and then on the scoring rubrics for the summative (performance) assessments. Incorporating Colorado Science Standards into the storyline and linking them to the embedded and summative assessments is a crucial step in aligning a kit's content and activities to how students were eventually assessed with the standardized tests in Colorado. Teachers who use the storylines, for example, remark that they now see why the entire kit should be taught—that is, so that children are amply prepared for the tests whose questions are also mapped to the Colorado Science Standards. Furthermore, teachers see how

Figure 20.3 "Wiring Our House"—Student Rubric

WIRING OUR HOUSE

Student Names: _____

_____

_____ Date: _____

UNSATISFACTORY	PARTIALLY PROFICIENT	PROFICIENT	ADVANCED
We "grab" the materials we need and start wiring without using our plan.	We are having difficulty following our plan to wire the house. We need help to use our plan to wire our house.	We are able to follow the plan to wire the house, and appropriately use materials to wire the house.	We are able to easily follow the plan to wire the house, using ALL tools and supplies effectively.
We have trouble wiring our house. Our materials are not in safe and convenient places. Sometimes our wiring works and sometimes it doesn't.	We can wire parts of our house that work. We can make a basic switch and use it to turn on our lights. Some of our materials are not always placed in safe and convenient places. Some of our lights work.	We can wire circuits (parallel or series) that work. We can make switches and use them to turn on our lights. We put our materials (batteries, bulbs, wires, and switches) in safe and convenient places. All of our lights work.	All of our lights work and we have added extra features*, to make the use of electricity easier in our house. (* i.e. an additional switch to control the same light, a more complex switch, a diode etc.),
If we have a problem, we do not know how to fix it.	If we have a problem, we have trouble fixing it.	If we have a problem, we can usually troubleshoot to fix it.	We can easily identify and troubleshoot problems. We change our plan to solve the problem.
We do not know why our wiring works or does not work. We could not do this again the same way.	We know why our wiring works or does not work, but we probably couldn't do this again the same way.	We know why our wiring works or does not work, and we could probably do this again the same way.	We know exactly why our wiring works or does not work and we know we could do this again the exact same way.

Electric Circuits Performance Assessment Task Rubric   Student Handout

41

©2003 STEP-uP. Reprinted with permission.

notebooks that include writing prompts embedded in the storylines prepare their students for the standardized tests.

Leadership Course: Phase Three

This part of the course focused on the development of formative assessments. Leadership and participants wrestled with many issues of formative assessment: What are the important attributes of a formative assessment? What should it assess? How tied should it be to the summative performance task? How does a formative assessment relate to the conceptual storyline? What type of formative assessment should be used (pen and pencil? performance? notebook related?)? Since the eighth-grade science state assessment included constructed-responses items, would it be important for elementary students to have experience with constructed responses?

Figure 20.4 "New Insect Project"—Teacher Rubric

NEW INSECT PROJECT: BODY STRUCTURE   For the Teacher

DIMENSIONS	UNSATISFACTORY	PARTIALLY PROFICIENT	PROFICIENT	ADVANCED
Conceptual Understanding, Use/Application of Evidence, Procedures, Communication				
(3.1b) Classify a variety of organisms according to selected characteristics • Sort a group of animals into two groups – insects and other	In my drawing of the new insect, I mostly used my imagination. My work does not represent an insect.	In my drawing of the new insect, I used some of the characteristics common to all insects.	In my drawing of the new insect, I used the characteristics common to all insects.	In my drawing of the new insect, I used the characteristics common to all insects. In addition, I included other characteristics.
(3.4a) Identify characteristics that are common to all individuals of a species • Know that adult insects have three body sections, six legs, and two antennae	My labels are missing or have a number of errors.	Some of the labels I used were scientifically accurate.	The labels I used were scientifically accurate.	The labels I used were scientifically accurate and demonstrate my extensive knowledge of insects.
(1c) Use data based on observations to construct a reasonable explanation • Describe things using appropriate senses • Draw, color, and label pictures showing main characteristics	I did not give purposes for the labeled body parts.	The scientific purposes I gave demonstrate a limited knowledge of insects.	The scientific purposes I gave demonstrate a basic knowledge of insects.	The scientific purposes I gave demonstrate a broad knowledge of insects.

Insects Performance Assessment Task   Rubric for the Teacher

20

As these questions were asked, the notion of an assessment storyline parallel to the conceptual storyline began to emerge. This "blinding glimpse of the obvious" drove the answers to our formative assessment questions. Assessing subconcepts that would lead to success on the final performance assessment seemed logical and productive in furthering student conceptual understanding as well as giving the teacher an opportunity to ensure that students were on the right conceptual track. Incorporating constructed responses in science notebooks would further the STEP-uP initiative of integrating standards and assessing in a way that benefited multiple content areas.

Rubrics were not necessary for the formative assessment work because the teacher only needed to know whether or not a student understood a particular subconcept. Moreover, meta-analysis research (NRC 2001) indicates the value of not grading formative assessments such as notebooks and of providing specific comments instead. In fact, teachers encouraged students to use a "line of learning" (the point at which a student recognizes "borrowed ideas" from others) on constructed responses so they could continue adding to their understandings as they proceeded through the unit.

Along with the constructed-response items, design teams determined the key elements showing evidence of understanding for each of the constructed responses and subconcepts. They used "Characteristics of Academic Prompts" (McTighe and Wiggins 1999) as a resource, along with Critical Friends protocols and "fishbowls" to refine constructed responses. The teams added a section to the User's Guide (discussed below) with prompts, the assessed subconcept, enabling skills, and "next steps" if a misconception or lack of understanding existed. "Next steps" ideally drive the classroom instruction of the teachers administering the science assessments.

Packaging of Materials: User's Guide and Implementation

The assessment design teams led the development of the User's Guide in the form of a binder that included multiple components: the conceptual storyline, assessment storyline, student writing prompts and exemplars of appropriate and inappropriate work, a revised summative performance assessment, and scoring rubrics for both teachers and students for the summative assessment. Usually five teachers, who were deemed highly skilled by the science resource teachers, piloted the assessment packages after they were trained in the conceptual storyline and assessment storyline. These teachers were observed administering the assessments, and they provided feedback regarding the performance task, rubrics, and constructed responses, as well as student exemplars. Once revisions were made based on the pilot, then approximately 10 volunteer elementary teachers were selected to field-test the materials. Once final revisions based on field-test feedback were made, the assessment binders were printed and distributed to districts to be included in their science kits. Typically, assessment packages were piloted the first year, field-tested the next, and implemented the following year.

Prior to implementation of the assessment packages, teachers in the five regional districts participated in a three-hour Assessment Implementation course. The sessions provided an overview of the assessment instruments and ways to administer the assessments. Additionally, help on using the feedback and determining "next steps" for instruction were addressed. Science resource teachers then mentored teachers as they implemented the assessments, and teachers were encouraged to score performance assessments in teams as part of their professional development. In this way, teachers clarified their own science conceptual understandings and identified ways in which they could further their students' understandings.

STEP-uP and Student Achievement

The comprehensive approach STEP-uP takes in professional development is paying off. While STEP-uP does not yet have research demonstrating the specific impact of our assessment professional development, the project has been able to show the power of specific aspects of its professional development, coupled with changes in classroom practices, that impacts student achievement. In one study (Revak et al. 2007), we examined the relationship between teacher professional development in elementary kit-based science, related teacher practices, and student achievement. The research was conducted in four of the five school districts (urban and suburban) in the Pikes Peak region of Colorado that are part of STEP-uP. The sample consisted of 2463 fifth-grade students during the 2003–2004 school year. Teacher participation in select professional development activities (e.g., use of notebooks) correlated with student achievement in reading, writing, and mathematics with effect sizes of 0.15 to 0.33. Select combinations of activities (e.g., notebook training *and* graphing in science) produced larger effect sizes of 0.32 to 0.56. Additionally, implementation of select teacher classroom practices produced still larger effect sizes, ranging from 1.0 to 1.75. Additional research studies are underway that will explore aspects of STEP-uP professional development linked with science achievement since Colorado began implementing the fifth-grade science CSAP (Colorado Student Assessment Program) in spring 2006.

Summary and Conclusions

The course offered during the school day provided the opportunity and time for the kit assessment design teams to meet for training in the development of the science kit assessments. Classroom teacher participants had permission from districts and principals to attend. The classroom substitute costs were covered by STEP-uP grant funds so teachers did not lose personal leave days for developing assessment products. Additionally, teachers were paid a stipend when work was completed and they received college or inservice credit for their work.

We encourage others to adapt our process to their own situations. It is the process as much as the product that provides powerful professional development for teacher-participants. Who can you identify as potential partners? How will you take the first step?

References

American Association for the Advancement of Science (AAAS). 1990. *Science for all Americans*. New York: Oxford University Press.

American Association for the Advancement of Science (AAAS). 1993. *Benchmarks for science literacy*. New York: Oxford University Press.

American Association for the Advancement of Science (AAAS). 2001. *Atlas of Science Literacy*. Washington, DC: AAAS and NSTA.

Appleby, J. 1998. *Becoming critical friends: Reflections of an NSRF [National School Reform Faculty] coach*. Providence, RI: Annenberg Institute for School Reform at Brown University.

Darling-Hammond, L., and M.W. McLaughlin. 1995. Policies that support professional development in an era of reform. *Phi Delta Kappan* 76 (8): 597–604.

Driver, R., A. Squires, P. Rushworth, and V. Wood-Robinson. 1993. *Making sense of secondary science: Research into children's ideas*. New York: Routledge.

Hazen, R., and J. Trefil. 1992. *Science matters: Achieving scientific literacy*. New York: Anchor Books.

Hein, G., and S. Price. 1994. *Active assessment for active science*. Portsmouth, NH: Heinemann.

Keeley, P. 2005. *Science curriculum topic study*. Thousand Oaks, CA Corwin Press.

Loucks-Horsley, S., P. W. Hewson, N. Love, and K.E. Stiles. 1992. *Designing professional development for teachers of science and mathematics*. Thousand Oaks, CA: Corwin Press.

Loucks-Horsley, S., N. Love, K. E. Stiles, S. Mundry, and P. W. Hewson. 2003. *Designing professional development for teachers of science and mathematics*. 2nd. ed. Thousand Oaks, CA: Corwin Press.

McTighe, J., and G. Wiggins. 1999. *The understanding by design handbook*. Alexandria, VA: Association for Supervision and Curriculum Development.

National Research Council (NRC). 1996. *National science education standards*. Washington, DC: National Academy Press.

National Research Council (NRC). 2001. *Classroom assessment and the national science education standards*. Washington, DC: National Academy Press.

O'Rourke, A., and W. O'Rourke. 2004. *A training manual for designing science assessments.* Private publication available from O'Rourke and O'Rourke, Colorado Springs, CO. E-mail: *or2@rmi.net*

Revak, M. A., S. Getty, P. J. Kuerbis, and L. B. Mooney. 2007. Linking professional development and classroom practice in science to student achievement in reading, writing, and math. (Submitted for publication and under review.)

Sparks, D., and S. Hirsh. 1997. *A new vision for staff development.* Alexandria, VA: Association for Supervision and Curriculum Development.

Wiggins, G. 1989. A true test: Towards more authentic and equitable assessment. *Phi Delta Kappan* 70(9): 703–714.

## CHAPTER 21

Using Formative Assessment and Feedback to Improve Science Teacher Practice

Paul Hickman
Science Education Consultant

Drew Isola
Allegan High School, Allegan, Michigan, and former PhysTEC (Physics Teacher Education Coalition) TIR (teacher in residence), Western Michigan University

Marc Reif
Ruamrudee International School, Bangkok, Thailand, and former PhysTEC TIR, University of Arkansas

In an *Education Week* article several years ago, James Hiebert and his colleagues made the case for moving educators away from a view of teaching as a solitary activity toward a view of teaching as a professional activity open to critical observation, study, and improvement.

Traditionally, classroom teaching in the United States has been viewed as a personal skill, invented and refined by each teacher during his or her career. Good teaching is considered to be the result of each teacher's doing his or her job behind the classroom door... To achieve small and continuing improvements in the average classroom requires a major shift in educators' thinking— from teachers to teaching. Rather than focusing only on evaluating the quality of teachers, the educational community must begin examining the quality of teaching. What kinds of methods are teachers using now and how could these methods be improved? Tackling this deep-seated problem begins with opening the classroom door. (Hiebert, Gallimore, and Stigler 2003)

The problem of teacher isolation is especially common among novice teachers who often see themselves working as solo practitioners expected to be prematurely expert and able to develop their craft without the support of a school-based professional network (Kardos and Johnson 2007).

Growth in knowledge, skills, and conceptual understanding should be the result of what teachers and students do in science classrooms. Teachers have to manage complex and demanding situations, shaping the personal, emotional, and social pressures of a group of students to help them learn now and become better learners in the future (Black and Wiliam 1998). Using formative assessment, the student's teacher or classmates provide frequent feedback that helps to plan future directions of the student's learning and the teacher's instruction. All teachers need to consider how their classroom activities, assignments, and tests support learning goals, allow the students to communicate what they know, and then use this information to improve their teaching and student learning (Boston 2002).

Formative assessment demands that teachers willingly redirect the plan (for the lesson) even when it is inconvenient, because learning is not about the teacher being ready; learning is about the students being ready. When assessment is regular and ongoing, teaching can adapt to help students develop deeper understanding and actively participate in their own learning. (Crumrine and Demers 2007)

There is much data, especially from the Trends in International Mathematics and Science Study (TIMSS), to support the idea that we teach the way we were taught (Stigler and Hiebert 1999). Recent education scholarship has shown that rather than continuing to use traditional, passive class formats, instructors K–20 should implement inquiry-based, problem-solving, and active-learning strategies coupled with ongoing formative assessment in their science courses. This means requiring that students develop hypotheses, design and conduct experiments, collect and interpret data, and discuss the results. Courses and instructional materials using these methods have been implemented and assessed in K–12 schools and in institutions of higher learning across the United States. They have been proven to spark student interest in science, help students—especially women and underrepresented minorities—learn more and get better grades, and lead more students to enroll in advanced science courses (Handelsman et al. 2004). However, Handelsman and colleagues also point out that classroom teachers and university faculty have been slow to abandon the traditional lecture format and adopt proven reform teaching techniques.

Now if we could only see what is happening behind the closed doors of our very private profession and had a useful tool to make sense of what we saw, we might be able to move current teachers and start new ones on a path to using more active learning and formative assessment strategies—strategies that have come to be called "reformed teaching"—in their classrooms.

Why Study Teaching Using the Reformed Teaching Observation Protocol (RTOP)?

For instructors at all levels to adopt reformed teaching strategies, they must both be convinced of the effectiveness of active learning and formative assessment (see it modeled in a number of content-specific settings) and have experienced these techniques (as learners themselves) in university courses or other professional settings. One impediment to this adoption is that standard evaluation instruments are not applicable to a reformed classroom because the structure, content, and delivery style for many reformed classes differ dramatically from the traditional lecture and demonstration approach. For example, in a reformed curriculum there is little emphasis on lecture, so the instructor may not have an opportunity to exhibit his or her level of content knowledge or class preparation that is measured on tradition evaluations.

New assessment instruments have been needed to evaluate reformed instruction, and one such instrument is the Reformed Teaching Observation Protocol (RTOP) (see Appendix, p. 437, for the complete RTOP). The RTOP was developed at Arizona State University as an observation instrument to provide a standardized means for measuring the quality of K–20 STEM (science, technology, engineering, and mathematics) classroom instruction. The instrument has high inter-rater reliability, and high scores measured by RTOP have been shown to correlate with increased student understanding in multiple studies (Lawson 2001). The use of this protocol was highlighted through a commissioned paper for and presented at the 2002 NRC Improving Undergraduate Instruction in Science, Technology, Engineering, and Mathematics Workshop.

Just as a science teacher might use an operational definition to initially get across a difficult concept like temperature (e.g., "Temperature is a measure of the hotness or coldness of an object. It is what we measure with a thermometer"), so

> RTOP operationally defines and assesses reformed teaching in the classroom—we henceforth explicitly reserve and define the term reformed teaching to mean those classroom practices that result in a high RTOP score. (MacIsaac and Falconer 2002)

RTOP classroom observations are made in five categories: Lesson Design and Implementation, Content (Propositional Knowledge), Content (Procedural Knowledge), Communicative Interactions, and Student/Teacher Relationships each with five items for a total of 25 items. The meaning of each of the 25 RTOP items is described carefully in the RTOP Training Guide included in the Appendix, page 437, to this chapter. Several of these items probe specifically for evidence of inquiry teaching and the teacher's level of pedagogical content knowledge. Other items are linked more strongly to the formative assessment of student learning, including

1. The instructional strategies and activities respected students' prior knowledge and the preconceptions inherent therein.

14. Students were reflective about their learning.

16. Students were involved in the communication of their ideas to others using a variety of means and media.

19. Student questions and comments often determined the focus and direction of classroom discourse.

22. Students were encouraged to generate conjectures, alternative solution strategies, and/or different ways of interpreting evidence.

24. The teacher acted as a resource person, working to support and enhance student investigations.

Each item is scored on a Likert scale from 0 (not observed) to 4 (very descriptive) of the classroom lesson. The RTOP protocol provides an interesting alternative to the traditional classroom evaluation and better reflects the values of reformed teaching as called for in current literature. Use of the protocol requires skilled observers with deep discipline-specific content knowledge who have completed training and co-observed classrooms or video to develop consistent use of the tool.

Using an observation tool or rubric like the RTOP has the potential to

- Build a common language and experience that reflects reformed teaching
- Offer models for providing students with meaningful feedback
- Be helpful for teacher self-evaluation of their classroom instruction
- Be a useful tool for mentors working with novice teachers
- Confront beliefs about teaching and learning
- Provide a focus for experienced teachers as they change to a more student-centered classroom
- Prepare for the Presidential Awards for Excellence in Mathematics and Science Teaching or state and national board certification

Why Study Teaching Using Video?

Film and video pervade our culture. Capturing and analyzing video records of classroom interactions have long been used for education research. More recently, observation tools have been used to provide evidence of changes in teaching practice, implementation of curricular innovations, and teacher growth over time. Both science and education researchers have long agreed that the process of observation tends to change the very things we wish to observe. There are ways, however, to minimize this effect in classrooms. As teachers and students become more comfortable with observers or cameras in the classroom, the impact of this intrusion into the usually private world of instruction is greatly reduced.

Traditionally, attempts to measure classroom teaching have used teacher questionnaires. Although questionnaires are relatively easy to administer on a large scale and can provide useful quantitative information, they rely on a teacher's memory and conscious behaviors. Moreover, questions are subject to a teacher's interpretation and cannot be revisited once the questionnaire has been completed. (Stigler et al. 1999)

Several states, including Arizona, Connecticut, and Indiana, require a videotaped lesson as part of the evidence provided to advance from an initial teaching credential to a more permanent one. Submission of a video record of teaching is a requirement for National Board for Professional Teaching Standards Certification and is part of the application process for the Presidential Awards for Excellence in Mathematics and Science Teaching. But most teachers have never seen a video of themselves teaching unless such an analysis was used in their teacher preparation program.

There may be some issues with getting institutional permission (from university, school, student, parents, and instructor) to conduct the classroom observation, especially if video is recorded, since the research is being performed on human subjects. That may make using live observations seem more attractive. But, many of the interactions and teachable moments that occur in active classrooms are fleeting and even skilled observers are challenged to take notice of more than a few elements of the complex instructional process. So, it is not a coincidence that large studies of classrooms, like TIMMS, used video records.

The benefits of a video record as compared to live observations are that video

- enables the study of complex processes,
- increases inter-rater reliability and decreases training difficulties,
- enables coding from multiple perspectives,
- stores data in a form that allows new analyses at a later time,
- facilitates integration of qualitative and quantitative information, and
- facilitates communication of the results.

Conclusion

There is no one right way to teach, no single instrument or method that can be used to provide evidence for improved student achievement and no single best form of professional development. But there are a small number of elements of accomplished teaching that can be observed and that should be present in all our lessons. Classroom practice is individual and quite complex, but the use of the RTOP tool and video records can help to provide the missing link between an instructor's behaviors and the growth of student understanding. It can also begin a constructive dialog among colleagues about science teaching. Having teachers study video of classroom practice can be a transformative professional development strategy for individuals, colleagues, or whole schools and joins other strategies like lesson study, topic study, looking at student work, and collaborative data inquiry as mechanisms to improve the quality of our students' learning.

Perhaps if more teachers were able to view a range of science teaching examples, including their own, and saw the power of getting feedback on the elements of reformed teaching in their own lessons, they would begin to include more opportunities for their students to receive frequent feedback to improve their learning. This might begin to increase the status of teaching from a performing art to a true clinical profession like law or medicine.

Acknowledgment

This chapter is an outgrowth of a short, awareness workshop to promote the use of RTOP within the PhysTEC* project. (The PhysTEC project was supported in part by the National Science Foundation [PHY-0108787].) The workshop has been given at the NSF/NSTA Assessment Conferences, AAPT (American Association of Physics Teachers) meetings, PhysTEC institutions, and at local science conferences. A special thank-you to each of those who provided concrete examples of how the RTOP tool can be used to support classroom science teachers.

References

Black, P., and D. Wiliam. 1998. Inside the black box: Raising standards through classroom assessment. *Phi Delta Kappan* 80(2).

Boston, C. 2002. The concept of formative assessment. Practical Assessment, Research & Evaluation 8(9). *http://PAREonline.net/getvn.asp?v=8&n=9*

Crumrine,T., and C. Demers. 2007. Formative assessment: Redirecting the plan. *The Science Teacher* 74(6): 28–32.

Kardos, M., and S. Johnson. 2007. On their own and presumed expert: New teachers' experiences with their colleagues. *Teachers College Record* 109(9).

Handelsman, J., D. Ebert-May, R. Beichner, P. Bruns, A. Chang, R. DeHaan, J. Gentile, S. Lauffer, J. Stewart, S. M. Tilghman, and W. B. Wood. 2004. Scientific teaching. Science 304: 521–522. *http://www.sciencemag.org/cgi/content/full/304/5670/521?ijkey=/Bhq mwmXov5vQ&keytype=ref&siteid=sci*

Hiebert, J., R. Gallimore, and J. Stigler. 2003. *Education Week*. The New Heroes of Teaching. November 5.

Lawson, A. 2001. Reforming and evaluating college science and mathematics instruction: Reformed teaching improves student achievement. *Journal of College Science Teaching* 31(6).

Lawson, A. 2002. Using the RTOP to evaluate reformed science and mathematics instruction. Commissioned for the NRC Improving Undergraduate Instruction in Science, Technology, Engineering, and Mathematics Workshop. *http://books.nap.edu/ books/0309089298/html/89.html#pagetop*

MacIsaac, D. L., and K. A. Falconer. 2002. Reforming physics education via RTOP. *The Physics Teacher* 40(8): 16–22.

Stigler, J., and J. Hiebert. 1999. *The teaching gap.* New York: The Free Press.

Stigler, J., P. A. Gonzales, T. Kawanka, S. Knoll, and A. Serrano. 1999. *The TIMSS Videotape Classroom Study: Methods and findings from an exploratory research project on eighth-grade mathematics instruction in Germany, Japan, and the United States.* Washington, DC: National Center for Education Statistics, U.S. Department of Education.

Useful Resource

Dan MacIsaac at Buffalo State College maintains an online version of the RTOP teacher workshop, traditionally offered at AAPT summer meetings. The site has edited and expertly scored video examples so that you can view a video clip, score it using the RTOP tool, and compare your score to an expert score. *http://physicsed.buffalostate.edu/AZTEC/ RTOP/RTOP_full/using_RTOP_1.html*

Applications of the Reformed Teaching Observation Protocol (RTOP) by Teachers at All Levels

To Build Collaboration

Marc Reif, Ruamrudee International School, Bangkok, Thailand

An experienced high school physics teacher, Marc mentored a recent college graduate who was going through an alternative certification process and had never taught high school before. To foster collaboration, Marc suggested that they observe each other using the RTOP. They were both teaching physics using the Modeling Instruction Program. Their RTOP scores were both reasonably high, which pleasantly surprised both of them. They had good discussions about what was meant by the items and what a high score on an item could mean for the students.

To Support Reflection on Teaching

Drew Isola, Allegan High School, Allegan, Michigan

When Drew, a high school math and science teacher, was first exposed to the RTOP, he was struck with its potential value as a tool to collect reliable data about teaching practices. He felt, however, that it was too artificial and not likely to capture the variability in any one teacher's classroom. Noting that the RTOP offers a detailed description of what it means to be an inquiry-based teacher, he edited the instrument—using more common, everyday terms—to make it more of a guide to personal reflection of his own teaching.

To Build Novice Teacher Effectiveness

Julia Olsen, University of Arizona

Since Julia has been mentoring middle and high school science teachers using RTOP, many have experienced tremendous growth. Novice teachers seem to find it relatively easy to take the feedback provided through the RTOP and to change elements of how they teach. Julia found that at least one teacher had tripled her initially low score within her first semester of teaching. That teacher worked in a very tough school, with a high number of English language learners, numerous behavior problems, and low student test scores. The teacher also went from having most of her students failing to all but a few passing, an improvement that can be directly attributed to the changes she made in her teaching since she hadn't changed her grading scale.

(continued)

(RTOP) *(continued)*

To Change Teacher's Ideas About Teaching and Learning Science
Dan MacIsaac and Kathleen Falconer, SUNY-Buffalo State College

At Dan and Kathleen's institution, a series of reform-centered physics courses are taught to preservice and inservice teachers. The courses feature extended student discourse with research-developed, guided discovery learning, and metacognitively supported curricula. At the start of these courses, many students have difficulty making sense of their experiences and show a lack of faith in their own ability to reason scientifically.

During the undergraduate courses for preservice teachers, students react to a single item from the RTOP and illustrate their response with an example directly from their own student teaching experiences. Experienced teachers write a pre-arrival paper in which they discuss what they think good teaching looks like, explain how their own teaching falls into this description, and provide an example from their own experience. At the end of the course, all teachers reflect on how their ideas of what good teaching looks like have changed during the course.

To Develop an Understanding of Inquiry
Joseph J. Bellina Jr., Saint Mary's College

Joe developed and led weeklong workshops designed to engage K–6 teachers in learning by grade-level-appropriate guided inquiry. Each morning teachers learned science by guided inquiry using materials designed for adults. Each afternoon they observed, practiced, and discussed teaching science by guided inquiry using materials designed for grades K–6. The RTOP instrument and training videos were used to bring closure to the week. The teachers read the guidelines and descriptions of the RTOP items during the evening before the final day of the workshop. Since the teachers had been immersed in guided inquiry instruction for the four previous days, the language and expectations presented were familiar.

On the afternoon of the last day, the teachers observed the training videos, rated the teacher's performance, discussed their ratings with a partner, and then reported scores orally. The RTOP experience was so successful that the school district used an edited list (to reduce the number of items) to evaluate the results of the district/college collaboration.

Appendix

Reformed Teaching Observation Protocol (RTOP)
TRAINING GUIDE

Daiyo Sawada   Michael Piburn
External Evaluator   Internal Evaluator
and

Jeff Turley, Kathleen Falconer, Russell Benford, Irene Bloom, and Eugene Judson

The Evaluation Facilitation Group

Arizona Collaborative for Excellence in the Preparation of Teachers
Arizona State University

ACEPT Technical Report No. IN00-2

The Reformed Teaching Observation Protocol (RTOP) is an observational instrument that can be used to assess the degree to which mathematics or science instruction is "reformed." It embodies the recommendations and standards for the teaching of mathematics and science that have been promulgated by professional societies of mathematicians, scientists and educators.

The RTOP was designed, piloted and validated by the Evaluation Facilitation Group of the Arizona Collaborative for Excellence in the Preparation of Teachers. Those most involved in that effort were Daiyo Sawada (External Evaluator), Michael Piburn (Internal Evaluator), Bryce Bartley and Russell Benford (Biology), Apple Bloom and Matt Isom (Mathematics), Kathleen Falconer (Physics), Eugene Judson (Beginning Teacher Evaluation), and Jeff Turley (Field Experiences).

The instrument draws on the following sources:

- National Council for the Teaching of Mathematics. *Curriculum and Evaluation Standards* (1989), *Professional Teaching Standards* (1991), and *Assessment Standards* (1995).
- National Academy of Science, National Research Council. *National Science Education Standards* (1995).
- American Association for the Advancement of Science, Project 2061. *Science for All Americans* (1990), *Benchmarks for Scientific Literacy* (1993).

It also reflects the ideas of all ACEPT Co-Principal Investigators, but especially those of Marilyn Carlson and Anton Lawson, and the principles of reform underlying the ACEPT project. Its structure reflects some elements of the *Local Systemic Change Revised Classroom Observation Protocol* , by Horizon Research (1997-98).

The RTOP is criterion-referenced, and observers' judgments should *not* reflect a comparison with any other instructional setting than the one being evaluated. It can be used at all levels, from primary school through university. The instrument contains twenty-five items, with each rated on a scale from 0 (not observed) to 4 (very descriptive). Possible scores range from 0 to 100 points, with higher scores reflecting a greater degree of reform.

The RTOP was designed to be used by trained observers. This *Training Guide* provides specific information pertinent to the interpretation of individual items in the protocol. It is intended to be used as part of a formal training program in which trainees observe actual classrooms or videotapes of classrooms, and discuss their observations with others. The *Guide*, in its present form, is also designed to solicit trainee thoughts and concerns so that they feel comfortable in using the instrument. For that reason, a space is provided after each item for trainee comments. Such input helps all those being trained to achieve a higher degree of consistency in using the instrument. Please keep this in mind in making comments.

March 2000 Revision
Copyright© 2000 Arizona Board of Regents
All Rights Reserved

I. BACKGROUND INFORMATION

This section contains space for standard information that should be recorded by all observers. It will serve to identify the classroom, the instructor, the lesson observed, the observer, and the duration of the observation.

comments:

II. CONTEXTUAL BACKGROUND AND ACTIVITIES

Space is provided for a brief description of the lesson observed, the setting in which the lesson took place (space, seating arrangements, etc.), and any relevant details about the students (number, gender, ethnicity, etc.) and instructor. Try to go beyond a simple description. Capture, if you can, the defining characteristics of this situation that you believe provide the most important context for understanding what you will describe in greater detail in later sections. Use diagrams if they seem appropriate.

comments:

The next three sections contain the items to be rated. Do not feel that you have to complete them during the actual observation period. Space is provided on the facing page of every set of evaluations for you to make notes while observing. Immediately *after the lesson*, draw upon your notes and complete the ratings. For most items, a valid judgment can be rendered only after observing the entire lesson. The whole lesson provides contextual reference for rating each item.

Each of the items is to be rated on a scale ranging from 0 to 4. Choose "0" if in your judgment, the characteristic *never* occurred in the lesson, not even once. If it did occur, even if only once, "1" or higher should be chosen. Choose "4" only if the item was very descriptive of the lesson you observed. Intermediate ratings do not reflect the number of times an item occurred, but rather the degree to which that item was *characteristic* of the lesson observed.

The remainder of this Training Guide attempts provides a clarification of each RTOP item and the subtest (there are five) of which it is a part.

March 2000 Revision
Copyright© 2000 Arizona Board of Regents
All rights reserved

III. LESSON DESIGN AND IMPLEMENTATION

1) The instructional strategies and activities respected students' prior knowledge and the preconceptions inherent therein.

A cornerstone of reformed teaching is taking into consideration the prior knowledge that students bring with them. The term "respected" is pivotal in this item. It suggests an attitude of curiosity on the teacher's part, an active solicitation of student ideas, and an understanding that much of what a student brings to the mathematics or science classroom is strongly shaped and conditioned by their everyday experiences.

comments:

2) The lesson was designed to engage students as members of a learning community.

Much knowledge is socially constructed. The setting within which this occurs has been called a "learning community." The use of the term community in the phrase "the scientific community" (a "self-governing" body) is similar to the way it is intended in this item. Students participate actively, their participation is integral to the actions of the community, and knowledge is negotiated within the community. It is important to remember that a group of learners does not necessarily constitute a "learning community."

comments:

3) In this lesson, student exploration preceded formal presentation.

Reformed teaching allows students to build complex abstract knowledge from simpler, more concrete experience. This suggests that any formal presentation of content should be preceded by student exploration. This does not imply the converse...that all exploration should be followed by a formal presentation

comments:

March 2000 Revision

3

ASSESSING SCIENCE LEARNING 439

4) This lesson encouraged students to seek and value alternative modes of investigation or of problem solving.

Divergent thinking is an important part of mathematical and scientific reasoning. A lesson that meets this criterion would not insist on only one method of experimentation or one approach to solving a problem. A teacher who valued alternative modes of thinking would respect and actively solicit a variety of approaches, and understand that there may be more than one answer to a question.

comments:

5) The focus and direction of the lesson was often determined by ideas originating with students.

If students are members of a true learning community, and if divergence of thinking is valued, then the direction that a lesson takes can not always be predicted in advance. Thus, planning and executing a lesson may include contingencies for building upon the unexpected. A lesson that met this criterion might not end up where it appeared to be heading at the beginning.

comments:

IV. CONTENT

Knowledge can be thought of as having two forms: knowledge of what is (Propositional Knowledge), and knowledge of how to (Procedural Knowledge). Both are types of content. The RTOP was designed to evaluate mathematics or science lessons in terms of both.

Propositional Knowledge

This section focuses on the level of significance and abstraction of the content, the teacher's understanding of it, and the connections made with other disciplines and with real life.

6) The lesson involved fundamental concepts of the subject.

The emphasis on "fundamental" concepts indicates that there were some significant scientific or mathematical ideas at the heart of the lesson. For example, a lesson on the multiplication algorithm can be anchored in the distributive property. A lesson on energy could focus on the distinction between heat and temperature.

comments:

4

7) The lesson promoted strongly coherent conceptual understanding.
The word "coherent" is used to emphasize the strong inter-relatedness of mathematical and/or scientific thinking. Concepts do not stand on their own two feet. They are increasingly more meaningful as they become integrally related to and constitutive of other concepts.
comments:

8) The teacher had a solid grasp of the subject matter content inherent in the lesson.

This indicates that a teacher could sense the potential significance of ideas as they occurred in the lesson, even when articulated vaguely by students. A solid grasp would be indicated by an eagerness to pursue student's thoughts even if seemingly unrelated at the moment. The grade-level at which the lesson was directed should be taken into consideration when evaluating this item.
comments:

9) Elements of abstraction (i.e., symbolic representations, theory building) were encouraged when it was important to do so.

Conceptual understanding can be facilitated when relationships or patterns are represented in abstract or symbolic ways. Not moving toward abstraction can leave students overwhelmed with trees when a forest might help them locate themselves.
comments:

10) Connections with other content disciplines and/or real world phenomena were explored and valued.

Connecting mathematical and scientific content across the disciplines and with real world applications tends to generalize it and make it more coherent. A physics lesson on electricity might connect with the role of electricity in biological systems, or with the wiring systems of a house. A mathematics lesson on proportionality might connect with the nature of light, and refer to the relationship between the height of an object and the length of its shadow.
comments:

March 2000 Revision
Copyright© 2000 Arizona Board of Regents
All rights reserved

5

Procedural Knowledge

This section focuses on the kinds of processes that students are asked to use to manipulate information, arrive at conclusions, and evaluate knowledge claims. It most closely resembles what is often referred to as mathematical thinking or scientific reasoning.

11) Students used a variety of means (models, drawings, graphs, symbols, concrete materials, manipulatives, etc.) to represent phenomena.

Multiple forms of representation allow students to use a variety of mental processes to articulate their ideas, analyze information and to critique their ideas. A "variety" implies that at least two different means were used. Variety also occurs within a given means. For example, several different kinds of graphs could be used, not just one kind.

comments:

12) Students made predictions, estimations and/or hypotheses and devised means for testing them.

This item does not distinguish among predictions, hypotheses and estimations. All three terms are used so that the RTOP can be descriptive of both mathematical thinking and scientific reasoning. Another word that might be used in this context is "conjectures". The idea is that students explicitly state what they think is going to happen before collecting data.

comments:

13) Students were actively engaged in thought-provoking activity that often involved the critical assessment of procedures.

This item implies that students were not only actively doing things, but that they were also actively thinking about how what they were doing could clarify the next steps in their investigation.

comments:

March 2000 Revision
Copyright© 2000 Arizona Board of Regents
All rights reserved

6

14) Students were reflective about their learning.

Active reflection is a meta-cognitive activity that facilitates learning. It is sometimes referred to as "thinking about thinking." Teachers can facilitate reflection by providing time and suggesting strategies for students to evaluate their thoughts throughout a lesson. A review conducted by the teacher may not be reflective if it does not induce students to *re-examine* or *re-assess* their thinking.

comments:

15) Intellectual rigor, constructive criticism, and the challenging of ideas were valued.

At the heart of mathematical and scientific endeavors is rigorous debate. In a lesson, this would be achieved by allowing a variety of ideas to be presented, but insisting that challenge and negotiation also occur. Achieving intellectual rigor by following a narrow, often prescribed path of reasoning, to the exclusion of alternatives, would result in a low score on this item. Accepting a variety of proposals without accompanying evidence and argument would also result in a low score.

comments:

V. CLASSROOM CULTURE

This section addresses a separate aspect of a lesson, and completing these items should be done independently of any judgments on preceding sections. Specifically the design of the lesson or the quality of the content should not influence ratings in this section. Classroom culture has been conceptualized in the RTOP as consisting of: (1) Communicative Interactions, and (2) Student/Teacher Relationships. These are not mutually exclusive categories because all communicative interactions presuppose some kind of relationship among communicants.

Communicative Interactions

Communicative interactions in a classroom are an important window into the culture of that classroom. Lessons where teachers characteristically speak and students listen are not reformed. It is important that students be heard, and often, and that they communicate with one another, as well as with the teacher. The nature of the communication captures the dynamics of knowledge construction in that community. Recall that communication and community have the same root.

March 2000 Revision
Copyright© 2000 Arizona Board of Regents
All rights reserved

7

16) Students were involved in the communication of their ideas to others using a variety of means and media.

The intent of this item is to reflect the communicative richness of a lesson that encouraged students to contribute to the discourse and to do so in more than a single mode (making presentations, brainstorming, critiquing, listening, making videos, group work, etc.). Notice the difference between this item and item 11. Item 11 refers to representations. This item refers to active communication.

comments:

17) The teacher's questions triggered divergent modes of thinking.

This item suggests that teacher questions should help to open up conceptual space rather than confining it within predetermined boundaries. In its simplest form, teacher questioning triggers divergent modes of thinking by framing problems for which there may be more than one correct answer or framing phenomena that can have more than one valid interpretation.

comments:

18) There was a high proportion of student talk and a significant amount of it occurred between and among students.

A lesson where a teacher does most of the talking is not reformed. This item reflects the need to increase both the amount of student talk and of talk among students. A "high proportion" means that at any point in time it was as likely that a student would be talking as that the teacher would be. A "significant amount" suggests that critical portions of the lesson were developed through discourse among students.

comments:

19) Student questions and comments often determined the focus and direction of classroom discourse.

This item implies not only that the flow of the lesson was often influenced or shaped by student contributions, but that once a direction was in place, students were crucial in sustaining and enhancing the momentum.

comments:

March 2000 Revision
Copyright© 2000 Arizona Board of Regents
All rights reserved

8

20) There was a climate of respect for what others had to say.

Respecting what others have to say is more than listening politely. Respect also indicates that what others had to say was actually heard and carefully considered. A reformed lesson would encourage and allow every member of the community to present their ideas and express their opinions without fear of censure or ridicule.

comments:

Student/Teacher Relationships

21) Active participation of students was encouraged and valued.

This implies more than just a classroom full of active students. It also connotes their having a voice in how that activity is to occur. Simply following directions in an active manner does not meet the intent of this item. Active participation implies agenda-setting as well as "minds-on" and "hands-on".

comments:

22) Students were encouraged to generate conjectures, alternative solution strategies, and/or different ways of interpreting evidence.

Reformed teaching shifts the balance of responsibility for mathematical of scientific thought from the teacher to the students. A reformed teacher actively encourages this transition. For example, in a mathematics lesson, the teacher might encourage students to find more than one way to solve a problem. This encouragement would be highly rated if the whole lesson was devoted to discussing and critiquing these alternate solution strategies.

comments:

March 2000 Revision
Copyright© 2000 Arizona Board of Regents
All rights reserved

9

ASSESSING SCIENCE LEARNING 445

23) In general the teacher was patient with students.

Patience is not the same thing as tolerating unexpected or unwanted student behavior. Rather there is an anticipation that, when given a chance to play itself out, unanticipated behavior can lead to rich learning opportunities. A long "wait time" is a necessary but not sufficient condition for rating highly on this item.

comments:

24) The teacher acted as a resource person, working to support and enhance student investigations.

A reformed teacher is not there to tell students what to do and how to do it. Much of the initiative is to come from students, and because students have different ideas, the teacher's support is carefully crafted to the idiosyncrasies of student thinking. The metaphor, "guide on the side" is in accord with this item.

comments:

25) The metaphor "teacher as listener" was very characteristic of this classroom.

This metaphor describes a teacher who is often found helping students use what they know to construct further understanding. The teacher may indeed talk a lot, but such talk is carefully crafted around understandings reached by actively listening to what students are saying. "Teacher as listener" would be fully in place if "student as listener" was reciprocally engendered.

comments:

VI. SUMMARY

The RTOP provides an operational definition of what is meant by "reformed teaching." The items arise from a rich research-based literature that describes inquiry-oriented standards-based teaching practices in mathematics and science. However, this training guide does not cite research evidence. Rather it describes each item in a more metaphoric way. Our experience has been that these items have richly intuitive meaning to mathematics and science educators .

Further information about the underlying conceptual and theoretical basis of the RTOP, as well as reliability and validity data and norms by grade-level and context, can be found in the *Reformed Teaching Observation Protocol MANUAL* (Sawada & Piburn, 2000).

Using Data to Move Schools From Resignation to Results: The Power of Collaborative Inquiry

Nancy Love
Research for Better Teaching

Just as the inquiry process can make the science classroom come alive with discovery, discourse, and deep learning, so *collaborative inquiry* among teachers can breathe new life into schools and classrooms. Teachers possess tremendous knowledge, skill, and experience. Collaborative inquiry creates a structure for them to share that expertise with each other, discover what they are doing that is working and do more of it, and confront what practices aren't working and change them. Collaborative inquiry is guided by the following simple questions:

- How are we doing?
- What are we doing well? How can we amplify our successes?
- Who isn't learning? What aren't they learning?
- What in our practice could be causing the lack of learning? How can we be sure?
- What can we do to improve? To deepen our knowledge of our content and of ways to teach it?

This chapter is largely based on material from *The Data Coach's Guide to Improving Learning for All Students: Unleashing the Power of Collaborative Inquiry* by Nancy Love, Katherine E. Stiles, Susan Mundry, and Kathryn Diranna (Corwin Press 2008) and was made available for this publication with permission from Corwin Press.

- How do we know if our new efforts worked?
- What do we do if students still don't learn?

The research base on the link between collaborative, reflective practice of teachers and student learning is well established (Little 1990; Louis, Kruse, and Marks 1996; McLaughlin and Talbert 2001). When teachers engage in ongoing collaborative inquiry focused on teaching and learning and making effective use of data, they improve results for students.

Inquiry Improves Student Learning

As staff of the National Science Foundation–supported Using Data Project, we have seen the power of collaborative inquiry firsthand. We worked with schools that are serving among the poorest children in this country—children from Indian reservations in Arizona, the mountains of Appalachia in Tennessee, and large and mid-size urban centers in the Midwest and West. A few years ago some of these children were simply passing time in school with "word search" puzzles or other time fillers; some were permanently tracked in an educational system that doled out uninspired, repetitive curriculum. Some of the schools in which we worked had not one single student pass the state test, and the vast majority was performing at the lowest proficiency level.

Today, students in these schools have a more rigorous curriculum and are experiencing significant and continuous gains in local and state assessments in mathematics, science, and reading. For example, in Johnson County, Tennessee, a poor, rural area with 70% of students on free-or-reduced lunch, the schools exceeded the growth rates of some of the wealthiest and highest-performing districts on the state assessment (see Figure 22.1). Most impressive were gains for students with disabilities. In science in grades 3 through 6, the percentage proficient for this group increased from 60% to 73% from 2004 to 2006. In grades 3, 5, and 8 mathematics, the percentage proficient for this group increased from 36% to 74% from 2004 to 2006 (Tennessee Department of Education 2005, 2006).

Other schools participating in the Using Data Project produced equally impressive gains in student achievement. For example, the Canton City, Ohio, middle schools tripled the percentage of African American students proficient in mathematics while making gains for all student groups (Ohio Department of Education 2005). The Canton City high schools, McKinley

Figure 22.1 Johnson County, Tennessee, Improved Performance in Science on State Assessment, 2004–2006

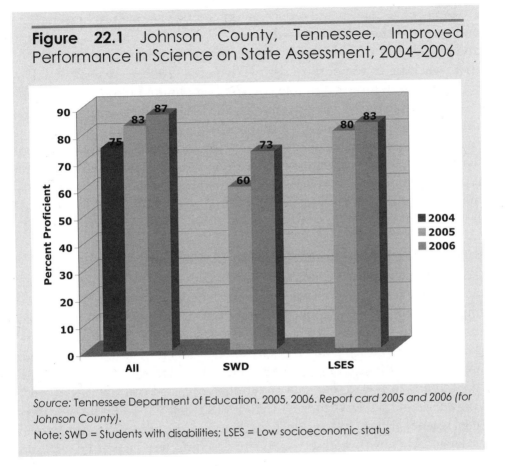

Source: Tennessee Department of Education. 2005, 2006. *Report card 2005 and 2006 (for Johnson County).*

Note: SWD = Students with disabilities; LSES = Low socioeconomic status

and Timkin, increased the percentage of students passing the Ohio Graduation Test by 23% and 22% respectively (Ohio Department of Education 2006). San Carlos Junior High School, serving all Native American children in rural Arizona, cut its failure rate in half (Arizona Department of Education 2005) while Wendall Williams Elementary School in Clark County (Las Vegas), Nevada, with 100% of students eligible for free-and-reduced lunch, increased mathematics performance by 24% on the Nevada state test in one year (Nevada Department of Education 2006).

According to Using Data Project external evaluators, schools implementing collaborative inquiry not only improved student achievement on state tests and other local measures; they changed their culture by increasing collaboration and reflection on practice among teachers. Teachers increased the frequency with which they used multiple data sources and engaged in data-driven dialogue; and they made improvements in their teaching in response to data (Love et al. 2008; Zuman 2006). Despite seemingly insurmountable barriers (e.g., limited resources, no common course or grade-level assessments, and historically low performance), these schools managed to solve one of the biggest problems science educators face: how to make effective use of the increasing amounts of school data now available to improve results for students. (A discussion of the types of school data that were used in the project—and that we recommend for use by other schools—begins on p. 453, "Data Use: What Kind? How Often?")

Building the Bridge Between Data and Results

Collaborative inquiry is the bridge that enables schools to connect the increasing amount of school data available to improved student learning. To implement collaborative inquiry, Using Data schools set out to build each step across the bridge and the foundation that supports it (Figure 22.2). First, they built the leadership and capacity of educators to engage in continuous improvement; through ongoing professional development, they gave educators training in how to use data ("data literacy") and collaborative inquiry knowledge and skills. The Using Data Project staff provided school leaders (e.g., teacher leaders, instructional coaches, building administrators), known as data coaches, with workshops, on-site coaching, and support materials in how to lead the process of collaborative inquiry in their schools and to develop data literacy and collaborative inquiry skills in others.

The second step across the bridge was collaboration. Teachers were organized into collaborative data teams—generally four to eight teachers and the building administrator or department chair—who worked together to use data to improve mathematics and science teaching and learning. At the elementary level, data teams were either grade-level teams or representatives of different grade levels who worked as a mathematics, science, or school improvement team. At the middle or high school, data teams were often organized by department, content area, or common courses taught. However configured, data teams met regularly, ideally weekly, during the school day.

Data teams used data frequently and in depth to guide instructional improvement. The most successful Using Data schools put in place benchmark

Figure 22.2 Collaborative Inquiry: Connecting Data to Results

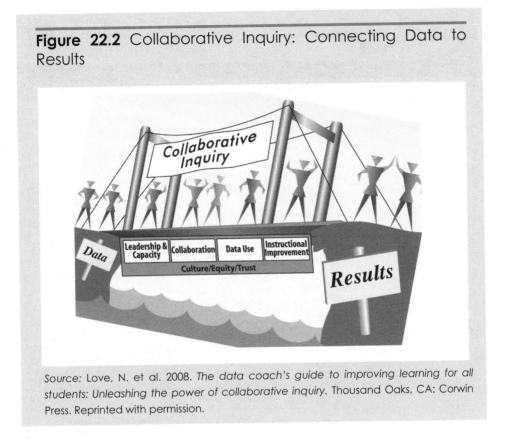

Source: Love, N. et al. 2008. *The data coach's guide to improving learning for all students: Unleashing the power of collaborative inquiry.* Thousand Oaks, CA: Corwin Press. Reprinted with permission.

common assessments and engaged teachers in regular analysis of item-level data and student work to identify and address student-learning problems. They learned to stop blaming students and their circumstances for failure and, instead, to use research and data about their instructional practices to generate solutions to identified gaps in student learning.

Data teams tried out new teaching strategies, such as graphing calculators, graphic organizers, and high-level questioning. They implemented new programs such as maximum inclusion for students with disabilities, inquiry-based science instructional materials, and use of school-based instructional coaches. And they frequently monitored results. In short, they made major shifts away from traditional data practices and toward those that build a high-performing, "Using Data" culture. These shifts are summarized in Table 22.1, page 452.

Table 22.1 Moving Toward a High-Performing, Using Data Culture

Element	Less emphasis on	More emphasis on
Leadership and capacity; data literacy and collaborative inquiry skills; content and pedagogical content knowledge; cultural proficiency; and leadership and facilitation skills	Individual charismatic leaders; data literacy as a specialty area for a few staff	Learning communities with many change agents; widespread data literacy among all staff
Collaboration	Teacher isolation; top-down, data-driven decision making; no time or structure provided for collaboration	Shared norms and values; ongoing data-driven dialogue and collaborative inquiry; time and structure provided for collaboration
Data Use	Use of data to punish or reward schools and sort students; rarely used by the school community to inform action	Use of data as feedback for continuous improvement and to serve students; frequent and in-depth use by entire school community
Instructional Improvement	Individually determined curriculum, instruction, and assessment; learning left to chance	Aligned learning goals, instruction, and assessment; widespread application of research and best practice; systems in place to prevent failure
Culture/	External accountability as driving force; focus on opportunities to learn for some	Internal responsibility as driving force; focus on opportunities to learn for all
Equity/	Belief that only the "brightest" can achieve at high levels; talk about race and class is taboo; culturally destructive or color-blind responses to diversity	Belief that all children are capable of high levels of achievement; ongoing dialogue about race, class, and privilege; culturally proficient responses to diversity
Trust	Relationships based on mistrust and avoidance of important discussions	Relationships based on trust, candidness, and openness

Source: Love, N. et al. 2008. *The data coach's guide to improving learning for all students: Unleashing the power of collaborative inquiry.* Thousand Oaks, CA: Corwin Press. Reprinted with permission.

Data Use: What Kind? How Often?

Let's focus further on the "data use" step across the bridge. The days of using data in schools once a year are over. If continuous improvement is the goal, then there is little point in examining only one source of data—state test results—which often become available only after students have moved on to the next grade. Data-literate teachers use a variety of different kinds of

Figure 22.3 The Data Pyramid: Different Types of Data Recommended for Use by Data Teams

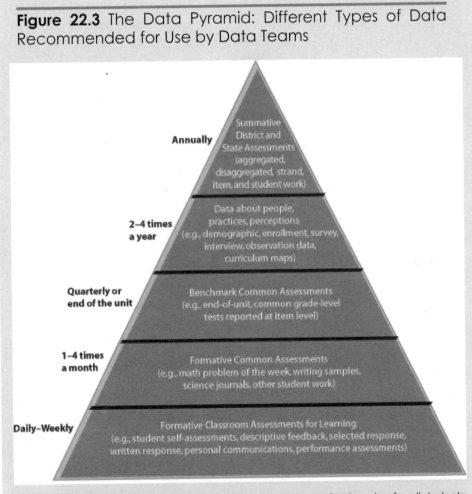

Source: Love, N. et al. 2008. *The data coach's guide to improving learning for all students: Unleashing the power of collaborative inquiry.* Thousand Oaks, CA: Corwin Press. Reprinted with permission.

data, some on a daily basis, some monthly or quarterly, and some annually, to continuously improve instruction and engage in collaborative inquiry. These include both formative and summative assessments. Formative assessments are assessments for learning and happen while learning is still under way and throughout teaching and learning to diagnose needs, plan next steps, and provide students with feedback. Summative assessments are assessments of learning and happen after learning is supposed to have occurred to determine if it did (Stiggins et al. 2004, p. 31). Figure 22.3, the Data Pyramid, illustrates the different types of data recommended for use by data teams.

The widest part of the pyramid shows the type of data that we suggest teachers spend the bulk of their time using—formative classroom assessments, done by teachers in their classrooms on an ongoing basis, including student self-assessments, descriptive feedback to students, rubrics, multiple methods of checking for understanding, and examination of student work such as science journals as well as tests and quizzes. These data inform teachers' instructional decisions—day-by-day, even minute-by-minute—and serve as the basis for feedback to help students improve their learning. For example, in Canton City, middle school data team members use handheld electronic devices—Texas Instruments Navigator and graphing calculators—to quickly assess student understanding of lessons while they are in progress. They then use this information to adjust their teaching, give specific feedback to students, and provide extra help for students who need it. Because of the strong research base indicating that these types of assessments improve student learning, we recommend that individual teachers spend the bulk of their data-analysis time developing, collecting, and analyzing these data (Black 2003; Black and Wiliam 1998; Bloom 1984; Meisels et al. 2003; Rodriguez 2004; Stiggins et al. 2004).

The next layer of the Data Pyramid represents formative common assessments, which the data team analyzes frequently (one to four times per month). These assessments include some of the same sources of data as the formative classroom assessments, the difference being that teams of teachers administer these assessments together and analyze them in their data teams. For example, teachers meet weekly to examine student entries in their science journals and brainstorm ideas for improving instruction. These formative common assessments are important in identifying student-learning problems, generating short cycles of improvement, and frequently monitoring progress toward the overall student-learning goal.

The next layer of the Data Pyramid illustrates benchmark common assessments, which teachers administer at the end of a unit or quarterly to assess the extent to which students have mastered the concepts and skills recently taught. These are administered together by teachers teaching the same content, either at the same grade level or in the same subject or course. The "common" feature makes them an ideal source of data for collaborative inquiry. In fact, they are among the most important sources of student-learning data the team has. They are timely, closely aligned with local curriculum, and available to teachers at the item level (i.e., results are reported on each individual item and the items themselves are available for the teachers' examination). Benchmark common assessments are most effective when they include robust performance tasks that provide evidence of student thinking and when multiple-choice items are analyzed item-by-item to uncover patterns in student choices and confusion underlying incorrect answer choices.

Benchmark common assessments can be used both formatively, to immediately improve instruction, and summatively, to inform programmatic changes in the future, such as increasing the amount of time a particular concept is taught or changing the sequence in which concepts are taught. Whether homegrown, included in science program materials, or purchased commercially, these tests must be of high quality: valid (they measure what is intended), reliable (they would produce a similar result if administered again), and as free of cultural bias as possible.

The next layer in the Data Pyramid, data about people, practices, and perceptions, is one that is often overlooked in schools, but it is extremely important. This type of data includes demographic data about student populations, teacher characteristics data, course enrollment data, and dropout rates. The data team analyzes demographic data to understand who the people are who make up the school community. This slice of data also includes (1) student enrollment in various types and levels of science and mathematics courses and (2) survey, observation, and interview data, which provide critical information about instructional practices, policies, and perceptions of teachers, students, administrators, and parents. These data become very important in exploring systemic causes of student-learning problems. They also help to ensure that diverse voices—by role (e.g., student, teacher, parent, administrator) and by race/ethnicity and economic, language, and educational status—are brought into the work of the data team. We recommend that data teams make use of these types of data two to four times per year to establish baseline data and monitor changes in practice.

The top of the Data Pyramid represents summative assessment data, including state assessments and annual district tests. These data are used to determine if student outcomes have been met and are for accountability purposes. The data team takes full advantage of these data, "drilling down" into them and analyzing them in as much detail as possible, including aggregated (largest group level) and disaggregated (broken out by student populations—e.g., race/ethnicity, gender, poverty, language, mobility, and educational status) data trends, strand (content domains), item-level data (student performance on each individual test item), and student work when available.

Along with other student-learning data sources described above, the summative assessment data become the basis for identifying a student-learning problem and setting annual improvement targets. These data, however, occupy a small part of the pyramid because they are only available annually and provide limited information about what to do to improve performance (especially if item-level data and released items are not available). In

Vignette: Improving Students' Graphing Skills Through Collaborative Inquiry

The analysis of the eighth-grade criterion-referenced test (CRT) science strand data for "investigation and experimentation" by the data team indicated that only 42% of the students scored at the proficient level. Item-level data revealed that three questions about plotting and interpreting graphs had the lowest percentage of correct answers. Even with this drill-down, the data team was left with lingering questions about why students were not able to answer these questions. It was evident to the team that analyzing only multiple-choice questions would not help them understand students' naive or alternate conceptions about graphing. The teachers knew that they needed to know exactly which concepts or content students were struggling to master.

(continued)

Vignette *(continued)*

The data coach brought the national and state science standards to the table for discussion. Using the documents helped the team to clarify their own content knowledge and build a common understanding of what eighth-grade students should know about charting, graphing, and summary statements. The list of concepts included: appropriate graphic representation (e.g., bar, line, pie); orientation of x- and y-axes; parallel and perpendicular lines; labeling of manipulated (independent) variables and responding (dependent) variables; and analysis of the relationship of manipulated and responding variables.

The team also discussed their experiences with teaching graphing and areas in which students always seem to struggle. Reviewing misconception research helped the team confirm that two of the most common misconceptions involve use of appropriate graphs to display the data and understanding the relationship between the variables.

Their interest was further piqued. What could they do to gather student work on this subject? The team decided to create an open-ended assessment prompt that asked students to graph data from a table that clearly labeled the variables and to make a summary statement from the graphic representation. They asked all eighth-grade science teachers to randomly select 10 students in their classes to take the open-ended assessment. This resulted in 50 pieces of student work (see one example on page 458).

To interpret the student work, the data team invited all teachers who gave the assessment to join in the analysis. First the teachers reviewed the scoring criteria (rubric) for expected student answers. Then they sorted the work into high-, medium-, and low-quality piles based on the scoring criteria and discussed the characteristics of each group. How was student understanding represented in the high-quality pile? What was lacking in a student's knowledge that indicated an intermediate level of understanding? What types of instructional interventions would be necessary to move a student from the low-quality to the medium-quality pile?

(continued)

Vignette *(continued)*

Student Work Example

Open-Ended Prompt

Please write (or draw) your answer directly on the lines or in the space provided

- You are the owner of a company that supplies local florists with tulips. Last year the tulips you produced tended to be smaller than usual and you wonder if it had something to do with the soil temperature in the winter.
- You recorded the ground temperature where the tulip bulbs were dormant and the average height of the plants when they sprouted. Your data chart looks like this:

HEIGHT OF TULIP PLANTS
ONE WEEK AFTER BREAKING THROUGH SOIL

	AREA A	AREA B	AREA C
Ground Temperature in Winter	7 °C	2 °C	0 C
Average Height of Plants	4 cm	8 cm	14 cm

1. Graph the data on the grid below. Remember to label the graph.

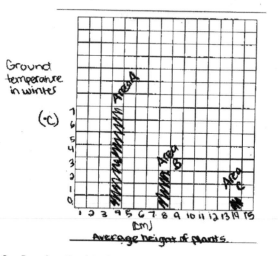

2. Based on the data from the graph, describe the relationship between ground temperature in winter and the height of tulip plants after a week of visible growth.

   The relationship is the more the temperature goes

   down the height gets taller.

(continued)

Vignette *(continued)*

The teachers made the following sorts of observations of the student work, without any interpretation or inference:

- Paper A uses a bar graph rather than a line graph.
- Papers A, B, and D have no title.
- Papers A and B have mixed up the variables, plotting the manipulated variable (ground temperature) on the y-axis instead of the x-axis.
- Most data points are plotted correctly.
- Papers A, B, and C don't use data from the graph to explain the changes, although they do state the change (colder temperature, taller plants).
- Paper D has a wrong relationship (warmer ground, taller plants).
- Paper E is the only one to use actual data numbers.

The team documented the following inference about student understanding: Students have difficulty understanding the difference between when to use a bar graph (discontinuous data) and a line graph (continuous data). They are also not using data as evidence when writing a summary statement of the data.

This led to a discussion of how graphing was taught. It soon was apparent that mathematics and science teachers don't articulate their content or their strategies. It was also clear that there were no common criteria for a quality graph or summary statements for grades 6–8. The team set out to implement the following changes: (1) meet with mathematics teachers to articulate content and strategies; (2) develop common criteria for quality graphing across content areas; (3) teach the students the criteria; (4) collect and analyze student work on graphing on a monthly basis; and (5) give students specific feedback on how to improve. Teachers were excited when their new samples of student work showed more students meeting the criteria for success.

Source: Love, N. et al. 2008. *The data coach's guide to improving learning for all students: Unleashing the power of collaborative inquiry.* Thousand Oaks, CA: Corwin Press. Reprinted with permission.

addition, these results often arrive too late for teachers who taught a group of students during the year of the test to respond to them. Finally, tests can be poorly constructed, culturally biased, inaccurate in science content, and lacking in rigor, underscoring the importance of using the rich array of data recommended in the pyramid.

Instructional Improvement—The Final Step to Results

The driving purpose for collecting all of the data described above is instructional improvement. There is no way to bridge the gap between data and results without changing what is taught, how it is taught, and how it is assessed. Instructional improvement is the last and essential step across the bridge linking data to results. The vignette "Improving Students' Graphing Skills Through Collaborative Inquiry" on page 456 illustrates several important features of using data for instructional improvement:

- Keep the conversation focused on improving instruction. Establish ground rules: No blaming students, their circumstances, other teachers, or factors outside of students' control.
- Use multiple data sources, including state and local test data at the strand and item level, to identify the specific knowledge and skills students may be having difficulty with.
- Use national and local standards and misconceptions research to deepen teachers' content knowledge about the particular content or skill students are struggling with. This enhances the teachers' ability to analyze the work for student thinking and misconceptions.
- Collect student work that will further elucidate student understanding relative to the learning problem being investigated.
- Clarify what quality student work looks like, using anchor papers and exemplars.
- When analyzing student work and other data, separate observations from inferences and further test inferences with additional data and research.
- When generating inferences, use the following questions to guide the dialogue:
  - Are our learning goals, instruction, and assessment aligned?
  - Did we teach this concept/skill? Did we teach it in enough depth? At the appropriate development level? In the best sequence?

- Did we use appropriate and varied instructional strategies to meet each student's needs?
- Did we use quality questions to extend student thinking?
- Did we use formative assessment data to give students feedback on their own learning and to identify student confusion and refocus our teaching?
- Did all students have access to this content and best practice?
- What content knowledge and pedagogical content knowledge will strengthen our ability to teach this content?
- Did we apply principles of cultural proficiency (knowledge and respect for diverse cultures) to ensure the best learning opportunities for culturally diverse learners?
- Use additional data (e.g., student and teacher surveys, classroom observations, student and teacher interviews, student enrollment in advanced science courses), and research to verify the causes of the student-learning problem and generate research-based solutions.
- Test solutions through ongoing monitoring of student learning in the problem area identified.

Building the Foundation

As illustrated in Figure 22.1, the foundation of the bridge of collaborative inquiry is a school culture characterized by collective responsibility for student learning and commitment to serve each and every child. In such cultures, educators trust each other enough to discuss "undiscussables" (e.g., race and racism, poor teaching), reveal their own practice and mistakes, root for one another, and face together the brutal facts that data often reveal (Barth 2006). Collaborative inquiry both thrives in such a culture and helps to establish it. Schools that want to unleash the power of collaborative inquiry

- take immediate steps to strengthen the collaborative culture, building a shared vision for science education, shared commitments to every child's learning, and norms of collaboration and risk-taking;
- build broad stakeholder support;
- provide teachers with high-quality and sustained professional development in data literacy and collaborative inquiry skills as well as in science content and pedagogical content knowledge;

- organize teachers into data teams;
- establish regular time for collaboration, ideally once a week for at least 45 minutes; and
- ensure timely access to robust local data sources, including common assessments at the item level and student work with evidence of student thinking.

References

Arizona Department of Education. 2005. Arizona school report card: San Carlos junior high school. Retrieved June 13, 2006, from *www.ade.az.gov/srcs/find_school.asp*

Barth, R. 2006. Improving relationships within the schoolhouse. *Educational Leadership* 63(6): 8–13.

Black, P. 2003. A successful intervention—Why did it work? Paper presented at American Educational Research Association annual meeting, Chicago (April 23).

Black, P., and D. Wiliam. 1998. Inside the black box: Raising standards through classroom assessment. *Phi Delta Kappan* 80(2): 139–48.

Bloom, B. S. 1984. The search for methods of group instruction as effective as one-to-one tutoring. *Educational Leadership* 41(8): 4–17.

Elmore, R. F. 2002. *Bridging the gap between standards and achievement: The imperative for professional development in education.* Washington, DC: Albert Shanker Institute.

Little, J. W. 1990. Teachers as colleagues. In A. Lieberman (Ed.), *Schools as collaborative cultures: Creating the future now* (pp. 165–193). New York: Palmer.

Louis, K. S., S. Kruse, and H. Marks. 1996. Schoolwide professional community. In F. Newmann and Associates (Eds.), *Authentic achievement: Restructuring schools for intellectual quality* (pp. 179–203). San Francisco: Jossey-Bass.

Love, N., K. Stiles, S. Mundry, and K. Diranna. 2008. *The data coach's guide to improving learning for all students: Unleashing the power of collaborative inquiry.* Thousand Oaks, CA: Corwin Press.

McLaughlin, M. W., and J. Talbert. 2001. *Professional communities and the work of high school teaching.* Chicago: University of Chicago Press.

Meisels, S. J., S. Atkins-Burnett, Y. Xue, J. Nicholson, D. D. Bickel, and S.-H. Son. 2003. Creating a system of accountability: The impact of instructional assessment on elementary children's achievement test scores. *Education Policy Analysis Archives* 11(9). Retrieved June 14, 2006, from *http://epaa.asu.edu/epaa/v11n9*

Nevada Department of Education. 2006. AYP (Annual Yearly Progress). Retrieved July 6, 2006, from *www.doe.nv.gov/accountability/ayp.html*

Ohio Department of Education. 2005. Local report card links (to 2004–05 data for Hartford, Lehman, and Souers Middle Schools, Canton, Ohio). Retrieved December 13,

2006, from *www.ode.state.oh.us/GD/Templates/Pages/ODE/ODEPrimary.aspx?page=2 &TopicID=1266&TopicRelationID=1266*

Ohio Department of Education. 2006. Local report card links (to 2005-2006 data for McKinley and Timkin High Schools, Canton, Ohio). Retrieved February 7, 2007, from *www.ode.state.oh.us/GD/Templates/Pages/ODE/ODEPrimary.aspx?page=2&Topic ID=1266&TopicRelationID=1266*

Rodriguez, M. C. 2004. The role of classroom assessment in student performance in TIMSS. *Applied Measurement in Education 17(1): 1–24.*

Stiggins, R. J., J. A. Arter, J. Chappuis, and S. Chappuis. 2004. *Classroom assessment for student learning: Doing it right—using it well.* Portland, OR: Assessment Training Institute.

Tennessee Department of Education. 2005. Report card 2005 for Johnson County. *http:// tennessee.gov/education/reportcard*

Tennessee Department of Education. 2006. Report card 2006 for Johnson County. *http:// tennessee.gov/education/reportcard*

Zuman, J. 2006. Using data project: Final evaluation report. Unpublished report. Arlington, MA: Intercultural Center for Research in Education.

Volume Editors

Janet Coffey is on the faculty of the Department of Curriculum and Instruction in the College of Education, University of Maryland, College Park. She is a former middle school science teacher.

Rowena Douglas is assistant executive director for professional development with the National Science Teachers Association, where she works to develop and implement a broad-based professional development component across all NSTA programs and products.

Carole Stearns is a science education consultant based in Portland, Oregon. Formerly a program officer at the National Science Foundation, she worked on initiatives serving K–12 teachers of science.

***

This is the second Research Dissemination Book developed under the leadership of the NSTA Professional Development Department and authored by presenters at the Research Dissemination Conferences held at the NSTA Conference on Science Education.

Contributors

Cari Herrmann Abell is a research associate with Project 2061 at the American Association for the Advancement of Science, Washington, DC.

Peter Afflerbach is a professor in the Department of Curriculum and Instruction, University of Maryland, College Park.

Alicia C. Alonzo is an assistant professor in the Department of Teaching and Learning, College of Education, University of Iowa.

Olga Amaral is associate dean, San Diego State University, Imperial Valley Campus.

Ruth Anderson is an education researcher at FACET Innovations, Seattle, Washington.

Matthew Anthes-Washburn is a physics teacher at Boston International High School, Boston, Massachusetts.

Carlos C. Ayala is an associate professor in the College of Education, Sonoma State University, California.

Alexandra Beatty is a senior program officer, Division of Behavioral and Social Sciences and Education, National Academy of Sciences, Washington, DC.

Meryl Bertenthal is a research scholar in the Office of the Vice Provost for Research, Indiana University, Bloomington.

Paul R. Brandon is a professor of education, Curriculum Research and Development Group, University of Hawaii, Manoa.

Barbara C. Buckley is a research scientist at Concord Consortium, Concord, Massachusetts.

Audrey B. Champagne is professor emerita of education and chemistry, University at Albany, State University of New York.

Rosalea Courtney is a senior research associate, Educational Testing Service, Princeton, New Jersey.

Charlene M. Czerniak is a professor of science education in the Department of Curriculum and Instruction and director of the Office of Research Partnership, University of Toledo.

Angela H. DeBarger is an educational researcher at the Center for Technology in Learning, SRI International, Menlo Park, California.

George E. DeBoer is deputy director of Project 2061 at the American Association for the Advancement of Science, Washington, DC.

Linda De Lucchi is codirector of FOSS, Lawrence Hall of Science, University of California, Berkeley.

Francis Q. Eberle is the executive director of the Maine Mathematics and Science Alliance, Augusta, Maine.

Arthur Eisenkraft is Distinguished Professor of science education and director of the Center of Science and Math in Context (COSMIC) at the University of Massachusetts, Boston.

Erin M. Furtak is an assistant professor in the School of Education at the University of Colorado, Boulder.

Linda Gentiluomo is a preK–6th grade science and math professional development specialist for the Schenectady, New York, school district.

Janice Gobert is a senior research scientist at Concord Consortium, Concord, Massachusetts.

Arhonda Gogos is a staff scientist (structural biology) with Sequoia Pharmaceuticals, Inc., Gaithersburg, Maryland.

Marian Grogan is project director, Center for Science Education, Education Development Center, Newton, Massachusetts.

Geneva Haertel is senior educational researcher, SRI International, Menlo Park, California.

Paul Hickman is a science education consultant in Andover, Massachusetts.

Paul Horwitz is a research scientist at Concord Consortium, Concord, Massachusetts.

Drew Isola is a math and physics teacher at Allegan High School, Allegan, Michigan.

Jacqueline Jones is assistant commissioner, Division of Early Childhood Education, New Jersey State Department of Education, Trenton.

Page D. Keeley is the senior science program director at the Maine Mathematics and Science Alliance, Augusta, Maine.

Thomas E. Keller is a program officer, Board of Science Education, National Academy of Sciences, Washington, DC.

Michael Klentschy is a faculty member in the College of Education, San Diego State University, Imperial Valley Campus.

Vicky L. Kouba is a professor of educational theory and practice in the School of Education, University at Albany, State University of New York.

Joseph S. Krajcik is a professor of science education in the School of Education, University of Michigan.

Pamela Kraus is a research scientist at Facet Innovations, Seattle, Washington.

Paul J. Kuerbis is a professor in the Education Department, Colorado College.

Okhee Lee is a professor in the Department of Teaching and Learning, School of Education, University of Miami.

Kathryn LeRoy is chief officer of curriculum services, Duval County Public Schools, Jacksonville, Florida.

Kathy Long is assessment coordinator, FOSS Project, Lawrence Hall of Science, University of California, Berkeley.

Nancy Love is the director of program development at Research for Better Teaching in Acton, Massachusetts.

Larry Malone is codirector of FOSS, Lawrence Hall of Science, University of California, Berkeley.

Katherine L. McNeill is an assistant professor of science education in the Lynch School of Education, Boston College.

An Michiels is a resident of Leuven, Belgium.

James E. Minstrell is director of operations and technology at FACET Innovations, Seattle, Washington.

Jim Minstrell is senior research scientist at FACET Innovations, Seattle, Washington.

Linda B. Mooney is principal investigator for the Science Teacher Enhancement Project—Unifying the Pikes Peak Region (STEP-uP) and administrative/principal coach in Colorado Springs, Colorado.

Marian Pasquale is senior research scientist, Center for Science Education, Education Development Center, Newton, Massachusetts.

Francis M. Pottenger III is a professor of education, Curriculum Research and Development Group, University of Hawaii, Manoa.

Edys S. Quellmalz is director of Technology-Enhanced Assessment and Learning Systems, Mathematics, Science, and Technology Program, WestEd, Redwood City, California.

Thomas Regan is a science test development specialist with the Education Assessment Group of the American Institutes for Research, Washington, DC.

Marc Reif is a physics teacher at Ruamrudee International School, Bangkok, Thailand.

Maria Araceli Ruiz-Primo is an associate professor in the Department of Education and Human Development, University of Colorado, Denver.

Patricia Schank is a computer scientist at the Center for Technology in Learning, SRI International, Menlo Park, California.

Richard J. Shavelson is the Margaret Jack professor of education and psychology in the School of Education at Stanford University.

Kathryn Show is a science coach, Instructional Services, Seattle (Washington) Public Schools.

Janet L. Struble is program coordinator for UToledo.UTeach.UTouch the Future (UT^3), University of Toledo.

Mark A. Templin is an associate professor of science education in the Department of Curriculum, University of Toledo.

Miki Tomita is a doctoral student in education at Stanford University.

Dylan Wiliam is deputy director and professor of educational assessment at the Institute of Education, University of London, UK.

Mark R. Wilson is a professor in the Graduate School of Education, University of California, Berkeley.

Paula Wilson is a resident of Kaysville, Utah.

Elaine Woo is science program manager, Instructional Services, Seattle (Washington) Public Schools.

Yue Yin is an assistant professor in the Department of Education at the University of Hawaii, Manoa.

Donald B. Young is director of the Curriculum and Research Group and professor of education, University of Hawaii, Manoa.

Index

*Page numbers in **boldface** type refer to figures or tables.*